STUDIES
SCIENTIFIC & SOCIAL

BY

ALFRED RUSSEL WALLACE

LL.D., D.C.L., F.R.S., ETC.

IN TWO VOLUMES.—VOLUME II

WITH NUMEROUS ILLUSTRATIONS

London
MACMILLAN AND CO., Limited
NEW YORK: THE MACMILLAN COMPANY
1900

Copyright © 2013 Read Books Ltd.
This book is copyright and may not be
reproduced or copied in any way without
the express permission of the publisher in writing

British Library Cataloguing-in-Publication Data
A catalogue record for this book is available from the
British Library

Alfred Russel Wallace

Alfred Russel Wallace was born on 8th January 1823 in the village of Llanbadoc, in Monmouthshire, Wales.

At the age of five, Wallace's family moved to Hertford where he later enrolled at Hertford Grammar School. He was educated there until financial difficulties forced his family to withdraw him in 1836. He then boarded with his older brother John before becoming an apprentice to his eldest brother, William, a surveyor. He worked for William for six years until the business declined due to difficult economic conditions.

After a brief period of unemployment, he was hired as a master at the Collegiate School in Leicester to teach drawing, map-making, and surveying. During this time he met the entomologist Henry Bates who inspired Wallace to begin collecting insects. He and bates continued exchanging letters after Wallace left teaching to pursue his surveying career. They corresponded on prominent works of the time such as Charles Darwin's *The Voyage of the Beagle* (1839) and Robert Chamber's *Vestiges of the Natural History of Creation* (1844).

Wallace was inspired by the travelling naturalists of the day and decided to begin his exploration career collecting specimens in the Amazon rainforest. He explored the Rio Negra for four years, making notes on the peoples and languages he encountered as well as the geography, flora, and fauna. On his return voyage his ship, Helen, caught fire and he and the crew were stranded for ten days before being picked up by the Jordeson, a brig travelling from Cuba to London. All of his specimens aboard Helen had been lost.

After a brief stay in England he embarked on a journey to the Malay Archipelago (now Singapore, Malaysia, and Indonesia). During this eight year period he collected more than 126,000 specimens, several thousand of which represented new species to science. While travelling, Wallace refined his thoughts about evolution and in 1858 he outlined his theory of natural selection in an article he sent to Charles Darwin. This was published in the same year along with Darwin's own theory. Wallace eventually published an account of his travels *The Malay Archipelago* in 1869, and it became one of the most popular books of scientific exploration in the 19th century.

Upon his return to England, in 1862, Wallace became a staunch defender of Darwin's landmark work *On the Origin of Species* (1859). He wrote responses to those critical of the theory of natural selection, including 'Remarks on the Rev. S. Haughton's Paper on the Bee's Cell, And on the Origin of Species' (1863) and 'Creation by Law' (1867). The former of these was particularly pleasing to Darwin. Wallace also published important papers such as 'The Origin of Human Races and the Antiquity of Man Deduced from the Theory of 'Natural Selection'' (1864) and books, including the much cited *Darwinism* (1889).

Wallace made a huge contribution to the natural sciences and he will continue to be remembered as one of the key figures in the development of evolutionary theory.

Wallace died on 7th November 1913 at the age of 90. He is buried in a small cemetery at Broadstone, Dorset, England.

CONTENTS

OF VOL. II

EDUCATIONAL.

CHAPTER		PAGE
I.	MUSEUMS FOR THE PEOPLE (*Macmillan's Magazine*, 1869)	1
II.	AMERICAN MUSEUMS (*Fortnightly Review*, September and October, 1887)	16
III.	HOW BEST TO MODEL THE EARTH (*Contemporary Review*, May, 1896)	59
IV.	EPPING FOREST AND TEMPERATE FOREST REGIONS	74
V.	WHITE MEN IN THE TROPICS (*The Independent*, New York, 1899)	99
VI.	HOW TO CIVILISE SAVAGES (*The Reader*, June 17, 1865)	107
VII.	THE EXPRESSIVENESS OF SPEECH, OR MOUTH-GESTURE AS A FACTOR IN THE ORIGIN OF LANGUAGE (*Fortnightly Review*, October, 1895)	115

POLITICAL.

VIII.	COAL A NATIONAL TRUST (*The Daily News*, September 16, 1873)	138
IX.	PAPER MONEY AS A STANDARD OF VALUE (*The Academy*, December 31, 1898)	145
X.	LIMITATION OF STATE FUNCTIONS IN THE ADMINISTRATION OF JUSTICE (*Contemporary Review*, December, 1873)	150
XI.	RECIPROCITY THE ESSENCE OF FREE TRADE, WITH A REPLY TO MR. LOWE (*Nineteenth Century*, April, 1879)	167
XII.	THE DEPRESSION OF TRADE, ITS CAUSES AND ITS REMEDIES (*Claims of Labour Lectures*, 1886)	188

CHAPTER		PAGE
XIII.	A REPRESENTATIVE HOUSE OF LORDS (*Contemporary Review*, June, 1894)	223
XIV.	DISESTABLISHMENT AND DISENDOWMENT, WITH A PROPOSAL FOR A REALLY NATIONAL CHURCH OF ENGLAND (*Macmillan's Magazine*, April, 1873)	235
XV.	INTEREST-BEARING FUNDS INJURIOUS AND UNJUST	254

THE LAND PROBLEM.

XVI.	HOW TO NATIONALISE THE LAND: A RADICAL SOLUTION OF THE IRISH LAND PROBLEM (*Contemporary Review*, November, 1880)	265
XVII.	THE "WHY" AND "HOW" OF LAND NATIONALIZATION (*Macmillan's Magazine*, August and September, 1883)	296
XVIII.	HERBERT SPENCER ON THE LAND QUESTION: A CRITICISM. (From an Address to the Land Nationalization Society, 1892)	333
XIX.	SOME OBJECTIONS TO LAND NATIONALIZATION ANSWERED. (From Tracts issued by the Land Nationalization Society)	345

ETHICAL.

XX.	A COUNSEL OF PERFECTION FOR SABBATARIANS (*Nineteenth Century*, October, 1894)	364
XXI.	WHY LIVE A MORAL LIFE? (*Agnostic Annual*, 1895)	375
XXII.	THE CAUSES OF WAR AND THE REMEDIES (*L'Humanité Nouvelle*, May, 1899)	384

SOCIOLOGICAL.

XXIII.	THE SOCIAL QUAGMIRE AND THE WAY OUT OF IT (*The Arena*, Boston, U.S.A., March and April, 1893)	394
XXIV.	ECONOMIC AND SOCIAL JUSTICE (*Vox Clamantium*, 1894)	432
XXV.	RALAHINE AND ITS TEACHINGS	455
XXVI.	REOCCUPATION OF THE LAND: THE ONLY IMMEDIATE SOLUTION OF THE PROBLEM OF THE UNEMPLOYED (*Forecasts of the Coming Century*, 1897)	478
XXVII.	HUMAN PROGRESS, PAST AND FUTURE (*The Arena*, January, 1892)	493
XXVIII.	THE INDIVIDUALISM: THE ESSENTIAL PRELIMINARY OF A REAL SOCIAL ADVANCE	510
XXIX.	JUSTICE NOT CHARITY, AS THE FUNDAMENTAL PRINCIPLE OF SOCIAL REFORM. (An Appeal to my Readers)	521

LIST OF ILLUSTRATIONS

IN VOL. II

FIG.		PAGE
1.	MUSEUM OF COMPARATIVE ZOOLOGY, CAMBRIDGE, MASS.	19
2.	ONE SIDE OF THE NORTH AMERICAN ROOM	27
3.	SOUTH AMERICAN FAUNA	29
4.	THE EUROPEAN FAUNA	33
5.	STONE IMPLEMENTS	41
6.	STONE SCRAPERS	41
7.	ARROW AND SPEAR HEADS	42
8.	ARROW AND SPEAR HEADS	43
9.	ARROW AND SPEAR HEADS	44
10.	TOOLS	44
11.	DISCS	45
12.	PLUMMETS	45
13.	STONE SPADE	46
14.	SPADE AND KNIVES	46

LIST OF ILLUSTRATIONS

FIG.		PAGE
15.	WEIGHTS	46
16.	CUPS AND SPOONS	48
17.	SHUTTLE AND REEL	48
18.	THREAD-WINDERS	48
19.	STONE DISCS FOR GAMES	49
20.	CEREMONIAL STONES	49
21.	CEREMONIAL STONES	49
22.	CEREMONIAL STONES	50
23.	ANIMAL SCULPTURES	51
24.	YOKES AND A CELT	52
25.	DIAGRAM OF EXPORTS AND IMPORTS 1864 TO 1883	192

STUDIES, SCIENTIFIC AND SOCIAL
VOL. II

CHAPTER I

MUSEUMS FOR THE PEOPLE

Museums of Natural History should be, one would think, among the most entertaining and instructive of public exhibitions, since their object is to show us life-like restorations of all those wonderful and beautiful animals, the mere description of which in the pages of the traveller, the naturalist, or the sportsman, are of such absorbing interest. Strange to say, however, such is by no means generally the case; and these institutions rarely appear to yield either pleasure or information at all proportionate to their immense cost. We can hardly impute this failure to anything in the nature of museums or of their contents, when we remember that good illustrated works on natural history are universally interesting and instructive; and that private collections of birds, shells, or insects are often very attractive even to the uninitiated, and at the same time of the highest value to the student. We must therefore seek for an explanation of the anomaly in the system on which public museums are usually constituted, in the quality of the specimens they exhibit, and in the mode of exhibiting them, all which, it is now generally admitted, are equally unsuited for the amusement and instruction of the public and for the purposes of the scientific student.

Public museums of natural objects being such entirely modern institutions, we can hardly wonder that no generally accepted principles have yet been laid down for their construction or arrangement. They most frequently originated with private collectors, whose plan was naturally followed in their enlargement; and when they outgrew their original domicile, an architect was called in, who, according to his special tastes, designed a temple or a palace for their reception. However inconvenient or unsuitable the original mode of exhibition might turn out, or however ill adapted to its purpose the new building might prove, it would, of course, be exceedingly difficult and expensive to alter either of them, more especially as the modified plan might be found, after trial, to have defects as great as that which it replaced.

Two eminent naturalists, Sir Joseph Hooker and the late Dr. J. E. Gray, both connected with great public museums, have made suggestions towards a more rational system; and as it is evident that museums will increase, and may be made an important agent in national education and the elevation of the masses of the people, it seems advisable that the subject should be brought forward for popular discussion.

Accepting as a basis the few essential principles that seem now to be agreed upon, I propose to follow them out into some rather important details.

I shall consider, in the first place, what should be the scope of a Typical Popular Museum, and then sketch out the arrangements best adapted to make it both entertaining and instructive to the young and ignorant, and a means of high intellectual culture and enjoyment to such as may be disposed to avail themselves fully of its teachings.

Museums are well adapted to illustrate all those branches of knowledge whose subject-matter consists mainly of definite movable and portable objects. The great group of the natural history sciences can scarcely be taught without them; while mathematics, astronomy, physics, and chemistry make use of observatories and laboratories rather than museums. The fine and mechanical arts, as well as history, can also be illustrated by extensive collections of objects; and we are thus led to a broad division of museums, according as they deal mainly with natural objects or with works of art.

A museum of natural objects appears, for a variety of reasons, best fitted to interest, instruct, and elevate the middle and lower classes, and the young. It is more in accordance with their tastes and sympathies, as shown by the universal fondness for flowers and birds, and the great interest excited by new or strange animals. It enables them to acquire a wide and accurate knowledge of the earth and of its varied productions; and if they wish to follow up any branch of natural history as an amusement or a study, it leads them into the pure air and pleasant scenes of the country, and is likely to be the best antidote to habits of dissipation or immorality. Such museums, too, offer the only means by which the mass of the working classes can obtain any actual knowledge of the wonderful productions of nature in present or past ages; and such knowledge gives a new interest to works on geography, travel, or natural history. Owing to the wide disconnection of these subjects from the daily pursuits of life, they are so much the better adapted for the relaxation of those who earn their bread by manual labour. The inexhaustible variety, the strange beauty, and the wondrous complexity of natural objects, are pre-eminently adapted to excite both the observing and reflective powers of the mind, and their study is well calculated to have an elevating and refining effect upon the character.

Works of art, on the other hand, though in the highest degree instructive and elevating to some minds, are not so universally attractive; and, what is more important, do not exercise so many faculties, and do not offer such wide and easily-reached fields of study for the working classes. Some previous training or special aptitude is required in order to appreciate them; and it may even be asserted with truth that the study of nature is a necessary preliminary to the appreciation of art. It does not seem improbable that, even if our object were to make artists and lovers of art, good museums of natural objects might be the most useful first step. We have further to consider that objects of art are already widely spread, and more or less accessible. Our great public buildings contain their art-decorations. The houses of the wealthy and the shops of our streets are full of art, and the artisan has frequent opportunities of seeing them, while local exhibitions of art are not uncommon, and will no doubt be come more and more frequent. The very young and the very ignorant would learn nothing in an art museum, while they would certainly gain both knowledge and pleasure in such an one as I am about to describe.

A Typical Museum of Natural History should contain a series of objects to illustrate all the sciences which treat of the earth, nature, and man. These are—1, Geography and Geology; 2, Mineralogy; 3, Botany; 4, Zoology; 5, Ethnology. I will briefly sketch what seems to be the best mode of illustrating these sciences in a museum for the people.

GEOGRAPHY AND GEOLOGY.—Some knowledge of the earth and its structure is so essential a preliminary to any acquaintance with natural history, and the working classes have so few opportunities of seeing large maps, globes, or models, that a good series of these should form a part of the museum. In particular, relief-maps, models, and maps to illustrate physical geography and geology, large sections and diagrams, and large globes, should be so exhibited that they could be conveniently examined and studied in detail.

The country around the museum should be shown on a large scale by a model or relief-map, in which the undulations of the ground and the hills, valleys, mountains and streams should be shown on a natural scale of heights, so as not to exaggerate the slopes to three or four times their actual steepness, as is usually done. The more important mountainous regions of our islands, as well as some portions of the Alps and Himalayas, should be shown in the same manner, and all on the same scale, so as to exhibit their true relations to each other. Only in this way can the erroneous ideas derived from maps on different scales and models on an exaggerated vertical scale, be counteracted.

Geological maps and sections should also be exhibited;

and some geological models on a large scale, built up in separate layers to show the actual position of the stratified rocks in some district not far removed from the museum, would give a more accurate notion of the geological structure of the earth than could be obtained by any number of maps and descriptions. In the same way the extent of the ice during the glacial period in the north of England might be shown, by movable layers of white cement or papier-maché, while very large-scale models of portions of the same area might show the results of the ice-action in the rounded and smoothed rocks, the moraines and the deposits of boulder-clays, as well as the erratic blocks coloured to show the parent-rock from which they had been brought by the ice and with lines to show, the course they had travelled. Ice-ground boulders from the moraines and slabs showing glacial striæ and polish should also be exhibited, with special models of remarkable perched blocks and erratics, moraines and extensive striated rocks. Clearly printed labels should explain how these various objects enable us to determine with considerable accuracy the height to which different areas were buried in ice and the directions in which it flowed. By these various illustrations the reality and the marvellous character of the Great Ice Age would be brought home to all intelligent visitors, and would so interest them that they would long to see the proofs of it in some glaciated country which they could reach by a holiday excursion; and the evidences of ice-action are so widely spread over all the northern and western portions of our islands that for most persons the opportunity could without difficulty be found.

MINERALOGY.—A series of the most important and best marked minerals should be exhibited, with tables and diagrams explaining the principles of their classification. Their number should not be too large, and every specimen should be accompanied by a label containing a brief account of all that is most interesting connected with it—its chemical constitution, its affinities, its distribution, and its uses. Combined with this collection there should be a series of specimens illustrating the mode of manufacture of the more important minerals, and their application to the arts and sciences. To give a local interest, all British specimens should be placed on tablets of one distinct colour, so as at once to catch the eye, and enable the student to form some idea of the comparative productiveness of his own country.

BOTANY.—The series of specimens to illustrate the science of botany in a popular museum may be of two kinds: such as show the main facts of plant-structure and classification; and others to teach something of the variety, the distribution, and the uses of plants.

By means of specimens, dissections, drawings, and models, the important radical differences of the great primary divisions of plants—cellular and vascular—acrogenous, endogenous, and exogenous—might be made clearly manifest. Alongside of the drawings and dissections there should be cheap fixed microscopes, showing the main structural differences, thus giving a reality and intensity to the characters which drawings or descriptions alone can never do.

Each of the most important natural orders of plants should next be illustrated by specimens of various kinds. Their structure and essential characters should first be shown, in the same way as the higher groups. Their geographical distribution should be marked out on small maps. Good dried specimens, and, if necessary, drawings or models of flowers or fruit, of the more characteristic and remarkable species, should then be exhibited; and along with these, samples of whatever useful products are derived from them. Where remarkable forest-trees occur in an order, good coloured drawings of them should be shown, as well as longitudinal and cross sections of their wood. In the same, or an adjoining case, specimens or casts of the most important fossil-plants of the same order may be exhibited, illustrating their range backward into past time.

By such a scheme as this, in a comparatively small space and with a small number of specimens, all that is of most importance in the vegetable kingdom would be shown. The attentive observer might learn much of the structure, the forms, and the varied modifications of plants: their classification and affinities; their distribution in space and time; their habits and modes of growth; their uses to savage and to civilized man. An outline of all that is most interesting and instructive in the science would be made visible to the eye and clear to the understanding; and it does not seem too much to expect that, so exhibited, Botany would lose much of its supposed difficulty and repulsiveness, and that many might be thereby induced to devote their leisure to this most useful and attractive study.

In order to assist those who are really students, a separate room should be provided, containing a Herbarium of British plants, as well as one illustrative of the more important exotic families and genera; and to this should be attached a collection of the more useful botanical works.

ZOOLOGY.—Owing to the superior numbers and greater variety of animals, their more complicated structure and more divergent habits, the higher interest that attaches to them, and their greater adaptability for exhibition, this department must always be the most extensive and most important in a Natural History Museum.

The general principles guiding the selection and exhibition of animals are the same as have been applied to plants, subject to many modifications in detail. The great primary divisions, or sub-kingdoms (Vertebrata, Mollusca, &c.), as well as the classes in each sub-kingdom (Mammalia, Birds, &c. and Cephalopoda, Gasteropoda, &c.), should be defined, by means of skeletons and anatomical preparations or models, so as to render their fundamental differences of structure clear and intelligible. At the head of each order (or subdivision of the class) a similar exposition should be made of essential differences of structure; and in every case the function or purpose of these differences should be pointed out by means of clearly-expressed tables and diagrams.

We now come to the specimens of animals to be exhibited, in order to give an adequate idea of their variety and beauty; their strange modifications of form and structure, their singular habits and mode of life, their distribution over the surface of the earth, and their first appearance in past time. To do this effectively requires a mode of exhibition very different from that which has been usually adopted in museums.

Throughout the animal kingdom, at least one or more species of every important family group should be exhibited; and in the larger and more interesting families, one or more species of each genus. The number of specimens is not, however, so important as their quality and the mode of exhibiting them. A few of the more important species in each order, well illustrated by fine and characteristic specimens, would be far better than ten times the number if imperfect, badly prepared, and badly arranged. Let any one look at an artistically mounted group of fine and perfect quadruped or bird skins, which represent the living animals in perfect health and vigour, and by their characteristic attitudes and accessories tell the history of the creature's life and habits; and compare this with the immature, ragged, mangy-looking specimens one often sees in museums, stuck up in stiff and unnatural attitudes, and resembling only mummies or scarecrows. The one is both instructive and pleasing, and we return again and again to gaze upon it with delight. The other is positively repellent, and we feel that we never want to look upon it again.

I consider it therefore an important principle, that in a museum for the people nothing should be exhibited that is not good of its kind, and mounted in the very best manner. Fortunately, specimens of a large number of the most beautiful and extraordinary animals are now exceedingly common, and every well-marked group in nature may be illustrated without having recourse to the rarer and more costly species. Carrying out these views, we should exhibit our animal in such a way as to convey the largest amount of information possible. The male, female, and young should be shown together, the mode of feeding or of capturing its prey, and the most characteristic attitudes and motions, should be indicated; and the accessories should point out the country the species inhabits, or the kind of locality it most frequents. A descriptive tablet should of course give further information; and in the immediate vicinity, specimens showing any remarkable points of its anatomy, and any useful products that are derived from it, should be exhibited.

Each group of this kind would be a study of itself, and should therefore be kept quite distinct and apart from every other group. It should be so placed that it could be seen from several points of view, and every part of each individual composing it closely examined. To encourage such examination and study, seats should be placed conveniently near it—a point strangely overlooked in most museums, where it seems to be taken for granted that visitors will pass on without any desire to linger, or any wish for a more close examination. It would add still further to the interest of these typical groups, if it were clearly shown how much they represented, by giving a list of all the well-known species of the genus or family, with their native country and proportionate size, and indicating, by means of a coloured line, which of them were exhibited in the museum. This would be an excellent and most intelligible guide to the collection itself, and would enable the visitor to judge how far it gave any adequate notion of the variety and exuberance of nature.

It would also, I think, be advisable, that as far as possible each well-marked and important group of any considerable extent should occupy one room or compartment only, where it would be separated from all others, where the attention could be concentrated upon it, and where the extent to which it was illustrated could be seen at a glance. This has not, I believe, been yet attempted in any museum; and when I come to speak of the building arrangements, I will explain how it can be easily managed. In this room, a department would also be devoted to the comparative anatomy of all the more important species and groups exhibited; and a large map should be suspended, showing in some detail their geographical distribution. Here, too, we should place specimens or casts of the fossil remains of the family, with restorations of some of the more important species;

and along with these, diagrams, showing the progress of development of the group throughout past time, as far as yet known.

This mode of attractive and instructive exhibition might be well carried out in the Mammalia, Birds, and Insects; less perfectly in the Reptiles and Fishes, whose colours can hardly be well preserved except in spirits. Even here, however, by using oblong earthenware vessels with glass fronts, instead of the usual bottles, many fishes and marine animals could be exhibited in life-like attitudes and with their colours well preserved. Mollusca may be well illustrated by means of models of the animals, as also may the marine and fresh-water Zoophytes. The more minute and delicate animals should be shown by means of a series of cheap microscopes or large lenses, fixed in suitable positions; and with a careful outline of the animal's history on a tablet or card, close by.

Connected with this, as with the botanical division of the museum, there should be a students' department, to which all should have free access who wished to obtain more detailed knowledge. Here would be preserved, in the most compact and accessible form, in cabinets or boxes, all specimens acquired by the museum and which were not required or were not adapted for exhibition in the popular department. Here, too, should be formed a complete local or British collection of indigenous animals, according to the extent and means of the institution, with the best zoological library of reference that could be obtained. In this department, donations of almost any kind would be acceptable; for, when not required for popular exhibition, an immense number of specimens can be conveniently and systematically arranged in a very limited space, and for purposes of study or for identification of species are almost sure to be of value. One of the greatest evils of most local museums is thus got rid of—the giving offence by refusing donations, or being forced to occupy much valuable space with such as are utterly unfit for popular exhibition.

ETHNOLOGY.—We now come to the last department of our ideal museum, and it is one to which a large or a small proportion of space may be devoted, according to the importance that may be attached to it. In accordance with the plan already sketched out for other departments, the following would be a fair representation of Ethnological science.

The chief well-marked races of man should be illustrated either by life-size models, casts, coloured figures, or by photographs. A corresponding series of their crania should also be shown; and such portions of the skeleton as should exhibit the differences that exist between certain races, as well as those between the lower races and those animals which most nearly approach them. Casts of the best authenticated remains of prehistoric man should also be obtained, and compared with the corresponding parts of existing races.

The arts of mankind should be illustrated by a series, commencing with the rudest flint implements, and passing through those of polished stone, bronze, and iron—showing in every case, along with the works of prehistoric man, those corresponding to them formed by existing savage races. Implements of bone and of horn should follow the same order.

Pottery would furnish a most interesting series. Beginning with the rude forms of prehistoric races, and following with those of modern savages, we should have the strangely-modelled vessels of Peru and of North America, those of Egypt, Assyria, Etruria, Greece, and Rome, as well as the works of China and of mediæval and modern Europe.

The art of sculpture and mode of ornamentation should be traced in like manner, among savage tribes, the Oriental nations, Greece, and Rome, to modern civilization. Works in metal and textile fabrics would admit of similar illustration. Characteristic weapons should also be exhibited; and painting might be traced in broad steps, from the contemporary delineation of a Mammoth up to the animal portraiture of Landseer.

This comprises a series of Ethnological illustrations that need not occupy much space, and would, I think, be eminently instructive. The clothing, the houses, the household utensils, and the weapons of mankind, can hardly be shown with any approach to completeness in a Popular Museum; and many of these objects occupy space quite disproportionate to their intrinsic interest or scientific value. They could in most cases be sufficiently indicated by drawings or models.

Situation and Plan of Museum.

The museum here sketched, beginning with illustrations of the earth and its component minerals, passing through the whole vegetable and animal kingdoms, and culminating in the highest art-products of civilized man, would combine a very wide range of objects with a clearly limited scheme, and would, I believe, well answer to the definition of a Typical Museum of Natural History. Although of such wide scope, it need not necessarily occupy a very large space; and I believe it might be instructively carried out in a building no larger than is devoted to many local museums. This brings me to say a few words on the kind of building best adapted to such an institution as is here sketched out.

In his President's address to the British Association at Norwich, Sir Joseph Hooker made some admirable remarks on the situation of museums. He observed:

"Much of the utility of museums depends on two conditions often strangely overlooked, viz. their situation, and their lighting and interior arrangements. The provincial museum is too often huddled away almost out of sight, in a dark, crowded, dirty thoroughfare, where it pays dear for ground rent, rates, and taxes, and cannot be extended. Such localities are frequented by the townspeople only when on business, and when they consequently have no time for sight seeing. In the evening, or on holidays, when they would visit the museum, they naturally prefer the outskirts of the town to its centre. . . . The museum should be in an open grassed square or park, planted with trees, in the town or its outskirts; a main object being to secure cleanliness, a cheerful aspect, and space for extension. Now, vegetation is the best interceptor of dust, which is injurious to the specimens as well as unsightly, whilst a cheerful aspect, and grass and trees, will attract visitors, and especially families and schools."

Evidently, then, the proper place for the museum is the centre of the park or public garden. This furnishes the largest and cleanest open space, the best light, the purest air, and the readiest access. With how much greater pleasure the workman and his family could spend a day at the museum, if at intervals they could stroll out on to the grass, among flowers and under shady trees, to enjoy the refreshments they had brought with them. They would then return to the building with renewed zest, and would probably escape the fatigue and headache that a day in a museum almost invariably brings on. The public park is the proper locality for the public museum.

In designing museums, architects seem to pay little regard to the special purposes they are intended to fulfil. They often adopt the general arrangement of a church, or the immense galleries and lofty halls of a palace. Now, the main object of a museum-building is to furnish the greatest amount of well lighted space, for the convenient arrangement and exhibition of objects which almost all require to be closely examined. At the same time they should be visible by several persons at once without crowding, and admit of others freely passing by them. None except the very largest specimens should be placed so as to rise higher than seven feet above the floor, so that palatial rooms and extensive galleries, requiring proportionate altitude, are exceedingly wasteful of space, and otherwise ill adapted and unnecessary for the real purposes of a museum. It is true that side-galleries against the walls may be and often are used to utilize the height, but these are almost necessarily narrow, and totally unadapted for the proper exhibition of any but a limited class of objects. By this plan, too, the whole upper-floor space is lost, which is of great importance, because a large proportion of objects are best exhibited on tables or in detached cases.

Following out this view, a simple and economical plan for a museum would seem to be, a series of long rooms or galleries, about thirty-five or forty feet wide, and twelve or fourteen feet high on each floor, the four or five feet below the ceiling on both sides being an almost continuous series of window openings, while at rather wide intervals large windows might descend to within three feet of the floor. At such distances apart as were found most convenient for the arrangement of the collections, movable upright cases might be placed transversely, leaving a central space of about five feet for a continuous passage; and the compartments thus formed might be completed by partitions and doors connecting opposite cases, wherever it was thought advisable to isolate any well-marked group of animals, or other division of the museum. By this means the proportion between wall-cases and floor space might be regulated exactly according to the requirements of each portion of the collection; and abundant light would be obtained for the perfect examination of every specimen.

Two of the great evils of museums are, crowding and distraction. By the crowding of specimens, the effect of each is weakened or destroyed; the eye takes in so many at once that it is continually wandering towards something more strange and beautiful, and there is nothing to concentrate the attention on a special object. Distraction is produced also by the great size of the galleries, and the multiplicity of objects that strike the eye. It is almost impossible for a casual visitor to avoid the desire of continually going on to see what comes next, or wondering what is that bright mass of colour or strange form that catches the eye at the other end of the long gallery. These evils can best be avoided, by keeping, as far as possible, each natural group of objects in a separate room, or a separate compartment of that room—by limiting as much as possible the number of illustrative groups of species, and at the same time making each group as attractive and instructive as possible. The object aimed at should be, to compel attention to each group of specimens. This may be done by making it so interesting or beautiful at first sight as to secure a close examination; by carefully isolating it, so that no other object close by should divide attention with it; and by giving so much information and interesting the mind in so many collateral matters connected with it, as to excite the observant and reflective as well as the emotional faculties.

The general system of arrangement and exhibition here pointed out does not at all depend on the building. It can be applied in any museum, and is, I believe, already to some extent adopted in our best local institutions. It has, however, never yet been carried out systematically; and till this is done, we can form no true estimate of how popular a Natural History Museum may become, or how much it may aid in the great work of national education.

NOTE.

The paper on American Museums, which follows this, was written eighteen years later, in Washington, immediately after a careful study of the two most remarkable museums in the United States. It will be seen, that one of these has carried out most of the suggestions of my early article, and has besides developed the idea of illustrating the Geographical Distribution of Animals which is, so far as I know, entirely new.

CHAPTER II

AMERICAN MUSEUMS

The Museum of Comparative Zoology, Harvard University.

THE immense energy of the American people in all that relates to business, locomotion, and pleasure, is to some extent manifested also in their educational institutions, and in approaching this great and all-important subject they possess some special advantages over ourselves. They are comparatively free from those old-world establishments and customs whose obstructiveness so often paralyzes the efforts of the educational reformer, and their originality of thought and action has thus freer scope; they are not afraid of experiments, and do not hastily condemn a thing because it is new; while, in all they undertake they are determined to have the best or the biggest attainable. Hence it is that colleges and universities for women, schools where the two sexes study together, institutes for the most complete instruction in technology and in all branches of experimental science, and the combination of manual with mental training as part of the regular school course, are to be found in successful operation in various parts of America, though, with rare exceptions, only talked about by us; while in most of the higher schools and colleges science and modern literature take equal rank with those classical and mathematical studies which still hold the first places in Great Britain.

The same originality of conception, and the same desire to attain the best practical results are manifested in some of the great American museums, which now rival, in certain special departments, the long-established national museums of Europe; although there is, of course, as yet, no approach to the vast accumulation of treasures of old-world natural history which is to be found at South Kensington. Notwithstanding the deficiency of material, however, the Harvard Museum is far in advance of ours as an educational institution, whether as regards the general public, the private student, or the specialist; and as it is probably equally in advance of every European museum, some general account of it may be both interesting and instructive, especially to those who have felt themselves bewildered by the countless masses of unorganized specimens exhibited in the vast and often gloomy halls and galleries of our national institution. Let us first consider, briefly, what are the usual defects of great museums, and we shall then be better able to appreciate both what has been aimed at, and what has been effected at Harvard.

Our British Museum, which may be taken as a type of the more extensive institutions of the kind, originated in the bequest of a private collector more than a century ago, and has since aggregated to itself most of the collections made by Government expeditions and explorations, while it has received extensive donations of entire collections made at great expense by wealthy amateurs, and has also of late years made large purchases from professional collectors. Such a museum began, of course, by exhibiting to the public everything it possessed, and with some exceptions this plan has been continued for the larger and more popular groups of animals. Large glazed wall-cases for stuffed quadrupeds and birds, with table cases for shells, starfish, insects, and minerals, were early in use; and while these were gradually improved in quality, size and workmanship, they have continued, till quite recently, to be almost the sole mode of arranging the collection. During the latter half of the present century the accession of fresh specimens has been so extensive that the task of naming, classifying, and cataloguing them has been beyond the power of the curators and their assistants. During the same period, while new species have been so rapidly added to the collections, the labours of anatomists and embryologists have led to constant and important changes in classification, and as it is quite impossible to be continually re-arranging scores of thousands of specimens, it necessarily follows that the museum cases have presented to the public an old and long-exploded arrangement, often quite at variance with the knowledge of the day as to the affinities of the different groups. A still further difficulty has been the overcrowding of the cases, because it was long the custom to exhibit to the public at least one specimen of every new species acquired by the museum; and the difficulty of finding room for the ever-increasing stores has rendered nugatory all attempts to group the specimens in varied ways, so as to convey the maximum of instruction and pleasure to the visitor.

Although the evils of this method of arranging a museum had been pointed out by many writers, notably by Sir Joseph Hooker, in his address as President of the British Association, at Norwich; by myself, in an article in *Macmillan's Magazine*, and by the late Dr. J. E. Gray, keeper of the zoological department of the British Museum, very little radical improvement has been effected in the new building at South Kensington. It is true that many of the large mammalia are more effectually exhibited in costly glazed floor-cases, and there is a great extension of the interesting series illustrating the habits and nesting of British birds; but the great bulk of the collection still consists of the old specimens exhibited in the old way, in an interminable series of over-crowded wall-cases, while any effective presentation on a large scale of the various aspects and problems of natural history, as now understood, is almost as far off as ever.[1] What may be done in this

[1] The late able Director of the Natural History Museum, Sir William Flower, utilized the entrance hall for educational purposes by means of a series of collections illustrating the comparative anatomy of animals, their protective colouring, and the phenomena of mimicry, thus showing a full appreciation of the true objects of a public museum. But the great bulk of the collection is still exhibited in the old manner,

direction, and how a museum should be constructed and arranged so as to combine the maximum of utility with economy of space and of money, will be best shown by an account of the Museum of Comparative Zoology at Harvard.

Origin of the Harvard Museum.

This museum originated in 1858, by a bequest of fifty thousand dollars from Mr. Francis C. Gray of Boston to

FIG. 1.—MUSEUM OF COMPARATIVE ZOOLOGY, CAMBRIDGE, MASS.

Harvard University, for the purpose of establishing a museum of comparative zoology; while the collections it contains were begun by the late Dr. Louis Agassiz, who had been for many years professor of zoology and geology. Owing to the exertions and influence of Professor Agassiz, the legislature of Massachusetts was induced to make a grant of one hundred thousand dollars, while over seventy thousand dollars were subscribed by citizens of Boston "for the purpose of erecting a fire-proof building in

the expense of altering which would be so great that it will probably be long before it is attempted. The building itself, though fine architecturally, is quite unsuited for such an educational museum as that described in the following pages.

Cambridge suitable to receive, to protect, and to exhibit advantageously and freely to all comers, the collection of objects in natural science brought together by Professor Louis Agassiz, with such additions as may hereafter be made thereto."

The general plan of the building and the arrangement of the contents were carried out in accordance with Professor Agassiz's views, while the collections have been greatly increased by the results of the great Thayer expedition to Brazil, by numerous gifts from private collectors, and especially by the many dredging expeditions carried out by Professor Alexander Agassiz, at his own cost, and by extensive purchases of specimens by the same gentleman, who, since his father's death, has occupied the post of curator of the museum, and has devoted his time and large private means to the development of the institution, so as to render it a worthy monument to his father's memory.

Plan of the Building.

The portion of the building already erected is about 280 feet long by 60 feet wide, inside dimensions. This forms the northern wing of the proposed museum, which, when completed, will consist of two such wings, connected by a front of 400 feet. A central partition wall runs lengthways through the building, dividing it into rooms, each 30 feet wide and 40 feet long, except in the centre of the wing, where a projection increases the width to about 70 feet, and this is left open on one floor, forming a room 70 feet by 40 feet for the exhibition of the larger mammalia. The angles connecting the wings with the front of the building are also somewhat larger, and are occupied by laboratories, professors' rooms, staircases, &c. The museum thus consists essentially of rooms of the uniform size of 40 feet by 30 feet, and from 10 to 12 feet high, each being well lighted by a row of windows on one of its sides, forming a building of five floors above the basement. In some of the public rooms the upper floor consists of a gallery, leaving the centre of the room open for the height of two floors.

This it will be seen is very different from what is usually considered the proper style of building for a great museum, which is characterized by lofty halls, magnificent staircases, and enormous galleries; but however grand and effective architecturally these may be, they are quite unsuited to the essential purposes for which a museum is constructed. Let us consider in the first place the supply of well-lighted cases on which the efficiency of a museum so much depends. A large gallery, such as is often seen in great museums, may be 200 feet long and 50 feet wide, giving 500 feet of wall. But if this is divided into five rooms, each 40 feet wide by 50 feet long, we shall have 900 feet of wall, the greater part of which, being opposite the windows and comparatively near to them, will be far better lighted. But the vast gallery must be proportionately lofty and would suffice for two floors of moderately sized rooms, so that, after allowing for the greater number of doors and windows in the smaller rooms, we have an economy of space of at least three to one in favour of the small-room plan, with an even greater proportionate saving of expense, owing to the smaller scale of all the ornaments and fittings.

But the chief advantage of this style of building consists in the facilities which it offers for subdivision and isolation of special groups of objects, and their arrangement so as to illustrate many of the most interesting and instructive problems of natural history. The galleries of a large museum, crowded with specimens arranged in a single series throughout the whole animal kingdom, confuse and distract the observer. As Professor Alexander Agassiz well says in one of his admirable reports as curator:

"The great defect of museums in general is the immense number of articles exhibited compared with the small space taken to explain what is shown. The visitor stands before a case which may be exquisitely arranged and the specimens carefully labelled, yet he does not know, and has no means of finding out, why that case is filled as it is; nothing tells him the purpose for which it is there. The use of general labels and a small number of specimens properly selected to illustrate the labels, would go far towards making a museum intelligible, not only to the average visitor, but often to the professional naturalist." . . . "The advantage, therefore, of comparatively small rooms, intended for a special purpose and for that purpose alone, will overcome at once the objections to be made to large halls where the visitor is lost in the maze of the cases, which, to him, seem placed without purpose and filled only for the sake of not leaving them empty."

Let us now see how these ideas have been carried out at the Harvard Museum.

The first thing to be noticed is the small proportion of the whole building open to the general public, as compared with that devoted to the preservation and study of the bulk of the collections. The existing portion of the building comprises seventy-four rooms, which are apportioned thus:—Ten rooms in the basement are filled with the vast collection of specimens preserved in alcohol, four rooms being occupied by the fishes, and the remainder by reptiles, mammals, birds, crustacea, mollusca, and other invertebrata. Four rooms are devoted to the entomological department. Seventeen rooms are devoted to storage and workrooms for the various departments. Four rooms are occupied by the libraries, and there are also seven laboratories for the students, an aquarium and vivarium, together with a large lecture-room. The remaining rooms are occupied by the curator and the professors in the several departments, except the seventeen exhibition rooms, which alone are open to the public. Before proceeding to describe these it will be well to notice the admirable manner in which space is economized and work facilitated throughout the building.

In all the storage and work rooms the side next the windows is wholly occupied by rows of tables, while the collections are preserved in cases running across the room in parallel rows, from front to back, and reaching from the floor to near the ceiling, with just space enough between them to get at the specimens conveniently. These cases are quite plainly constructed to hold series of drawers or trays of a uniform size and depth, but which will admit drawers of two or three times the depth where the size of the specimens require it. The drawers run loosely in open frames so as to be freely interchangeable, and the whole case is enclosed by well-fitting glass doors. Every drawer or tray is distinctly labelled to show its contents, while a part of the room (or of an adjacent one) is devoted to a library of books specially treating of the groups stored in it. In such a room the student or specialist finds, close at hand, all that he requires, with ample light, and table-room on which to arrange and compare the specimens he may be studying. The general library is arranged on a similar plan, on tiers of shelves running across the room, with just space to walk between them, the cases being enclosed by open wirework doors; and it is a striking proof of the purity of the atmosphere in this suburb of Boston, that there was not the least visible accumulation of dust on books which had not been removed or dusted for several years. The fine trees which surround the museum for some distance no doubt greatly assist in preserving a dust-free atmosphere. The vast number of specimens thus conveniently stored can only be realized by seeing the tiers of cases in room after room, the collection being especially rich in fishes, radiate animals, and marine organisms generally. The advantages of the uniform interchangeable drawers are enormous, as they admit of the growth of the collection in any department and the re-arrangement of the several groups with the least possible amount of labour. To admit of this growth and re-arrangement, a case is here and there left empty; while even the transference of a large part of the collection from one room to another would be effected with ease and rapidity.

Rooms devoted to the Public.

Having thus seen the general character of the arrangements for students and specialists, let us proceed to examine the rooms devoted to the instruction and amusement of the general public. On entering the building the visitor finds opposite to him an open room, over which is painted in large letters, "Synoptic Room—Zoology," and, when inside he finds, on several blank spaces of wall, an intimation that this room contains a

Synopsis, by means of typical examples, of the whole animal kingdom. Two large wall-cases are devoted to the Mammalia; each Order being represented by three or four of its most characteristic forms, from the monotremes and marsupials up to the apes and monkeys. The rodents, for example, are illustrated by means of stuffed specimens and skeletons of an agouti, a porcupine, a rabbit, a squirrel, and a jerboa; the ungulates by a small tapir and a young hippopotamus, always accompanied by their skulls or skeletons. The birds are similarly represented, in one wall-case, by stuffed specimens and skeletons of all the chief types. Another case is filled with reptiles—fine examples of lizards and snakes in spirits, tortoises, alligators, toads, &c., while the fossil forms are shown by a small but very perfect oolitic crocodile, a Plesiosaurus, a beautiful slender lizard of Jurassic age, and a cast of the Pterodactyle with its wings. Another case contains some striking specimens of fishes, both in spirits and stuffed, with their skeletons, as well as some beautifully-preserved fossil fishes. The worms, sponges, and insects are exhibited in three more wall-cases, while the crustacea, radiata, and mollusca occupy two cases in the centre of the room, and over these is suspended a model of a gigantic cuttle-fish twenty feet in diameter.

The special features to be noted in this room are, that its contents and purpose are clearly indicated to every visitor, each group and each specimen being also well and descriptively labelled; that every specimen is good and perfect, well mounted, and beautiful or interesting in itself; that skeletons exhibiting the differences of structure, and fossils exhibiting some of the strange forms of earlier ages of the world, are placed along with stuffed specimens; and, lastly, that the specimens are comparatively few in number, not crowded together, and so arranged and grouped as to show at the same time the wonderfully varied forms of animal life, as well as the unity of type that prevails in each of the great primary groups under very different external forms. We here see that a room of very moderate dimensions is capable of exhibiting all the chief types of form and structure that prevail in the animal kingdom, and of thus teaching some of the most important lessons to be derived from the study of nature. It constitutes of itself a typical museum of animal life, and is more really instructive, as well as more interesting, than many museums which contain ten times the number of specimens and occupy far greater space. It may serve as a model of the kind of room which should form part of every local museum of Natural History, leaving all the remaining available space for the purpose of giving a complete representation of the local fauna and flora.

The visitor now ascends to the third floor, which is wholly devoted to exhibition rooms. He first enters the largest room in the building (about seventy feet by forty), in which is arranged a systematic collection of mammalia, of sufficient extent to exhibit all the chief modifications of form and structure without confusing the spectator by a vast array of closely allied species or badly preserved specimens. A large gallery surrounds this room, devoted to the systematic collection of reptiles, and on a level with this gallery is suspended a very fine skeleton of the Finback whale, about sixty feet long, in a position to be thoroughly inspected both from below and above. The other prominent objects are fine specimens, with skeletons, of the American bison, the giraffe, and the camel; skeletons of each of the five great races of man, and of the three chief types of anthropoid apes; and some casts of the large extinct Australian marsupials in the same cases with the skeletons of their comparatively small modern representatives. Four other rooms, each of the standard size—forty feet by thirty—are devoted to a similar representative collection of birds, fishes, mollusca, and polyps, respectively; while in galleries over these rooms are the collections of batrachians, crustacea, insects and worms, echinoderms, acalephs, and sponges. The most striking objects here are, perhaps, in the bird room, a grand skeleton of the *Dinornis maximus*, as compared with that of an ostrich; in the molluscan room, a model of the giant squid or calamary of Newfoundland, about twenty feet long, with two arms thirty feet in length, their dilated ends armed with powerful suckers; and among the lower forms the beautiful glass models of the sea-anemones and polyps.

This systematic collection differs from the usual collections exhibited in public museums in the following important points. It is strictly limited to a series of typical species, which may be from time to time improved by the substitution of better or more representative specimens, by alterations of arrangement, &c., but which are never to be extended, because they are already quite as numerous as the average intelligence even of well-educated persons can properly understand. The skeletons and fossil types are all exhibited in juxtaposition with the stuffed specimens. Each class of animals is exhibited by itself, with ample explanatory labels to teach the spectator what he is examining, and what are the main peculiarities of the different groups. Of course, in a comparatively new institution, the best and most illustrative species have not always been obtained, or the best and most instructive methods of exhibiting them hit upon. In all these matters improvements will be constantly made, while the space devoted to each class and the number of specimens exhibited will undergo no material alteration.

Illustrations of Geographical Distribution.

We will now pass on to the special feature of the museum and that which is most to be commended, the presentation to the public of the main facts of the geographical distribution of animals. This is done by means of seven rooms, each one devoted to the characteristic animals of one great division of the earth or ocean, which we will now proceed to describe.

Beginning with a room devoted to the North American fauna, we at once note its general characteristics, in its wolves, foxes, bears, and seals; its numerous deer and squirrels, its noble bison now approaching extinction, while a grand skeleton of the mastodon exhibits its most prominent mammal of the immediately preceding age. A closer examination shows us its more special peculiarities, its prong-horn antelope; its raccoon, skunk, and prehensile-tailed porcupine; with its numerous small carnivora and rodents. Several of these types are shown in the illustration (Fig. 2) from a photograph of one side of this room. Among its birds we notice the wild turkey, the black vulture or "turkey-buzzard," the fine ruffed grouse and crested quail, as characteristic features; while among the smaller birds its numerous woodpeckers, its

FIG. 2.—ONE SIDE OF NORTH AMERICAN ROOM.

tyrants, and its prettily coloured thrushes, warblers, and finches are most prominent. Its reptiles and amphibia are characterized by numerous fresh-water tortoises, many curious lizards, the rattlesnakes, and other striking forms; many varieties of frogs, some of large size; and its very curious and interesting salamanders and other tailed batrachia. Its fishes are rich in fine and characteristic forms, and we notice specimens of the siluroid cat-fish, the garpike, and the mud-fish, belonging to the extremely ancient type of the ganoids, the huge devil-fish of South Carolina, one of the most gigantic of the rays; with many others. Among its shells, the fresh-water Uniodæ are prominent; and, in the insect collection, the number of large and brilliantly-coloured butterflies is very striking as compared with those of Europe.

The next room takes us into South America, and here we are at once struck with many remarkable contrasts. First, there is the comparative scarcity of large mammalia, the higher groups being represented by the llama, the tapir, a few small deer, and the jaguar, which is common to North America; while such low and ancient types as the sloths, ant-eaters, and armadillos abound, together with an unusual number and variety of large rodents, and many peculiar forms of monkeys. Some of these are shown in the accompanying illustration (Fig. 3) from a photograph taken in a corner of this room. The extinct mammals are well represented by a fine skeleton of the Megatherium or giant sloth of the Pampas. The birds exhibit a wonderful richness and variety, with a similar preponderance of low types of organization. The blue and claret-coloured chatterers, the many-coloured little manikins, the strange white bell-birds, the wonderfully-crested umbrella-bird of the Upper Amazonian islands, the brilliant crested cock-of-the-rock, and the innumerable tyrants, bush-shrikes, and ant-thrushes, all belong to a type of perching birds in which the peculiar singing-muscles of the larynx have not been developed, and which are but scantily represented in any other part of the world. The metallic trogons, with yellow or rosy breasts; the ungainly but finely-coloured toucans, with their huge but exquisitely-tinted bills; the green and gold jacamars; as well as the hundreds of species of those winged gems, the humming-birds, represent a yet lower and more archaic type of bird life nowhere so strongly developed as in this marvellous continent. The beautiful crested curassows are also a low form perhaps allied to the Australian mound-makers. Reptile life is abundantly represented, but except, perhaps, the iguanas, there are none to strike the ordinary observer as being especially characteristic. The insects, however, at once attract attention; the grand blue morpho butterflies; the exquisite catagrammas, with their fantastic markings beneath; the immense variety of the Heliconoid butterflies, with their elongated wings and antennæ and striking colouration, and the wonderful variety and beauty

FIG. 3.—SOUTH AMERICAN FAUNA.

of the little Erycinidæ, a family almost confined to South America. Among other insects we notice the strangely-formed and fantastically-coloured harlequin-beetle; the huge rhinoceros-beetle; the large lanthorn-fly, and many others, as being equally peculiar.

Passing next to the room which illustrates the opposite continent of Africa, we are presented with a contrast in the forms of life at once marvellous and interesting. From the poorest continent in mammals we pass to the richest, our eyes being at once greeted by the elephant, rhinoceros, and hippopotamus, the buffalo, the giraffe, and the zebra, with a vast array of antelopes, the lion, and the great man-like apes. The most cursory inspection of these two rooms will teach the visitor a lesson in natural history that he will not learn by a dozen visits to our great national storehouse at South Kensington—the lesson that each continent has its peculiar forms of life, and that the greatest similarity in geographical position and climate may be accompanied by a complete diversity in the animal inhabitants.

When we examine the birds, the difference between the two continents is almost equally great, although not so conspicuous to any one but an ornithologist. The great bulk of the South American groups have no representatives whatever in Africa. Instead of toucans we have hornbills and turacos; instead of humming-birds we have the totally different group of sunbirds; instead of the tyrants, hang-nests, and chatterers, we have flycatchers, starlings, and orioles; instead of bush-shrikes and ant-thrushes we have true shrikes and caterpillar-catchers—in almost every case a high grade of organization in Africa in place of the low grade in South America. Passing over the reptiles and fishes, as not presenting forms sufficiently well known or whose external characteristics are sufficiently distinctive, we find in the insects equally marked differences. The African butterflies have a peculiar style of form and colouring, distinguishing them from those of most other parts of the world, sober greens and blues or rich orange browns being common. The Heliconidæ of America are here replaced by the allied but distinct subfamily of the Acræidæ, while among beetles the huge goliaths and the monstrous tiger-beetles are altogether peculiar.

The next room we enter is the Indian, or Indo-Malayan; and here the scene again changes, though not so radically as we found to be the case in passing from South America to Africa. There are still many great mammalia, but of distinct characteristic forms; the tiger replaces the lion, deer and bears are abundant groups, which are entirely unknown in Africa, the orangs and the long-armed apes replace the gorilla and the chimpanzee, true wild cattle are found as well as buffaloes, while the musk-deer, the strange flying lemur, and the gigantic fox-bats are characteristic forms unknown elsewhere. Among birds, the most typical group is that of the pheasants, which reach their highest development in the peacock and many-eyed argus; the hornbills are of a different type and more varied forms than those of Africa; the cuckoo family is abundant and varied, while the gorgeously-coloured broadbills and ground-thrushes belong to the low type of perchers so abundant in South America. Among the insect tribes we especially notice the glorious yellow and green-winged ornithopteræ, the princes of the butterfly world; the huge atlas moth, the largest of lepidoptera and probably the largest-winged of all insects; the three-horned rhinoceros beetle; the grand buprestidæ, and the strange leaf-insects of Java and Ceylon.

We now enter the room devoted to the Europe-Siberian fauna, the chief object in it being a fine skeleton of the great Irish elk, while its most representative living mammals are deer, wolves, wild boars, bears, wild oxen, wild sheep and goats, the chamois, and some peculiar forms of antelopes. Its most prominent birds are its partridges, grouse, bustards and pheasants, but it is deficient in gay-coloured perching-birds as compared with all other regions. Its reptiles are few and insignificant, as are its fresh-water fishes. In insects its chief characteristic is the abundance of beetles of the genus Carabus, its dung-feeding lamelliscorns and its fritillary butterflies.

Lastly, the Australian room brings us into an altogether distinct world of life. All the conspicuous mammals are of the marsupial type, from the giant kangaroos down to the diminutive kangaroo-rats and flying-opossums; and these comprise representatives of all the chief types of the higher mammalia in the form of herbivorous, carnivorous, rodent, and insectivorous marsupials. Among the birds we have such peculiar forms as the emu, the mound-making brush-turkeys, the lyre-birds and bower-birds, the birds of paradise, the cockatoos and lories, the brush-tongued honey-suckers, and the varied and beautiful forms of the kingfishers and fruit-pigeons—an assemblage of peculiar and brilliant developments of bird life hardly to be equalled except in South America. The recently extinct forms—the colossal kangaroos and wombats of Australia, and the huge dinornis of New Zealand—were equally remarkable.

The six rooms now briefly described complete the exposition of the geographical distribution of land animals, and the visitor who makes himself thoroughly acquainted with their contents by repeated inspection and comparison, will obtain a conception of the general aspects of animal life in each of the great divisions of the globe which hardly any amount of reading or of visits to ordinary museums would give him. It is a remarkable thing that so interesting and instructive a mode of arranging a museum, and one so eminently calculated to impress and educate the general public, has never been adopted in any of the great collections of Europe, in all of which ample materials exist for the purpose. It is a striking proof of the want of any clear perception of the true uses and functions of museums that pervade the governing bodies of such institutions, and also perhaps, of the deadening influence of routine and red-tapeism in rendering any such radical change as this almost impossible. But we have yet to see some further applications of the same principle at the Harvard Museum.

Two rooms not yet opened to the public are being prepared to illustrate the fauna of the Atlantic and Pacific Oceans respectively. Here will be exhibited specimens of the peculiar forms of whales and porpoises, seals, walruses, and sea-lions, the oceanic birds, the fishes and mollusca characteristic of each ocean, while separate cases will illustrate the land fauna of the more remarkable of its oceanic islands. On my suggesting to Professor Agassiz that the northern and southern portions of these faunas were usually distinct, he thought that these might be perhaps exhibited at opposite ends of each room.

By the kindness of Prof. Agassiz and of Mr. Samuel Henshaw, his representative at the museum, I have been able to give a view of a corner of the European room, showing the goats, deers, bears, rabbits, and other characteristic

animals; a similar corner of the South American r
showing the llamas, armadillos, and sloths; and one si
the North American room with its bison, elk, &c. (Figs.
but any general picture of the assemblage of anima

FIG. 4.—THE EUROPEAN FAUNA.

impossible in a photograph, owing to the distribution of
separate cases, which stand out between the windows.

It might perhaps be better in any future attempt
show the geographical distribution of animals in a muse
that a different method should be adopted. The two lon
sides of a large room lighted from above or from the e

should have cases about eight or ten feet wide, wholly glazed in front with as few bars as possible. If the floor of the cases were raised two feet above the floor of the room, sheets of glass eight or ten feet high might be fitted edge to edge, the joints being filled with Canada balsam, or some similar material, to render them-dust tight, the openings to the cases being at the back or the two ends. Each case should represent a scene characteristic of the Region represented. In that illustrating the Neotropical Region for instance, one case would represent a Brazilian forest, with, say, a tapir, some agoutis, ant-eaters, and sloths, all in natural attitudes and surroundings. A troop of spider-monkeys; some macaws, toucans, chatterers, trogons, and curassows, would be seen perched upon the branches; an iguana, some ground lizards, and the great harlequin and elephant beetles would also appear in the foreground; while sitting upon leaves or on the ground, or flying in the air, would be a score or two of the most characteristic butterflies—the blue morphos, the lovely catagrammas, the brilliant heliconii and ithomiæ, &c. There should be no crowding, no attempt to show too many species, but just that amount of characteristic life and that variety of form, structure, and colour, which might, under the most favourable conditions, be witnessed by a concealed observer.

The other side of the same room might be fitted to show the south temperate plains and the highlands of the Andes; and here would be seen the llamas and huanacos, the rheas, the condor, the vischaca, the chinchilla, the crested screamer, the puma, armadillos, and many humming-birds; with the characteristic vegetation and insects of the district.

If the six great regions of the globe were thus illustrated in the best possible manner, in some cases two rooms being devoted to a region, such a museum would be at once so attractive and so instructive, that comparatively little space would be required for a general collection to be exhibited to the public. In fact what is termed a typical collection, illustrating all the more important families, would be quite sufficient.

Extinct Forms of Life.

Four other rooms are also being prepared to exhibit the geological succession of animal life. In the first room the visitor will find illustrations of the mollusca, the trilobites, and the strange and often gigantic fishes of the palæozoic era down to the Devonian age. The next will contain the same groups as exhibited in the Carboniferous period, with the earliest forms of amphibia and reptiles, and their later developments in the Jurassic period when the first small mammals made their appearance. Here will be exhibited models of the huge reptile (Atlantosaurus) discovered by Professor Marsh, by far the largest of all terrestrial animals. Then will come a room devoted to the Cretaceous deposits, the wonderful giant Ammonites, and the abundant reptilian and bird forms which have been discovered in America. The last room of the series will be devoted to the Tertiary deposits, and will show the many curious lines of modification by which our most highly-specialized animals have been developed. If some of the preceding rooms contain the most marvellous products of remote ages, here assuredly will be the culminating point of interest in seeing the curious changes of form by which our existing cattle and horses, sheep, deer, and pigs, our wolves, bears, and lions, have been gradually modified from fewer and more generalized ancestral types.

Of all the great improvements in public museum-arrangement which we owe to the late Professor Agassiz and his son, there is none so valuable as this. Let any one walk along the vast palæontological gallery at South Kensington, and note the crowded heaps of detached bones and jaws and teeth of fossil elephants and other animals, all set up in costly mahogany and glass cases for the public to stare at, with here and there a more complete specimen or a restoration; but all crowded together in one vast confusing series from which no clear ideas can possibly be obtained, except that numbers of strange animals, which are now extinct, did once live upon the globe, and he will certainly admit the imperfections of this mode of exhibition, as profitless and puzzling to the general public as it is wasteful of valuable space and inconvenient to the student or the specialist. In a proper system of arrangement all these fragments would be treated as material for study, not as specimens to be exhibited to the public. Casts and models of bones and other fossils can now be cheaply and easily made of paper, which when carefully coloured are to the ordinary eye indistinguishable from the specimen itself; and the materials already existing in the museums of Europe and America are so vast that nearly complete skeletons can be obtained of a great number of the more interesting extinct animals.

What ought to be exhibited to the public is a typical series of such skeletons or models, so arranged as to show the progression of forms and the evolution of the more specialized types as we advance from the earlier to the later geological periods. Instead of one huge gallery, a series of moderate-sized rooms should be constructed, each to illustrate one geological epoch, with subsidiary rooms where necessary to show the successive modifications which each class or order of animals has undergone. Where only fragments of an important type have been obtained, these might be exhibited with an explanation of *why* they are important, and an outline drawing showing the probable form and size of the entire animal. A museum of this kind, utilising the palæontological treasures of the whole world, would be of surpassing interest, and would probably exceed in attractiveness and popularity all existing museums. It would offer scope for a variety of groupings of extinct and living animals, calculated, as Professor Agassiz intended his museum to do, "to illustrate the history of creation, as far as the present state of scientific knowledge reveals that history." It is surely an anomaly that the naturalist who was most opposed to the theory of evolution should be the first to arrange his museum in such a way as best to illustrate that theory, while in the land of Darwin no step has been taken to escape from the monotonous routine of one great systematic series of crowded specimens, arranged in lofty halls and palatial galleries, which may excite wonder, but which are calculated to teach no definite lesson.

A grand opportunity is now afforded for a man of great wealth, who wishes to do something for the intellectual advancement of the masses. Let him build and endow a "Museum of Comparative Palæontology," for the purpose of carrying out Agassiz's idea on a scale worthy of it. Such a museum, built on the plan of that at Harvard, but with rooms of a larger average size, would easily accommodate the far larger number of spectators that would certainly visit it, and would tend more than anything else could do to raise the sciences of palæontology and zoology in popular estimation, and to clear away the clouds of misunderstanding which still enshroud the grand theory of evolution. It would enable the general public to appreciate for the first time the marvellous story presented by the sequence of animal life upon the globe, and would at once instruct and elevate the mind by exhibiting the comparative insignificance of existing animals, in variety and often in size, to those which have preceded them, and by demonstrating the innumerable and startling changes of the forms of life upon the globe during the long series of ages which preceded the advent of man. Such a museum would certainly become the most popular, as it would be the most instructive, of all the great scientific exhibitions yet established, while its founder would secure to himself an amount of honourable fame rarely accorded to those who devote money to public purposes.

Museums of American Pre-historic Archæology.

Few Englishmen have any adequate idea of the present condition of the study of prehistoric archæology in America, or are at all aware of the vast extent and interesting character of the collections which illustrate the early history of that continent. The recognition of the antiquity of man in Europe, and the establishment of the successive periods characterized by the palæolithic and neolithic implements, are events within the memory of many of us; while even at the present day the existence of man before the glacial period is vehemently denied by some geologists, and all the evidence brought forward to establish the fact is sought to be explained away with as much misspent ingenuity as was exerted in the case of the early finds of McEnery and Boucher de Perthes. Notwithstanding that almost every fact of the early discoveries has now been proved to have been a reality, every new fact which goes to show that man is only a little older than we have hitherto supposed, is still received with incredulity or neglect, although it is universally admitted that not only is there no antecedent improbability in these new discoveries, but that the theory of evolution if it is worth anything, demands that the origin of man be placed very far back in the tertiary period.

While such has been the frame of mind with which each new discovery in Europe has been met, it was natural that comparative ignorance should prevail as to the course of discovery across the Atlantic; more especially as there was a common notion that America was really a new world as regards man, and that except a few puzzling facts, like the ruined cities of Central America, Mexico, and Peru, its native races were comparatively recent immigrants from Asia by the north-western route, and that their prehistoric history was brief, simple, and altogether unimportant as compared with that of early Europe. The facts, however, point to an exactly opposite conclusion, the prehistoric remains of North America being at least equally abundant, equally varied, and offering as numerous and as interesting problems for solution as are met with in the European continent. In no other part of the world has the use of stone for all the purposes of savage and barbarous life been so extensive and so highly elaborated; nowhere else has a race which has many features in common, and which was long held to be perfectly homogeneous, been found to present more diversities in customs, in arts, in language, and in physical characteristics.

The study of prehistoric archæology and of man's antiquity has run almost a parallel course in America and in Europe. The early discoveries of Schmerling and Godwin-Austen compare with those of the Natchez human bones in the Mississippi loess, and of arrowheads, pottery, and burnt wood in close connection with skeletons of the mastodon. The kitchen-middens of Denmark are far less extensive than the shell-heaps of New England, Florida, and Alaska; while the discoveries in the lake-dwellings, peat-bogs, and tumuli may be compared with the still more extensive finds in the "mounds" of the great valley of the Mississippi. Even the mysterious structures at Stonehenge, on Dartmoor, and in Brittany, are not more mysterious than some of the animal mounds or extensive systems of earthworks of Ohio and Illinois, nor offer more difficult problems than the sculptures and hieroglyphics of Central American and Mexican temples.

Before giving a brief sketch of the varied specimens which illustrate the history of early man in America, it may be well to state the character of the museums in which they may be best studied—the Peabody Museum of American Archæology and Ethnology at Cambridge, Massachusetts, and the Museum of Prehistoric Archæology at the Smithsonian Institution in Washington. These two museums illustrate very distinct methods of arrangement, each of which has its advantages. At Cambridge the collections are arranged according to localities or areas. Everything found in one mound, or group of mounds, is kept together, so as to illustrate, as far as possible, the life history of the constructors. Surface finds are grouped according to States or districts; the instruments, bones, shells, &c., of the shell-heaps are similarly arranged; the same is done with objects found in caves, in stone-graves, in the old Pueblo villages, &c. In the words of the curator, Mr. F. W. Putnam:

"A natural classification has been attempted, grouping together objects belonging to each people. By this method is brought out the ethnological value of every object in the museum, so that in the mind of the student each is put into the great mosaic of human history. Thus it is that throughout the arrangement of the museum the chip of stone and the polished instrument are side by side. There is no forcing into line, no selection of material, in order to illustrate a theory. Every object falls into its place with its own associates, and tells its part of the story of the efforts of man and the results which he has reached at different times and in different places. By this method of arrangement nothing is forced, and misconception is impossible. Separate the objects and classify them by their kind, independently of their source, and the result is simply a series of collections illustrating the development of the *arts* of man; and although such collections will find appropriate places in a museum like this, they should be secondary to the main collection, and be formed of duplicate material. Upon these principles and methods the arrangement of the collections in the present building has been carried on."

The great collection in the National Museum at Washington, on the other hand, is arranged to illustrate the development of prehistoric industry and arts. First we have cases filled with the rudest chipped implements, many quite as rude as the palæolithic flints of Europe, and closely resembling them in form. These are of the most varied materials—calcite, chalcedony, obsidian, quartzite, slate, sandstone, or trap. Many are scrapers, rude knives, spears, &c., and come from every part of the continent. In other cases we find leaf-shaped, arrow-shaped, and spear-shaped stones; passing on successively to all the varied uses to which stone has been applied, through a long gallery containing probably a hundred large floor-cases. Besides this progressive series there are some special cases containing the whole of the contents of certain mounds or graves, or the weapons and implements from some specially interesting locality or island. This method of arrangement has the advantage of enabling a visitor more easily to appreciate the endless variety in the forms of each class of articles, and to compare the development of the stone age in America with that of Europe. As in the case of zoological collections, a great national museum should combine both methods of arrangement; and it is therefore fortunate that in the present progressive condition of the study the two great museums of American prehistoric archæology should have adopted different systems.

The first thing that strikes the visitor is the immense number and variety of forms of stone weapons, implements, and ornaments, far exceeding anything known in Europe. First we have ovoid or leaf-shaped stones, often of flint quartzite or other hard material, and probably used as scrapers. Fig. 5, b, is a common type, often shorter and rounded with the broad end worked to a fine edge. These scrapers are usually better shaped and more carefully worked than the flint scrapers so commonly found in England.

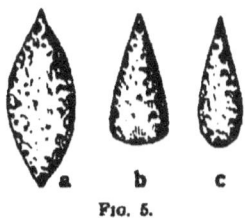

Fig. 5.

Diverging from these we have a great variety of shapes evidently adapted for special purposes, such as scraping down the hafts of spears and arrows, as in the strange forms shown in Fig. 6, b, c, and d. Borers, probably used for making holes in skins for lacing them together, are shown in Fig. 6, a, and these too vary greatly, some being very slender and delicate, and all are formed of flint, quartzite, or other hard stone. It may be noted that very rude tools and weapons are found in certain deposits in America as in Europe, some in New England and others in Utah closely resembling the palæolithic implements of the high gravels of our own country and France.

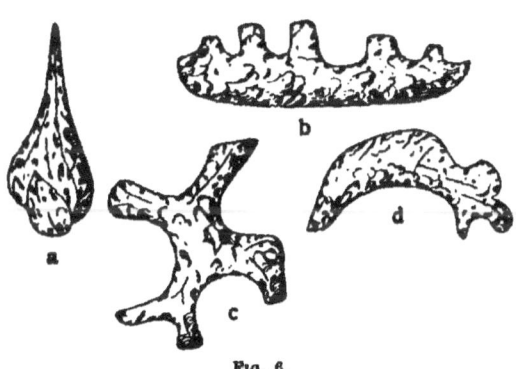

Fig. 6.

Passing on to the more decided weapons we notice a number of very distinct types. Fig. 7, h, shows a transition from the well-formed scraper to the spear-head. This takes a more definite form in g, and j; while in i, and k, we have the base narrowed and elongated evidently to fix into a notched haft and to fasten by a binding of thongs. Another development is into the triangular thin arrow-head a, sometimes with the angles elongated and

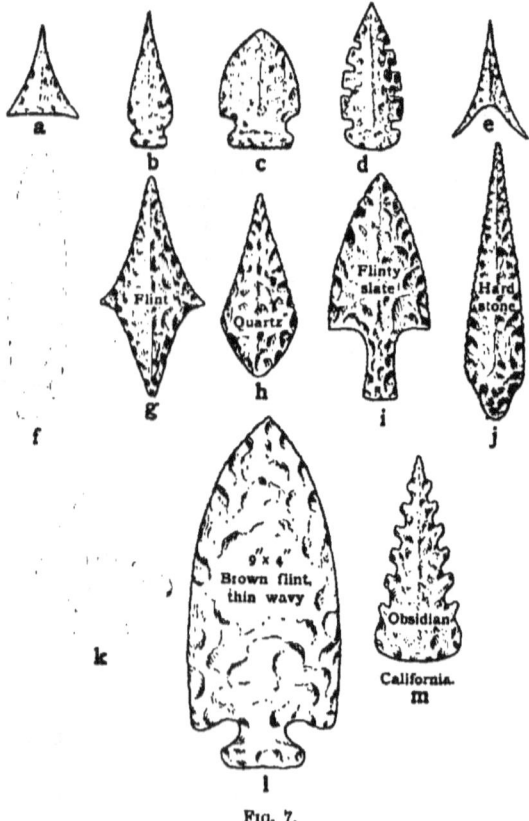

FIG. 7.

delicately, pointed, as in e. These carry us on to the more definite spear head with a notched base, as in b, c, f, and l, an improvement which would render the fastening to the shaft more secure. Many of these are large and beautiful weapons, like that here figured (1), which is nine inches long by four inches wide.

There remain two other weapons figured above, d and m, characterized by the deep square serrations. These are formed of obsidian and are from California, to which State they appear to be confined. In a Californian magazine, *The Land of Sunshine* (Oct. 1899), Mr. H. C. Meredith gives figures of a number of very remarkable forms of these deeply serrated obsidian weapons, some curved like a broad bladed knife, others shaped like an Australian boomerang, while some are bent at right angles. They are about two or three inches long, with one end formed to be fastened to a handle, and would then be a very formidable weapon to throw in an enemy's face. There are also some fine obsidian spear-heads, four and a half inches long. Mr. Barr, of Stockton, has a very fine collection of these obsidian weapons of varied shapes, especially the curious "curves." They have been found only in the central valley of California.

From these we pass to more perfectly formed barbed arrow and spear-heads, such as the three examples in Fig. 8, the exceedingly broad and delicate type on the right being from California. These very small arrow-heads are often formed of agate, jasper, cornelian or other gem-like stones.

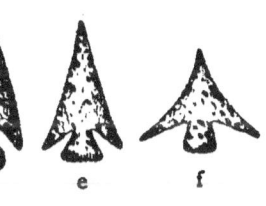

Fig. 8.

Among the very interesting forms of arrow-head are those represented in Fig. 9, in which the two faces are bevelled off on opposite sides, so as to produce the effect of a slight screw or twist. In some cases the whole arrow-head appears to be twisted, probably owing to a favourable grain in the stone. These specimens have been found in the ancient mounds of Wisconsin, Ohio, Georgia, and Alabama, and seem to have been designed for the purpose of giving a rotation to the arrow about its axis, thus counteracting any slight curvature in the shaft and producing a straighter flight.

The next set shown in Fig. 10 are flat and finely chipped tools of uncertain use. The upper one, a, was of a black flinty material from Oregon and was fourteen inches long,

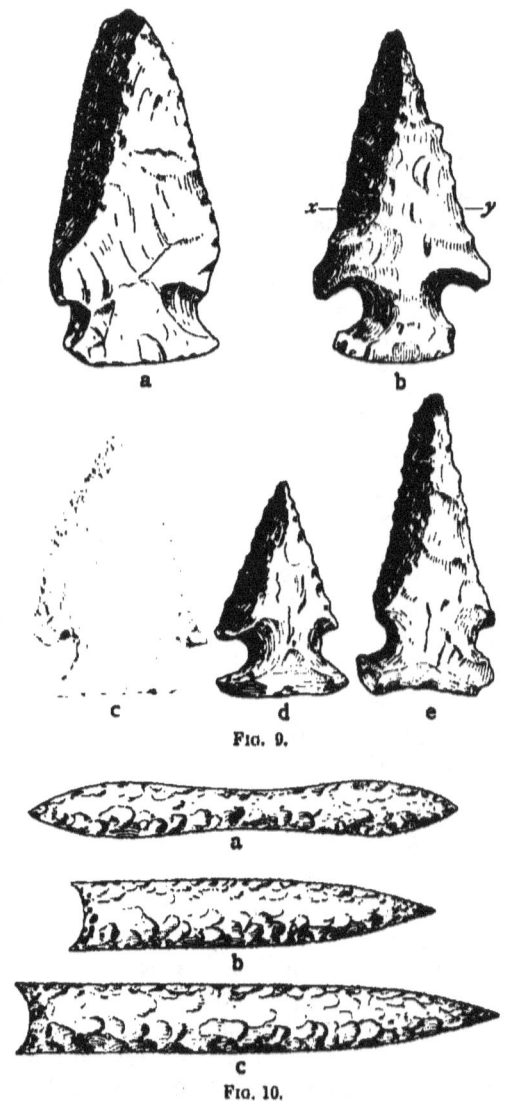

Fig. 9.

Fig. 10.

but there was a still slenderer one of the same type that was no less than two feet long. The others, b and c, were probably the heads of ceremonial spears and were also from a foot to two feet long. They were found in California.

Fig. 11 shows a series of flat discs of flint or shale of

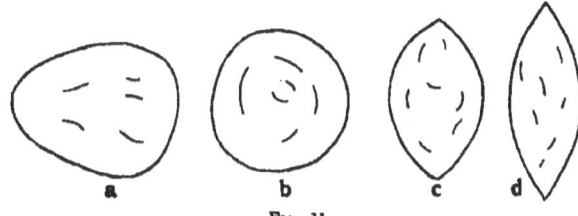

Fig. 11.

various round oval or elongate shapes, perhaps used in some game.

The set of seven articles in Fig. 12 have the common feature of being nearly circular in section, and having a constriction at one end or projection at both ends, so as to facilitate suspension by a string. They are very numerous,

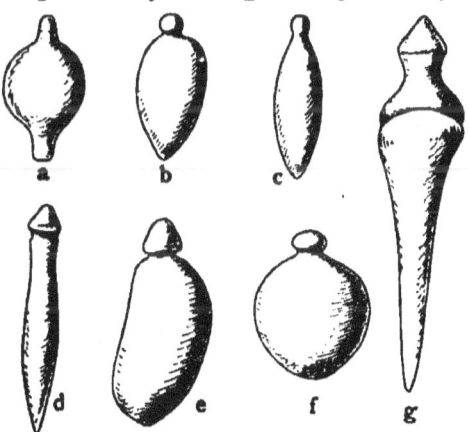

Fig. 12.

of many curious forms, often very rude, yet adapted for the same purpose, and they were found in many localities. Some may have been sinkers for nets or fishing lines, while the more symmetrical may be weights used in twisting thread.

Numbers of flat flint hoes have been found of considerable size, having a projection like that of the best-formed arrow-heads for fastening to a handle; and some of these have the outer edge highly polished, evidently by use as a hoe in fine alluvial soil. Fig. 14, b, shows one of these hoes or spades about fifteen inches long with a hole near the base evidently for the purpose of tying it more firmly to the handle; while Fig. 13 is a curious tool notched on each side at the base, perhaps for the purpose of cutting through roots, and with a long projection, giving a very firm attachment to the handle. This was formed of a tough black stone, perhaps basalt, and was found in Louisiana.

FIG. 13.

Among other tools we find numbers of hammers, pounders, grinders, pestles and mortars, rubbing stones, crushers, club-heads, weight-stones for diggers, as now used by some Indians to assist in piercing the earth, and many others of unknown use. Fig. 14, a, and c, are some kind of knife or cutting tool, the lower edge being finely ground, and the base fastened to a wooden handle to allow of great pressure. Fig. 15, a, and b, shows two small oval pebbles carefully grooved round the longer diameter, to be used apparently as well secured weights.

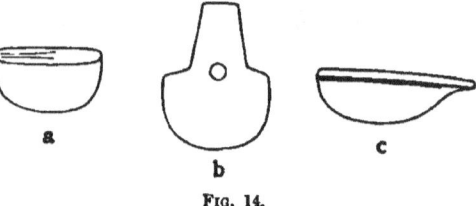

FIG. 14.

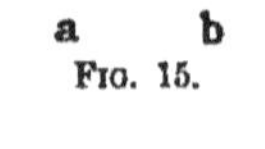

FIG. 15.

The great variety of cutting tools, such as axes, adzes, chisels and gouges, are often beautifully worked, of the hardest fine-grained rocks, such as syenite or hæmatite, and are sometimes highly polished. The gouges especially are often deeply hollowed out and ground to a perfect cutting edge.

The tools and household implements here noted show

that those who made them were an agricultural people, cultivating the ground largely and with skill, as did most of the tribes on the eastern coast and in California when Europeans first encountered them. We have heard so much of late years of the warlike and nomad character of the American Indian tribes that we are apt to forget that many of the more peaceful agricultural peoples have been exterminated. Yet the early settlers in the north-eastern States were often saved from famine by the stores of maize of the peaceful Indians.

The extensive use of roots, nuts, acorns, maize, &c. as food required facilities for cracking, crushing, or grinding; and hence some of the most common implements, both of modern Indian tribes and throughout all prehistoric ages, are hammers, grinders, pestles, and mortars, of varied sizes, forms, and workmanship. The pounding, crushing, and grinding stones are of very varied forms, from the unworked pebble up to the most elaborate grinder with a broad handle, something like a tailor's iron, but carved out of solid stone. Corresponding to these are the grinding-stones and mortars, of equally varied forms and sizes. Some are flat, some slightly hollowed; some have numerous small pits or cups in them, probably to hold nuts of various kinds, so as to prevent them from flying away when being cracked. From these we pass on gradually to shallow basins and large deep mortars, some of the latter found in California being a foot or eighteen inches wide, and having corresponding stone pestles, some of which are two and a half feet long.

We now come to a series of implements or articles for domestic use of varied form and size, but often involving a large amount of labour.

In Fig. 16, we have representations of five types of stone spoons or cups with handles, some of the former are only two inches diameter, while the latter are often six or eight inches. These are very nicely finished. Others were somewhat ruder, and there are many of larger size used as plates, or bowls, up to ten or twelve inches across. These are all from California, where stone work attained a high degree of perfection.

In Fig. 17 we have examples of curious boat- or cup-shaped stones of doubtful use, but c, and d, have grooves with a hole through the stone near each end, suggesting a thread reel or a shuttle for weaving.

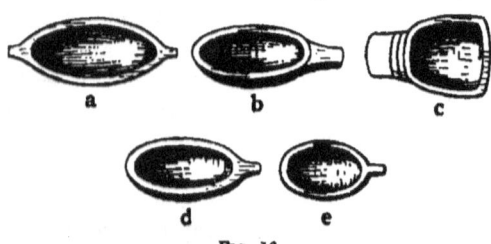

Fig. 16.

In Fig. 18 we have ten different forms of flat smooth stones symmetrical in shape, and with one or two holes through them. These look much as if used for winding thread for weaving, and perhaps the different shapes may

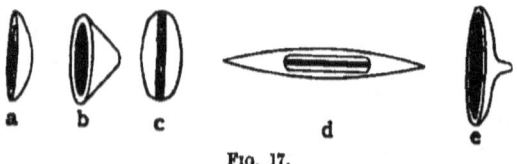

Fig. 17.

have been used for different colours, so that patterns could be woven correctly in a dark hut or at night, the colours being known by the shape of the winder. These indicate a considerable amount of skill in the weaving of textile

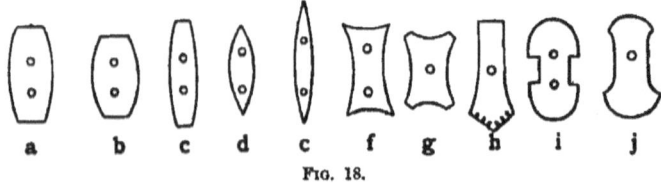

Fig. 18.

fabrics, remains of which have been found in some of the mounds.

Besides these varied implements and weapons, whose uses are known from observation of modern savages, or may be fairly conjectured, there are many others which appear to be either personal ornaments or objects used in favourite games, or for ceremonial purposes. Of the former class are small stones of various forms, and more or less decorated with pits or incised lines, some of which were probably ear ornaments, others gorgets (Fig. 19). Great numbers of stone discs have been found, of various sizes, from two or three up to eight inches in diameter, some of which are worked beautifully true and smooth. They are usually hollowed on one or both surfaces, and many have a central perforation. Some are formed of hard quartzite, three or four inches diameter, and must have required an enormous amount of labour to cut and polish them without a lathe or any of the appliances of the modern lapidary.

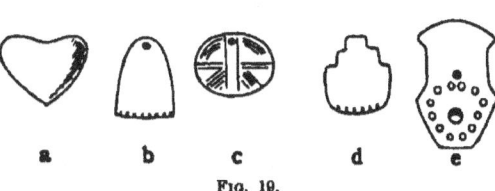

FIG. 19.

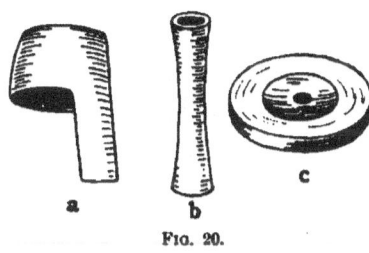

FIG. 20.

These were probably used in a game called chungke, practised among some Indian tribes, and resembling a combination of bowls and spear-throwing; and the Creek Indians had chungke yards kept smooth and level on purpose for the game (Fig. 20 c). The supposed ceremonial stones have been found from Connecticut to Florida, mostly in mounds, and are of very varied symmetrical forms, and all have a

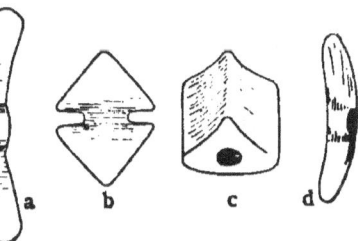

FIG. 21.

central hole sufficiently large to admit a small stick. One has a form closely resembling the "key" of the maple (Fig. 21 a), others are cylindrical, but slightly curved (Fig. 21 d), some are like triangles joined by a narrow connecting bar at the centre of their opposite bases (Fig. 21 b); others, again, like the longitudinal section of a dice-box (Fig. 22 a), with many more, only a few of which are here figured.

Sculptured objects are numerous, and some have considerable artistic merit. Among the modern Indians the Sioux carve animal and human figures on pipes of catlinite or red pipe stone, some of which are well executed and of fanciful design. The Haida Indians, of Queen Charlotte Island, are celebrated for their skilful carving

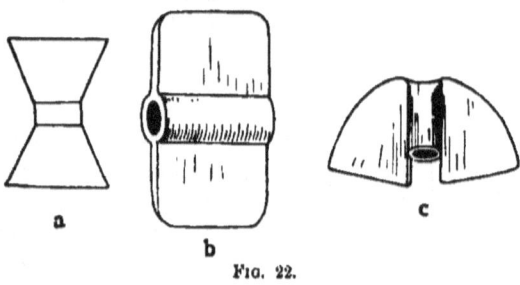

Fig. 22.

in wood and slate, the latter being very elaborate, highly polished, and having the appearance of black marble. These are grotesquely idealised into more or less symmetrical designs, and bear a considerable resemblance to some of the Mexican sculptures, while in language and physiognomy these tribes differ from all the Indians of the adjacent regions. It is, however, in the mounds that the greatest variety of sculptures have been found, and among them are some of a very remarkable character.

The pipes from the mounds of Ohio and Illinois are often carved into the form of human heads, some of which have Indian characteristics, while others seem quite distinct. Animal forms are also abundant, and among them are seen the dog, bear, otter, prairie-dog, beaver, tortoise, frog, serpent, hawk, heron, coot, duck, woodpecker, owl, &c.

The supposed tropical animals carved by the mound builders, such as the manatee and the parrot, are errors of identification. There is, however, a curious carving repre-

Fig. 23.

senting some form of llama or camel found on the site of a mound in Ohio.[1] (Fig. 23, a.) Many carvings of animals,

[1] The history of this remarkable piece of sculpture is as follows. Mr. J. F. Snyder, M.D., purchased it along with a few other prehistoric relics, flint arrow points, stone axes, &c., of a typical backwoodsman, who was migrating from Marion Co., Ohio, to the west, with his family and household goods. The man was rough and uneducated, and profoundly ignorant of archæology, but attached some value to the specimens, partly because others did, but chiefly because he had himself found them. He stated that he had ploughed up the llama, together with many Indian bones, and two of the stone axes, and some of the flints, from a low flat mound in his field, while preparing the ground for corn-planting. He sold the specimens because he needed money to prosecute his journey. These facts were communicated in a letter to myself from Dr. Snyder, in answer to an inquiry as to the history of the "llama." There can, I think, be no reasonable doubt of the genuineness of the find. A number of similar objects have been found in Peru, and several of them are figured in *The U.S. Naval Astronomical Expedition to the Southern Hemisphere during the Years* 1845-52, but none of these *exactly* correspond with the Ohio specimen. It has been suggested that this relic was brought to Florida by one of De Loto's men, who had obtained it in Peru while engaged there under Pizarro, and that it reached Ohio from Florida by Indian conquest or by trade and barter. This purely hypothetical explanation seems highly improbable and quite unnecessary. There are many proofs of widespread intercommunication among most savages, and there can be no doubt that it existed among such ancient and comparatively advanced peoples as the inhabitants of Peru and Mexico and the mound builders. In an interesting paper published in the *Proceedings of the American Antiquarian Society*, in 1886, Mr. F. W. Putnam shows that jade ornaments have been found in a mound in Michigan, and also in burial mounds in many localities in Central America, which have evidently been formed by cutting up jade celts; and further, that the same material is nowhere found *in situ* in America, while it exactly corresponds with Asiatic jade, some of the specimens exactly matching the material of the jade celts of New

not on pipes, some rude (Fig. 23, b), others more delicate, have been found in New York and other States. In Iowa two pipes, with rude carvings of an elephant or mastodon but with neither tusks nor tail, have been found by two separate individuals; but suspicion has been thrown on their genuineness because they both passed through the hands of the same person, and because they resemble in general form the well-known elephant mound of Wisconsin. It is, however, absolutely demonstrated, by bones pierced with stone arrows and others burnt with fire, that the mastodon was coeval with man in America, and there is therefore no antecedent improbability in its being represented both in mounds and carvings.

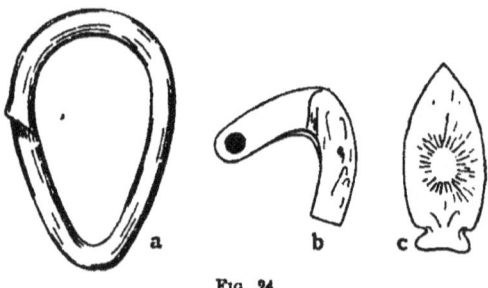

Fig. 24.

Very strange are the stone collars, or "sacrificial yokes," found in great abundance in the island of Porto Rico, and more rarely in Mexico. These are in shape and size like small horse-collars, but carved out of single blocks of hard volcanic rock. They all have a curious ornamental projection on one side, as if to represent the junction of the material out of which the type collar was formed. Some are slender and comparatively light, while others are so massive that they would be a heavy load for a man. They are said to be found in surface deposits, and along with them are many finely worked and polished celts and axes (Fig. 24, a, b, yokes; c, celt).

The long-continued use of stone in America for the

Zealand. These specimens, as well as the carved llama, may therefore be considered to prove the widespread intercommunication between distant peoples at a very remote epoch.

most varied purposes, and the occupation of the country by Indian tribes down to comparatively recent times, is the obvious cause of the extreme abundance of stone weapons and implements all over the country. As indications of this abundance, the case of Dr. Abbott's farm at Trenton, New Jersey, may be mentioned. This gentleman has obtained, on a very limited area, about twenty thousand stone implements and several hundreds of associated objects made of bone, clay, and copper, besides numerous pipes and carved stone ornaments. In a small field on the banks of the Potomac, near Washington, arrow-heads of quartz and quartzite have been collected for many years, and are sometimes still so abundant that hundreds may be found in a few days. This is on the site of an Indian settlement abandoned about two hundred years ago. In California, the large stone mortars used for pounding the acorns, which seem always to have formed the food of the indigenes, are scattered over the country by thousands; while the beautiful little arrow-heads of jasper and chalcedony found abundantly in some districts, are systematically collected to be set in gold and used as ornamental jewellery.

Next in interest and extent to the stone weapons and implements are the articles of pottery found abundantly in the various classes of mounds and sites of villages. These consist chiefly of cooking vessels, water jars, drinking cups, and mortuary urns, extremely varied in form, size, and ornamentation, and often exhibiting a considerable amount of artistic skill. In a group of mounds in New Madrid, in Missouri, over a hundred such vessels were found, exhibiting about thirty distinct types of form, from flat dishes to long-necked jars, vessels with or without handles or feet, and with the handles greatly varied in number, form, and position. Many of these are moulded above into the form of human heads or busts, and some of them are in strange attitudes, recalling the fantastic Peruvian pottery. Similar pottery has been found in the mounds of Tennessee, Kentucky, and Ohio, as well as in the curious stone graves found extensively in the Southern States; but their various peculiarities can only be understood by examining the specimens or a good series of figures. Very numerous tools and utensils of shell have also been found in the mounds, a moderate quantity in copper, with many ornaments of mica and some of silver and of gold.

Ancient Mounds and Earthworks.

The general character of the mounds and earthworks of various parts of the United States, and which are more especially abundant in the great valley of the Mississippi and its tributaries, is sufficiently known, though their vast numbers and the great variety of form and structure which they present is hardly understood in England. A voluminous memoir will shortly be published by the Bureau of Ethnology, which will give most important information on the entire subject. In some parts of Indiana and Kentucky a hundred mounds have been found in a hundred acres. The enclosed area of the ancient earthworks at Aztalan, Wisconsin, is more than fourteen hundred feet long and near seven hundred wide. The great mound of Cahokia, St. Louis, was ninety feet high, and covered an area of seven hundred feet by five hundred feet, with an inclined road up one side to reach the flat platform on the top. Another almost equally large mound exists at Seltzertown, Mississippi. In Louisiana are some curious platform mounds, in the form of squares or parallelograms, connected by terraces. Besides the wonderful Fort Ancient in Ohio, containing five miles of embankment, now, sad to relate, being gradually destroyed by cultivation, there are in Georgia and other southern states several fortified mountain-tops, recalling, in their inaccessibility, the hill-forts of India.

Ash-pits, Cemeteries and House sites.

Another curious class of works are the ash-pits, discovered a few years since near Madisonville, Ohio. Mr. Putnam, curator of the Peabody Museum, has opened no less than one thousand of these pits, and has obtained from them a large amount of implements, ornaments, pottery, and other articles. They are found on a plateau which is covered with a remnant of the virgin forest. There is a surface deposit of twelve to eighteen inches of leaf-mould, below which is hard clay. These pits are found to be circular in form, from three to four feet in diameter, and from four to seven feet deep. At the bottom there is often a small circular excavation, either in the centre or at one side. They are usually filled with ashes, in more or less defined layers, the bottom portion being very fine grey ashes, while the upper part may be more or less mixed with gravel or sand, with occasianal layers of charcoal. Throughout the whole mass of ashes and sand, from the top of the pit to the bottom, are bones of fishes, reptiles, birds, and mammals. Those of the larger species of mammalia, such as the elk, deer, and bear, are generally broken, and appear to have been those of animals used for food. Half a bushel of such bones are sometimes taken out of one pit. Shells of many species of Unio are also found. There is also much broken pottery, but rarely any entire vessels. Numbers of implements of bone or horn are found, some of large size and apparently used for digging, as well as awls, beads, harpoon points, and small whistles. Arrow points, drills, scrapers, and other stone instruments are common, with some polished celts and rough hammer-heads. Stone pipes and copper beads and finger-rings are also found. In some of the pits a considerable quantity of charred corn has been found, together with nuts and other articles of food, and in one case only a human skeleton was found at the bottom of a pit. A considerable area, including that occupied by the pits, seems to have been used as a cemetery, both before and since they were constructed. A great number of skeletons are found buried just beneath the layer of leaf mould, and in some cases these skeletons lie across a pit, while in others skeletons already buried have been evidently disturbed by digging the pit.

In the same district, but at a little higher elevation, are a number of earth-circles, from forty-three to fifty-eight feet in diameter, which prove to be sites of houses, with a central fire-place of clay, and with implements and utensils agreeing with those found in the pits. After an extensive and most laborious investigation of this locality, the only explanation of the peculiar feature of the pits is, that at certain times or on certain special occasions the whole contents of a house were burned, and the remains and all the ashes buried in a pit, while the quantity of bones found indicates that the ceremony was accompanied by feasting. The thick layer of leaf-mould covering the pits, graves, and house-sites would indicate an antiquity much greater than that of the large forest-trees which grow on the present surface, while the enormous number of the pits and the extent of the cemetery, covering over fifteen acres of ground and from which over five hundred skeletons have been obtained, indicates that the place was permanently occupied by a large population.

American Shell-banks.

Another class of remains, the shell-banks, are far more numerous and extensive than the kitchen-middens of Europe. They are found from Nova Scotia to the Gulf of Mexico, and on the west coast they have been discovered in Alaska and in California, while similar mounds, composed entirely of fresh-water shells, occur in the valleys of the Mississippi, Ohio, and other rivers. These accumulations are often of great extent. One on the coast of Georgia covers ten acres to a depth of from five to ten feet. In Florida, on Amelia Island, a shell-heap extends a quarter of a mile inland by a hundred and fifty yards along the shore; and many others are found thickly scattered over a district a hundred and fifty miles long. An immense number of works of art and animal and human remains have been found in them, some of which indicate a considerable antiquity.

Cave-dwellings.

America also has its cave dwellings, with characteristic remains of their human inhabitants; its cliff-houses, forts, and towns, partly excavated and partly built up with good stone walls, so as to resemble mediæval castles or eastern rock-cities; and its ruined towns of the Zuni and Pueblo Indians scattered over the vast desert-regions of Arizona and New Mexico. Some of these are highly interesting and remarkable. The ruined pueblo of Penasca Blanca in the Chaga Cañon, New Mexico, forms a regular oval of about five hundred by four hundred feet, the houses being symmetrically placed around the outside so as to enclose an open area, which contains a depression, probably a pond for storing water. The walls of the houses are regularly and solidly built of stone. Equally remarkable is a large round tower about forty feet in diameter with double walls, the space between which is divided into numerous small rooms. This is in ruins, but was evidently well constructed of good stone masonry. Accurate models of these and many other structures exist in the National and Smithsonian Museums.

The preceding brief outline of the materials which exist in American Museums for the study of prehistoric man are sufficient to show that they are not inferior in extent, variety, and interest to those of Europe; while if we extend our survey to the marvellous prehistoric remains of Mexico, Central America, Peru, and Bolivia, their pyramids and temples, their ruined cities, their cemeteries, their highways and aqueducts, their highly characteristic sculpture, their fantastic pottery, and their still undeciphered hieroglyphics, we may claim for the American continent a position, as regards the early history and development of the human race, hardly inferior to that of the whole of the Eastern hemisphere. A body of earnest and painstaking students are now engaged in the collection, preservation, and study of these various classes of remains; and at the same time a vast mass of most valuable material is being brought together relating to the manners and customs, the tools, weapons, and ornaments, the tribal relations, the migrations, the folk-lore, the religions, and the languages of the aboriginal inhabitants. Already much light has been thrown on the prehistoric remains by their comparison with objects still in use in some parts of the continent; and this study has resulted in the formation of two schools of American anthropologists. The one school, impressed by the very numerous resemblances to be found between existing Indians and the mound-builders, maintain the practical identity of race and continuity of habitation from the epoch of the earliest prehistoric remains down to the date of the European discovery. The other school, laying more stress on the differences between the remains left by the mound-builders and other prehistoric races and the works of modern Indians, and being convinced, further, that there are indications of great antiquity and successive occupation in many areas, believe that there has been a long series of changes in America as in the old world, that each group of remains and each area has its characteristic features, that there have been higher grades of civilisation succeeded by lower as well as lower by higher, and that the facts, no less than the probabilities, are all in favour of successive displacements of tribes or races, of which the displacement of the mound-builders by the ancestors of the historic "red men" was perhaps the latest.

This divergence of opinion is probably the very best security for the ultimate discovery of the truth, since it assures us that no important evidence on either side will be neglected. The whole inquiry is in good hands; fresh material is continually being obtained and elaborated; and we may look forward with some confidence to a final consensus of opinion which shall disperse, by the light of accurate knowledge, some portion at least of the obscurity which has hitherto overshadowed the early history of the American continent.

CHAPTER III

HOW BEST TO MODEL THE EARTH

M. ELISÉE RECLUS, the well-known geographer, in a pamphlet printed at Brussels,[1] has elaborated a startling and even sensational proposal for the construction of a huge globe, on a scale of one hundred thousandth the actual size of our earth. This is only about one-third smaller than the maps of our own one-inch Ordnance Survey; and the magnitude of the work will be appreciated when it is stated that the structure will be 418 feet in diameter, so that the London Monument, if erected inside it, would not reach to its centre, while even the top of the cross of St. Paul's Cathedral would fall short of its highest point by fourteen feet. This enormous size is considered to be necessary in order to allow of the surface being modelled with minute accuracy and in true proportions, so as to show mountains and valleys, plateaux and lowlands, in their actual relations to the earth's magnitude. Even on this large scale the Himalayas would be only about three and a half inches high, Mont Blanc, about two inches, the Grampians half an inch, while Hampstead and Highgate would be about one-sixteenth of an inch above the valley of the Thames. It may be thought that these small elevations would be quite imperceptible on the vast extent of a globe

[1] Elisée Reclus, *Projet de Construction d'un Globe Terrestre à l'échelle du Centmillième.* Edition de la Société nouvelle. 1895.
More recently (1898) a paper on the same subject was read before our Royal Geographical Society, in which the same eminent geographer explained the advantages of his plan, even if the globe were constructed on a smaller scale than he first proposed.

which would be a quarter of a mile in circumference; but the visibility of inequalities of surface depends not on their actual magnitude so much as on their steepness or abruptness, and most hills and mountains rise with considerable abruptness from nearly level plains. All irregularities of surface are appreciated by us owing to the effects of light and shade produced by them; and by a proper arrangement of the illumination the smallest deviations from a plane can be easily rendered visible. Again, the slopes of mountains are always much broken up by deep valleys, narrow gorges, or ranges of precipitous cliffs, which give a distinct character to mountainous countries, thus producing striking contrasts with lowlands and plateaux, which, when brightened by appropriate colouring and brought to view by a suitable disposition of the sources of light, would give them any desired amount of distinctness.

It is proposed that the globe shall always be kept up to the latest knowledge of the day, by adding fresh details from the results of new explorations in every part of the world; so that, by means of photography, maps of any country or district could be formed on any scale desired; and for a small fee the globe might be available to all map-makers for that purpose. Such maps would be more accurate than those drawn by any method of projection, while the facility of their construction would render them very cheap, and would thus be a great boon to the public, especially whenever attention was directed to any particular area.

M. Reclus states the scientific and educational value of such a globe as due to the following considerations—(1) its accuracy of proportion in every part, as compared with all our usual maps, especially such as represent continents or other large areas; (2) the unity of presentation of all countries, by which the erroneous ideas arising from the better-known countries being always given on the largest scale will be avoided; and (3), that the true proportions of all the elevations of the surface will be made visible, and thus many erroneous ideas as to the origin, nature, and general features of mountain ranges, of valleys, and of plateaux, will be corrected. He has fixed upon the scale

of one hundred thousandth for several reasons. In the
first place, it gives the maximum size of a globe that, in
the present state of engineering science, can probably be
constructed, or that would be in any case advisable;
secondly, it is the scale of a considerable number of important maps in various parts of the world; and, thirdly,
it is the smallest that would allow of very moderate
elevations being modelled on a true scale. He considers
that even Montmartre at Paris, and Primrose Hill at
London, would be distinctly visible upon it under a proper
oblique illumination.

When, however, we consider the size of such a globe,
nearly 420 feet in diameter, it is evident that both the
difficulties and the cost of its construction will be
very great; and both are rendered still greater by
the particular design adopted by M. Reclus—a design
which, in the opinion of the present writer, is by
no means the best calculated to secure the various
objects aimed at. I will therefore first briefly describe
the exact proposals of M. Reclus as set forth in his
interesting and suggestive pamphlet, and will then
describe the alternative method, which seems to me to be
at once simpler, less costly, and more likely to be both
popular and instructive.

The essential features of the proposed globe are said to
be as follows. Nothing about it must destroy or even
diminish its general effect. It must not therefore rest
upon the level ground, but must be supported on some
kind of pedestal; and it must be so situated as to be seen
from a considerable distance in every direction without
any intervening obstruction by houses, trees, &c. But, in
our northern climate, the effects of frost and snow, sun
and wind, dust and smoke, rain and hail, would soon
destroy any such delicate work as the modelling and
tinting of the globe; it is therefore necessary to protect
it with an outer covering, which will also be globular, its
smooth outer surface being boldly and permanently
coloured to represent all the great geographical features
of the earth, so as to form an effective picture at a considerable distance. In order to allow room for the various

stairs and platforms which will be required in order to provide for access to every part of the surface of the interior globe, and to afford the means of obtaining a view of a considerable extent of it, there is to be a space of about fifty feet between it and its covering, so that the latter must have an inside diameter of about 520 feet. It is also to be raised about sixty feet above the ground, so that the total altitude of the structure will be not far short of 600 feet.

M. Reclus adds to his general description a statement furnished by a competent engineer giving a general estimate for the erection of the globe, with some further constructive details which are, briefly, as follows: Both the globe and the envelope are to be built up of iron meridians connected by spiral bands, leaving apertures nowhere more than two metres wide. The envelope is to be covered with thick plates of glass, and either painted outside on a slightly roughened surface, or inside with the surface remaining polished, either of which methods are stated to have certain advantages with corresponding disadvantages. The envelope being exposed to storms and offering such an enormous surface to the wind would not be safe on a single pedestal. It is therefore proposed to have four supports placed about 140 feet apart, and built of masonry to the required height of sixty feet. The globe itself is to have a surface of plaster, on which all the details are to be modelled and tinted, the oceans alone being covered with thin glass. In order to provide access to every part of the surface of the globe it is proposed to construct in the space between the globe and its covering, but much nearer to the former, a broad platform, ascending spirally from the South to the North Pole in twenty-four spires, with a maximum rise of one in twenty. The balustrade on the inner side of this ascending platform is to be one metre (three feet three inches) from the surface of the globe, and the total length of the walk along it will be about five miles. But as the successive turns of this spiral pathway would be about twenty feet above each other, the larger portion of the globe's surface would be at too great a distance, and would be seen too obliquely, to permit of the details being studied. It is therefore proposed that the globe should rotate on its polar axis, by which means every part of the surface would be accessible, by choosing the proper point on the platform and waiting till the rotation brought the place in question opposite the observer. But as such an enormous mass could only be rotated very slowly, and even more slowly brought to rest, this process would evidently involve much delay and considerable cost. Again, as the facility of producing accurate maps by photography is one of the most important uses which the globe would serve, it is clear that the spiral platform, with its balustrade and supporting columns, would interfere with the view of any considerable portion of the surface. To obviate this difficulty it is stated that arrangements will be made by which every portion of the spiral platform may be easily raised up or displaced, so as to leave a considerable portion of the globe's surface open to view without any intervening obstruction. In order that this removal of a portion of the roadway may not shut off access to all parts of the globe above the opening, eight separate staircases are to be provided by means of which the ascent from the bottom to the top of the globe may be made.

Objections to the Plan of M. Reclus.

This account of the great earth-model proposed by M. Reclus clearly indicates the difficulties and complexities in the way of its realisation. We are required to erect, not one globe, but two, the outer one, to serve mainly as a cover for the real globe, being very much larger, and therefore much more costly, than the globe itself. Then we have the eight staircases of twenty-four flights each, and the five or six miles of spiral platform, wide enough to allow of a pathway next the surface of the globe and a double line of road outside for the passage of some form of auto-motor carriages. Then, again, the greater part of this huge spiral platform is to be in movable sections, which can be either swung aside or lifted up in order to allow of an uninterrupted view of any desired portion of the globe's surface. But even this will not suffice to get an adequate view of the globe in all its parts, and this enormous mass is to be rendered capable of rotating on a vertical axis. It is suggested that this rotation shall be continuous in the space of a sidereal day, and it is thought that it will be so slow as not to interfere with any photographic operations that may be desired.

But a little consideration will show us that, even with all these complex constructions and movements, and supposing that they all work with complete success, the main purposes and uses of the globe, as laid down by M. Reclus himself, would be very imperfectly attained. His first point is that such a globe would correct erroneous ideas as to the comparative size and shape of different regions due to the use of Mercator's or other forms of projection. But in the globe as proposed no comparison of different countries, unless very near together, would be possible; and even if considerable portions of the platform could be removed, and the observer could be placed near the outer covering, at a distance of, say, forty feet from the globe, only a comparatively small area could be seen or photographed in its accurate proportions. If we take a circle of forty feet diameter as our field of view it is evident that all the marginal portion would be seen very obliquely (at an angle of 30° from the perpendicular if the surface were flat, but at a somewhat greater angle owing to the curvature of the surface), and would also be on a smaller scale owing to their greater distance from the instrument, so that the central portions only would be seen in their true proportionate size and shape. For ordinary views this would not much matter, but when we have to produce maps from a globe which is estimated to cost somewhere about a million sterling, and one of whose chief uses is to facilitate the production of such maps, a high degree of accuracy is of the first importance. In order to attain even a fair amount of accuracy comparable with that of a map on any good projection, we should probably have to limit the portion photographed to about ten feet square, equal to 190 or 200 miles, so that even such very restricted areas as Scotland or Ireland would be beyond the limits of any high degree of accuracy. Larger areas, such as the British Isles, France, or Germany, would be quite beyond the reach of any accurate reduction by means of photography. As affording exceptional facilities for accurate map-making the globe would be of very limited service.

The second advantage to be derived from the proposed globe is stated to be the correction of erroneous ideas as to the comparative sizes of various countries and islands, owing to the fact of their representation in atlases on very different scales, while each country gives its own territories the greatest prominence. But a large part of this advantage would be lost owing to the fact that distant countries could never be seen together. That Texas is much larger than France would not be impressed upon the spectator when, after losing sight of the one country several hours might pass before he came in sight of the other, while even the various States of Europe, such as Great Britain and Italy, or Portugal and Turkey, would never be in view at the same time. For this special purpose, therefore, the globe would not be so instructive as the large wall maps of continents at present used in every schoolroom.

The third advantage, that the globe would admit of the varied contours of the surface being shown in their true proportions, does undoubtedly exist, and is very important; but even as regards this feature, its instructiveness would be very largely diminished by the impossibility of seeing the contours of any considerable area in its entirety, or of comparing the various mountain ranges with each other, or even the different parts of the same mountain range. It may be doubted whether the relief-maps now made do not give as useful information as would be derived from a globe of which only so limited a portion could be seen at one view.

Alternative Plan.

It thus appears that the gigantic earth-model proposed by M. Reclus would very imperfectly fulfil the purposes for which he advocates its construction. But this defect is not at all inherent in a globe of the dimensions he proposes, but only in the particular form of it which he appears to consider to be alone worthy of consideration. I believe that such a globe can be made which shall comply with the essential conditions he has laid down, which shall be in the highest degree scientific and educational, which shall be a far more attractive exhibition than one upon his plan, and which could be constructed for about one-third the amount which his double globe would cost. It would only be necessary to erect *one* globe, the outer surface of which would present a general view of all the great geographical features of the earth, while on the inner surface would be formed that strictly accurate model which M. Reclus considers would justify the expense of such a great work, and which, as I shall presently show, would possess all those qualities which he postulates as essential, but which the globe described by him would certainly *not* possess.

I make no doubt that the eminent geographer would at once put his veto upon this proposal as being wholly unscientific, unnatural, and absurd. He would probably say that to represent a convex body by means of a concave surface is to turn the world upside-down, or rather outside-in, and is fundamentally erroneous; that it must lead to false ideas as to the real nature of the earth's surface, and that it cannot be truly educational or scientifically useful. But these objections, and any others of like nature, are, I venture to think, either unsound in themselves or are wholly beside the question at issue. M. Reclus has himself declared the objects of the gigantic earth-model, and the educational and scientific uses it should fulfil. I take these exactly as he has stated them, and I maintain that if the plan proposed by me can be shown to fulfil all these requirements, then it can not be said to be less scientific, or less instructive, than one which can only fulfil them in a very inferior degree.

Before showing the overwhelming advantages of the concave over the convex globe for all important uses, I would call attention to two strictly illustrative facts. Celestial globes have been long in use, and I am not aware that it has ever been suggested that they are unscientific and deceptive, and that they ought to be abolished. Positions seen on such a globe can be, and are, easily transferred to the apparently concave sky; while many problems relating to the motions of the earth and the planets are clearly illustrated and explained by their use. A concave surface suspended from the ceiling of a schoolroom would, no doubt, show more accurately the position of the heavenly bodies, but would probably not be so generally useful as the unnatural convex globe.

The representation of the earth's surface on the inside of a sphere has been tried on a considerable scale by Wyld's globe in Leicester Square, and was found to be extremely interesting and instructive. Before seeing it I was prejudiced against it as being quite opposed to nature; but all my objections vanished when I entered the building and beheld the beautiful map-panorama from the central gallery. I visited it several times, and I never met with any one who was not delighted with it, or who did not find it most useful in correcting the erroneous views produced by the usual maps and atlases. It remained for twelve years one of the most instructive exhibitions in London, when it was removed owing to the lease of the ground having expired. This globe was sixty feet in diameter, and it showed how grand would be the effect of one many times larger and admitting of greater detail, while variety would be obtained by the view at different distances and under various kinds of illumination.

One other consideration may be adduced in this connection, which is, that even the outer surface of a huge globe has its own sources of error and misconception. It would perpetuate the idea of the North-pole being up and the South-pole down, of the surface of the earth being not only convex but sloping, while for the whole southern hemisphere we should have to look upwards to see the surface, which we could never do in reality unless we were far below that surface. Again, we all know how the sea-horizon seen from an elevation, and especially from a balloon, appears not convex but concave. A convex globe, therefore, will not represent the earth as we see it, or as we can possibly see it; and to construct such a globe with all the details of its surface clearly manifest, while at the same time we *see* the convexity and have to look *up* to some parts of the surface and *down* upon others, really introduces fresh misconceptions while getting rid of old ones. We cannot reproduce in a model all the characteristics of the globe we live on, and must therefore be content with that mode of representation which will offer the greater number of advantages and be, on the whole, the most instructive and the most generally useful. This, I believe, is undoubtedly the hollow globe, in which, however, the outer surface would be utilised to give a general representation of the earth as proposed by M. Reclus, and which would no doubt be itself a very interesting and attractive object.

Advantages of a Concave Globe.

I will now proceed to show, in some detail, how the concave surface of a hollow globe is adapted to fulfil all the purposes and uses which M. Reclus desires.

We should, in the first place, be able to see the most distant regions in their true relative proportions with a facility of comparison unattainable in any other way. We could, for instance, take in at one glance Scandinavia and Britain, or Greenland and Florida, and by a mere turn of the head could compare any two areas in a whole hemisphere. Both the relative shape and the relative size of any two countries or islands could be readily and accurately compared, and no illusion as to the comparative magnitude of our own land would be possible. In the next place, the relief of the surface would be represented exactly as if the surface were convex, but facilities for bringing out all the details of the relief by suitable illumination would be immensely greater in the hollow globe. Instead of being obliged to have the source of illumination only fifty feet from the surface, it could be placed either at the pole or opposite the equator at a distance of 200 or 300 feet, and be easily changed in order to illuminate a particular region at any angle desired, so as to bring out the gentlest undulations by their shadows. Of course, electric lighting would be employed, which by passing through slightly tinted media might be made to represent morning, noon, or evening illumination.

It is, however, when we come to the chief scientific and educational use of such a globe—the supply of maps of any portion of the earth on any scale by means of photography—that the superiority of the concave model is so overwhelming as to render all theoretical objections to it entirely valueless. We have seen that on the convex surface of a globe such as M. Reclus has proposed, photographic reproductions of small portions only would be possible, while in areas of the size of any important European State, the errors due to the greater distance and the oblique view of the lateral portions would cause the maps thus produced to be of no scientific value. But, in the case of the concave inner surface of a sphere, the reverse is the case, *the curvature itself being an essential condition of the very close accuracy of the photographic reproduction.* A photograph taken from anywhere near the centre of the sphere would have every portion of the surface at right angles to the line of sight, and also at an equal distance from the camera. Hence there would be no distortion due to obliquity of the lateral portions, or errors of proportion owing to varying distances from the lens. We have, in fact, in a hollow sphere with the camera placed in the centre, the ideal conditions which alone render it possible to reproduce detailed maps on the surface of a sphere with accuracy of scale over the whole area. For producing maps of countries of considerable extent the camera would, therefore, be placed near the centre, but for maps of smaller areas on a larger scale, it might be brought much nearer without any perceptible error being introduced, while even at the smallest distances and the largest scale the distortion would always be less than if taken from a convex surface. It follows that only on a concave globular surface would it be worth the expense of modelling the earth in relief with the greatest attainable accuracy, and keeping it always abreast of the knowledge of the day, since only in this way could accurate photographic reproductions of any portions of it be readily obtained. For absolute accuracy of reduction the sensitive surface would have to be correspondingly concave, and this condition could probably be attained.

I will now point out how much more easily access can be provided to every part of the surface of a concave than to that of a convex globe. Of course, there must be a tower in the position of the polar axis. This would be as small in diameter as possible consistent with stability, and with affording space for a central lift; and it would be provided with a series of outside galleries supported on slender columns, at regular intervals, for affording views of the whole surface of the globe. This general inspection might be supplemented by binocular glasses with large fields of view and of varying powers, by means of which all the details of particular districts could be examined. For most visitors this would be sufficient; but access to the surface itself would be required, both for purposes of work upon it, for photographing limited areas at moderate distances, and for close study of details for special purposes. This might be provided without any permanent occupation of the space between the central tower and the modelled surface, in the following manner.

Outside the tower and close to the galleries will be fixed, at equal distances apart, a series of three or four circular rails, on which will rest by means of suitable projections and rollers, two vertical steel cylinders, exactly opposite to each other and reaching to within about ten feet of the top and bottom of the globe, with suitable means of causing them slowly to revolve. Attached to these will be two light drawbridges, which can be raised or depressed at will, and which also, when extended, will have a vertical sliding motion from the bottom to the top of the upright supports. The main body of this drawbridge would reach somewhat beyond the middle point from the tower to the globular surface, the remaining distance being spanned by a lighter extension sliding out from beneath the main bridge and supported by separate stays from the top of the tower. When not in use, the outer half would be drawn back and the whole construction raised up vertically against the galleries of the tower. The two bridges being opposite each other, and always being extended together, would exert no lateral strain upon the tower.

By means of this arrangement, which when not in use would leave the whole surface of the globe open to view, access could be had to every square foot of the surface, whether for purposes of work upon it or for close examination of its details; and, in comparison with the elaborate and costly system of access to the outer surface of a globe of equal size, involving about five miles of spirally ascending platform and more than a mile of stairs, besides the rotation of the huge globe itself, is so simple that its cost would certainly not be one-twentieth part that of the other system. At the same time, it would give access to any part of the surface far more rapidly, and even when in use would only obstruct the view of a very small fraction of the whole globe.

A Suggested Mode of Construction.

A few words may be added as to a mode of construction of the globe different from that suggested in the project of M. Reclus. It seems to me that simplicity and economy would be ensured by forming the globe of equal hexagonal cells of cast steel of such dimensions and form that when bolted together they would build up a perfect oblate spheroid of the size required. As the weight and strain upon the material would decrease from the bottom to the top, the thickness of the walls of the cells and of the requisite cross struts might diminish in due proportion, while the outside dimensions of all the cells were exactly alike. At the equator, and perhaps at one or two points below it, the globe might be encircled by broad steel belts to resist any deformation from the weight above. A very important matter, not mentioned by M. Reclus, would be the maintenance of a nearly uniform temperature, so as to avoid injury to the modelling of the interior by expansion and contraction. This might be secured by enclosing the globe in a thick outer covering of silicate or asbestos packing, or other non-conducting material, over which might be formed a smooth surface of some suitable cement, or papier-mache, on which the broad geographical features of the earth might be permanently delineated. With a sufficiency of hot-water pipes in and around the central tower, and efficient arrangements for ventilation, the whole structure might be kept at a nearly uniform temperature at all seasons.

It has now, I think, been shown that the only form of globe worth erecting on a large scale is one of which the inner surface is utilised for the detailed representation and accurate modelling of the geographical features of the earth's surface, while on the outside, either by painting or modelling or the two combined, all the grander features could be so represented as to be effectively seen at considerable distances. But as to the dimensions of such a globe there is room for much difference of opinion. I am myself disposed to think that the scale of $\frac{1}{100000}$, proposed by M. Reclus, is much too large, and that for every scientific and educational purpose, and even as a popular exhibition, half that scale would be ample. The representation of minute details of topography due to human agency, and therefore both liable to change and of no scientific importance—such as roads, paths, houses, and enclosures—would be out of place on such a globe, except that towns and villages and main lines of communication might be unobtrusively indicated. And for adequately exhibiting every important physiographical feature—the varied undulations of the surface in all their modifications of character, rivers and streams with their cascades and rapids, their gorges and alluvial plains, lakes and tarns, swamps and peat-bogs, woods, forests, and scattered woodlands, pastures, sand dunes and deserts, and every other feature which characterises the earth's surface, a scale of $\frac{1}{200000}$th, or even one of $\frac{1}{250000}$th, would be quite sufficient. And when we consider the difficulty and expense of constructing any such globe, and the certainty that the experience gained during the first attempt would lead to improved methods should a larger one be deemed advisable, there can, I think, be little doubt that the smaller scale here suggested should be adoped. This would give an internal diameter of 167 feet, and a scale of almost exactly a quarter of an inch to a mile, and would combine grandeur of general effect, scientific accuracy, and educational importance, with a comparative economy and facility of construction which would greatly tend to its realisation. It is with the hope of showing the importance and practicability of such a work that I have ventured to lay before the public this modification of the proposal of M. Reclus, to whom belongs the merit of the first suggestion and publication. Now that Great Wheels and Eifel Towers are constructed, and are found to pay, it is to be hoped that a scheme like this, which in addition to possessing the attractions of novelty and grandeur, would be also a great educational instrument, may be thought worthy of the attention both of the scientific and the commercial worlds.

CHAPTER IV

EPPING FOREST, AND THE TEMPERATE FOREST REGIONS

Introductory Note

THIS article is reprinted in its original form for two reasons. It describes the actual condition of Epping Forest from personal observation, when it was first secured for the public in 1877, and will thus enable residents or visitors to know how much has been done in the way of re-afforestation of the bare portions of it, and the improvement of the unsightly waste of gravel-pits near Whip's Cross and the dreary Flats of Wanstead. It also describes some of the most interesting phenomena of the distribution of the forest trees of the temperate zone, a matter of permanent interest to all lovers of nature, and makes suggestions for the formation of illustrative forests of three or four distinct types, which, though not adopted at Epping Forest as here proposed, might still be undertaken in the New Forest, where there are several thousand acres of open pastures or boggy heaths of dreary aspect, which could thus be rendered at once interesting and beautiful.

Epping Forest in the Past.

Our greatest legal authorities will not admit that the people of England have any right whatever to enjoy the beautiful scenery of their native land, beyond such glimpses as may be obtained of it from highways and footpaths. Legally there is no such thing as a " common," answering to the popular idea of a tract of land over which anybody has a right to roam at will[1] Every supposed common is said by the lawyers to belong absolutely to some body of individuals, to a lord or lords of the manor and the surrounding owners of land who have rights of common over it; and if these parties agree together, the said common may be enclosed, and the public shut out of it for ever. The thousands of tourists who roam every summer over the heathy wastes of Surrey or the breezy downs of Sussex, who climb the peaks or revel on the heather-banks of Wales or Scotland, are every one of them trespassers in the eye of the law; and there is, perhaps, no portion of these favourite resorts of our country-loving people that it is not in the power of some individual or body of individuals to enclose and treat as private property.

How far this legal assumption accords with justice or sound policy, it is not our purpose now to inquire; that question having been treated by many able pens, and being one which will assuredly not become less important or less open to discussion as time goes on. We have now a far pleasanter task, that of calling attention to one of our ancient woodland wastes, Epping Forest, which, in the words of an Act of Parliament passed at the end of last session, is to be for ever preserved as "an open space for the recreation and enjoyment of the public." Here at length every one will have a right to roam unmolested, and to enjoy the beauties which nature so lavishly spreads around when left to her own wild luxuriance. We shall possess, close to our capital, one real forest, whose wildness and sylvan character is to be studiously maintained, and which will possess an ever-increasing interest as furnishing a sample of those broad tracts of woodland which once covered so much of our country, and which play so conspicuous a part in our early history and national folk-lore. Unfortunately the spoilers have been at work, and much of the area now dedicated to the people has been

---

[1] "Although the public have long wandered over the waste lands of Epping Forest without let or hindrance, we can find no legal right to such user established in law." (*Preliminary Report of the Epping Forest Commissioners*, 1875, p. 12.)

more or less denuded of its woodland covering and otherwise deteriorated. Before, however, we describe the present state of the forest, and discuss the important question of how best to restore its beauty and increase its interest, it will be well to give our readers some notion of its former extent and of the circumstances that have led to its preservation.

It appears by the Reports of the Epping Forest Commission (1875 and 1877) that in the reign of Charles I. the Forest of Essex, or of Waltham, as it was then called, comprised the whole district between the rivers Lea and Roding, extending southward to Stratford Bridge, thus including the site of the great Stratford Junction Station, and northward to the village of Roydon, a distance in a straight line of sixteen miles. Much of this wide area was, however, even at that early date, only forest in a legal sense, for it included many towns and villages and much cultivated land, and these seem to have left the actual unenclosed forest not much larger than in the first half of the present century. We are told, for example, that during the two centuries from 1600 to 1800 only eighty acres of the forest were enclosed, and that even up to 1851 barely 600 acres had been enclosed. The unenclosed forest at that date is estimated by the Commissioners at 5,928 acres. Then came the development of our railway system, and the discovery of Californian and Australian gold. The wealth of the country began to increase at an unprecedented rate; the growth of London became more rapid than ever, and its citizens more and more acquired the habit of residing in the country. Land everywhere rose in value; the wastes of Epping were temptingly near at hand; and illegal enclosures went on at such a pace that during the twenty years between 1851 and 1871 they amounted to almost exactly half the entire area, leaving only 3,001 acres still open.

This wholesale process of enclosure, which, if quietly submitted to, would soon have left nothing of Epping Forest but the name, roused the indignation of many who dwelt near the forest or felt an interest in it, and a powerful agitation was commenced, in which the Corporation of the City of London and many members of the Legislature took a prominent part. In 1871 the Epping Forest Commissioners were appointed by Act of Parliament, and they gave in their final report only in the spring of last year. But in the meantime a most important case had been decided in the courts. At the request of the Corporation of London, which supplied all the necessary funds, the Commissioners of Sewers (as freeholders in the forest) commenced a suit in Chancery against the lords of manors and persons to whom they had granted lands, claiming a right of common over all the waste lands of the forest, and that all enclosures made since 1851 should be declared illegal. The Master of the Rolls decided (on the 24th November, 1874) in favour of the plaintiffs, and against this decision the defendants did not appeal. It has therefore been made the basis of legislation in the Act just passed, which declares, that the whole 5,928 acres which the Commissioners found to have been open waste of the Forest in 1851 are to be treated as common lands, and (the lords of manors or their grantees being first duly compensated for their manorial rights and property in the soil) that the whole of this extensive area, with the exception of lands built upon before 1871, gardens, and pleasure-grounds, is to be preserved " uninclosed and unbuilt upon as an open space for the recreation and enjoyment of the public."

Large sums of money were, however, required to buy up the manorial rights, and although this might possibly have been done by public subscription, the necessity for this course was obviated by the liberality and public spirit of the City of London, which offered to supply all the needful funds, not only for this purchase, but also for all work that might be found necessary for the preservation, management, and replanting of the forest. This munificent offer was accepted, and the very reasonable desire of the Corporation to have the chief voice in the management of the newly acquired domain in trust for the public, was acceded to by the Legislature; and the Act accordingly declares that Epping Forest is to be managed by a committee consisting of twelve members of the Corporation of London, and four verderers, chosen by the commoners of the twelve parishes in which the forest is situated.

Condition of the Forest when Acquired.

Let us now take a brief glance at the present state of the land thus dedicated to the public, before proceeding to discuss the question—how it may be made the most of. First, and nearest to London, we have the open expanse of Wanstead Flats, not half a mile from the Forest Gate Station of the Great Eastern Railway, and which, together with some illegally enclosed ground northwards towards the village of Wanstead, comprises an area of nearly five hundred acres. Crossing it from north to south opposite Lake House is an avenue of lime-trees, never very fine, and now rapidly dying from the combined effects of want of shelter and the smoky atmosphere. With this exception almost the whole of the Flats is denuded of trees, and offers a drear expanse of wiry grass interspersed with a few tufts of broom, stretching for more than a mile in length and not far short of half a mile wide. On the northern side considerable excavations have been made for brickfields, and here, where the ground rises somewhat, there is a very nice turf, with fern, broom, and even heather, in considerable patches. Northwestward is a large piece of recovered land, about fifty acres in extent, dotted over with oaks and bushes, and intersected by a fine double avenue of limes a third of a mile long, but many of the trees, in the part nearest London, are rapidly dying. Planes are probably the only trees which would now thrive well here. This is, on the whole, a rather pretty piece of half-wild woodland, well worth careful preservation for the use of the dense population surrounding it.

To the west of Wanstead and Snaresbrook, and northward towards Woodford, is a fine expanse of unenclosed land, nearly a mile long, and from a quarter to half a mile wide; and when some illegal enclosures are thrown open, this will be continued uninterruptedly to Woodford Green. The southern portion of this tract between Wanstead Orphan Asylum and Whip's Cross has been utterly devastated by gravel-digging, the whole surface being a succession of pits and hollows with stagnant pools of water, and a few miserable oaks left standing on mounds where the gravel has been dug away around them. One would think that here the lords of the manors had infringed on the rights of the commoners, by destroying the pasture and even the surface soil on which any herbage can grow; and that in equity they should be called on to pay damages instead of receiving payment for their alleged property in the soil, which they have here succeeded in rendering almost wholly worthless either for use or enjoyment. North-westward, towards Woodford Green, is a rather pretty piece of wild forest-land, with open grassy glades, intervening thickets, and ponds swarming with interesting aquatic plants. There are, however, very few ornamental trees, the oaks being mostly small, with a quantity of miserable pollard-beeches hardly more sightly than so many mops.

Passing Higham Park we come upon a large extent of illegally enclosed land, now to be thrown open, and much of it already given up. Between Woodford Green and Chingford Hatch there are about sixty acres of poor grass and fallow-land adorned with a few bushes and one fine oak-tree, but sloping gently towards the north-west, and with extensive views over the wooded country beyond. Further north there are more than a hundred acres of small enclosures—rough pasture, fallow-land, or cultivated fields, dotted with a few poor trees, and at present far from picturesque, but with an undulating surface offering considerable opportunity for improvement. To the west these fields are bounded by Chingford brook, by the side of which are some very handsome willow-trees growing in stiff clay and indicating what this part of the land is adapted for. A little to the north-east is the new village of Buckhurst Hill, to the south-east of which is a fine piece of enclosed forest, about a hundred acres in extent and called the Lodge Bushes.

We now enter the northern and grandest division of the Forest, which stretches away for a distance of five miles from Queen Elizabeth's Lodge to near the town of Epping. North and west of the Lodge are nearly three hundred acres of illegally enclosed fields, now dreary fallows and poor pastures, but with fine slopes affording opportunity for producing new effects of forest-scenery. To the west and south of Loughton village are more extensive enclosures of several hundred acres of land, much of it arable or pasture land of good quality; and further north, near Theydon Church and on towards Epping, are other enclosures of less extent, and almost all of this will again be thrown open to the forest.

To the north of the road from Loughton to High Beech there is a vast extent of rough forest-land, nearly three miles long and from half a mile to a mile wide, which has all been recovered after having been illegally enclosed by the lords of the manors, but not before they have denuded large portions of it of everything deserving the name of a tree, and left it a scrubby waste without any pretensions to sylvan beauty. Here are square miles of land, once as luxuriant as the unenclosed portions further west, but now presenting a hideous assemblage of stunted mop-like pollards rising from a thicket of scrubby bushes.

From this brief sketch of the present condition of Epping Forest (1878), with more especial reference to the newly recovered portions of it, we find, that probably not much less than a thousand acres, which are now or have recently been enclosed and cultivated fields, will soon be thrown into the forest; while, in addition to this, there are considerably more than a thousand acres which are almost entirely denuded of trees and in a generally unsightly condition. The question at once arises—How can these wide tracts of land be *best* dealt with for the future recreation and enjoyment of the public? The Act of Parliament, it is true, empowers the conservators to form playgrounds and cricket-grounds in suitable places, and some portion of these lands may be so applied. But a very few acres will serve for this purpose, or indeed are at all suitable for it; and there will remain by far the larger portion to be otherwise dealt with. After all the agitation, all the arduous legal struggles, all the liberal, nay, lavish, expenditure of money to secure this land to the people, it cannot surely be left as it is. Some steps must be taken to make it beautiful and picturesque in the future, and at least as well adapted for the recreation and enjoyment of coming generations as the old forest was for those which have passed away. The obvious course, and that which will at once occur to every one, is to plant this ground in some way or other. It was once all forest. It is as a forest that the whole domain is dedicated to the public; and it is the forest scenery which has always given to the entire district its peculiar charm. Our country still has wide tracts of common and of open wastes, as well as extensive enclosed woods, and parks, and plantations; but our genuine forests are few and far between. Undoubtedly, therefore, as forest or woodland of some kind this land should be restored; and the question we have to decide is—Of what kind?

Some may say, restore it as much as possible to its ancient state; plant it with oaks and beeches, with a sprinkling of elm, birch, and ash. This may be the easiest and the simplest, but it is certainly the least advantageous mode of dealing with the land. While these trees were growing—for a couple of generations at least—they would be utterly uninteresting woods, and even in the far-distant future would hardly surpass many other parts of the forest, while they would increase the monotony which is its chief defect. Another plan would be, to make a mixed planting of choicer trees, shrubs, and evergreens, which would be more beautiful while growing, and would in time form a forest of a more diversified character. Or again, a regular arboretum might be formed, a great variety of trees, and especially choice pines and firs, being planted so as to form specimens. Either of these plans would at once possess some interest; but they would be utterly deficient in novelty, or in that special and peculiar interest we should aim at, when we have to deal with such an extensive and varied area as the recovered portions of Epping Forest. We have already fine mixed plantations and woods, and many splendid arboretums; and at Kew we have in process of formation a magnificent collection of specimen trees which it would be out of place to attempt to imitate, while the expense would be far greater than almost any other kind of planting.

Proposed illustration of Temperate Forests.

The plan I have now to propose is very different from all these. It is one which would be perfectly novel, perfectly practicable, intensely interesting as a great arboricultural experiment, attractive alike to the uneducated and to the scientific, not more expensive than any other plan, and perfectly in harmony with the character of the domain as essentially "a forest." It is, briefly, to form several distinct portions of forest, each composed solely of trees and shrubs which are natives of one of the great forest regions of the temperate zone.

In order to understand how interesting and how instructive this would be, and, especially, to how great an extent it would add to the variety and beauty of the scenery, while retaining to the fullest extent its character as a wild and picturesque woodland district, it will be necessary to give a brief sketch of the great forests of the north temperate zone, to point out their comparative richness, their distinctive characters, and their different styles of beauty; and in doing this I shall avail myself largely of the writings of the greatest authority on the subject, the late Professor Asa Gray, who has made the relations and origin of the various forest regions of the Northern Hemisphere the study of his life.

The two northern continents, America on the one side, Europe and Asia on the other, have each two great and contrasted forest regions, an eastern and a western; and in both cases the eastern is very rich, while the western is comparatively poor. The trees of our own country belong to the western or European forest region, which includes also the adjacent parts of Western Asia. That region contains about eighty-five different kinds of trees (seventeen being conifers, or firs and pines), and of these only twenty-eight are really natives of Britain, about twenty being tolerably common, and forming the wild trees of our woods and wastes, with which we are all more or less familiar.

If we compare the European set of trees with that of the forest region of Eastern America we find a wonderful difference. Instead of a total of eighty-five, we have there no less than 155 different kinds of trees, and a large number of these are very distinct from those of Europe, constituting altogether new types of vegetation, many of which, however, we have long cultivated for ornament. Among these are magnolias, tulip-trees, red and yellow horse-chestnuts, the locust or common acacia, the honey-locust (a far handsomer tree), the liquidambar, the sassafras, the hickories, the catalpa, the butter-nut and black walnut, many fine oaks, the hemlock spruce, the deciduous cypress, and a host of others less generally known. Most of these differ from our native trees by their more varied and beautiful foliage, by many of them being flowering trees often of the most magnificent kind, and, what is equally important, by the glorious tints which a large proportion of them assume in autumn. Every one has heard of the rich autumnal tints in Canada and the United States as something of which our woods, beautiful as they are, give hardly any idea. Instead of the yellows and browns of our trees, there is in the American forest every tint from the richest scarlet and crimson to yellow, which, combining in endless varieties, give a splendour to the autumnal landscape which is worth a journey across the Atlantic to behold. The Virginian creeper, which drapes our houses with a crimson mantle even amid the smoke of London, the red maple and the sumach of our shrubberies, give us some notion of these tints, but hardly any idea of the effect they produce when their colours are lavishly spread over a varied landscape. Most of the trees which acquire these brilliant hues grow as well with us as in their native country. Some American trees, strange to say, seem to grow even better, for the beautiful ash-leaved Negundo is a small tree in its native country, rarely exceeding thirty feet high, while Loudon tells us that it grows to forty feet in England; the white maple reaches only forty feet in America and fifty feet here; and a similar difference occurs with many other trees. So favourable, indeed, is our climate to the growth of trees generally, that, according to Professor Asa Gray, we "can grow double or treble the number of trees that the United States can," although their native species are five times as numerous as ours!.

There is therefore really no difficulty in producing in England an almost exact copy of a North American forest, with all its variety of foliage, with its succession of ornamental flowers, and with its glorious autumnal tints; yet this has never been attempted either in this country or in any part of Europe. That many of these trees will reach noble dimensions there is no doubt whatever. A honey-locust (*Gleditschia triacanthos*) in Professor Owen's garden at Richmond Park was, in 1872, a magnificent tree nearly eighty feet high, and was then sixty years old. There is at Dorking a tulip-tree about the same size; while the many beautiful American oaks, maples, birches, and poplars, form noble forest trees in many of our parks and pleasure-grounds. Were such trees planted in masses, they would grow upwards more rapidly and produce a forest-like effect in from twenty to forty years; while from their varied foliage and general novelty of aspect, they would be both beautiful and interesting at a far earlier period.

Here, then, we may do something which has never been done before, which is sure to succeed (since it is only growing trees in masses which have already been grown singly), and which will ultimately produce a real addition to our landscape, while the individual trees will be a constant source of gratification and delight. As yet we have only mentioned the different kinds of trees, but North America is not less rich in beautiful shrubs to form an underwood to the forest or open patches here and there in its recesses. The rhododendrons, azalias, and kalmias, will grow as underwood wherever there is peat or loam, while the well-known snowberry, the aloe-like yuccas, several fine spiræas, American blackberries, and many others, would grow anywhere.

Now let us suppose one of the most suitable of the open tracts recovered at Epping to be thus converted into an American forest, in which as many trees and shrubs peculiar to Eastern North America as we know to be hardy, are planted in masses and variously intermingled. Such an experiment would excite interest at every stage of its growth. The paths and open glades intersecting it would be visited year after year to see how it was thriving, and this would necessarily lead many of its visitors to acquire an intelligent interest in the trees, and shrubs, and flowers of other lands. And as time rolled on, and one kind of tree after another arrived at its period of blossoming, and displayed each succeeding year in greater perfection its glowing autumnal tints, the "American forest" would become celebrated far and wide, and would attract visitors who would never think of going to see the more homely beauties of a native woodland, and still less a young plantation of common trees.

Before proceeding to describe the other characteristic "forest pictures" which might be produced in the waste lands of Epping, it will be well to answer an objection sure to be made, that the kind of planting here proposed, consisting wholly of foreign, and largely of rare trees and shrubs, would be very expensive. This, however, is a complete error. Many of the trees in question are certainly rather expensive when large specimens are purchased of nurserymen; but this is chiefly because there is so little demand for them, and they occupy ground and require attention for many years unprofitably. But nearly all these American trees could be raised from seed almost as cheaply as the very commonest kinds. The seeds could be obtained from their native country at a mere nominal cost; and by forming a nursery-ground, small at first, and increased year by year, in which to raise them, their removal at the most suitable age and season to the places which they were permanently to occupy would ensure rapid and vigorous growth. The great item of expense in forming any extensive plantation is labour, and this would be little if any more in growing one kind of tree than another, supposing both to be raised from seed and to be equally hardy. The question of expense cannot, therefore, be of importance, as compared with the vast difference in permanent results between the plan here advocated and that of the ordinary English wood, the mixed plantation, or the systematic arboretum. The latter, indeed, would be very much more expensive, because, few specimens being wanted, it would not be worth while raising them from seed, while an arboretum would require more weeding and pruning, as well as some amount of permanent gardening, which in a forest is unnecessary.

Another important feature of such a forest would be, that it would furnish reliable information as to what valuable timber trees may be profitably grown in this country. Among American trees the sugar-maple, hickory, tulip-tree, redwood, and locust, are well-known as producing valuable timbers for special purposes; and there are many trees of Eastern Europe and Asia equally valuable, which it might be profitable to grow largely. As, however, they have been hitherto almost always grown singly for ornament, we have been unable to test, either the rapidity of their growth under more natural conditions, or the quality of their timber at different ages; all which points would be determined, were they grown in quantity as here proposed, by the mere periodical thinnings-out necessary to encourage the free development of those that were to remain and form the permanent forest.

Passing now to the western or Californian coast of North America, we find another forest region, remarkably different from that of the Eastern States. It is characterized at once by extreme richness in coniferous trees, and what Professor Asa Gray terms its "desperate poverty" in deciduous kinds, of which it has only one-fourth as many as Eastern America, and one-half as many as Europe.[1] Almost all the trees which are especially characteristic of Eastern America are wanting, their place being chiefly supplied by peculiar species of oaks, maples, ashes, birches, and poplars, groups which are equally abundant on both sides of the Atlantic. When we turn to the coniferous trees, however, Western America stands pre-eminent, possessing nearly twice as many different kinds as the Eastern States, and nearly three times as many as all Europe, while it exhibits the grandest, tallest, and most beautiful firs, pines, and cypresses in the world.

[1] Deciduous trees, 34 species; conifers, 44 species!

Here we find the giant Wellingtonia and redwood, the magnificent Douglas fir, the exquisitely beautiful piceas-*nobilis* and *lasiocarpa*, such fine cypresses as *Lawsoniana* and *Lambertiana*, such unequalled pines as *insignis* and *macrocarpa*, the well-known handsome thujas, *gigantea* and *Lobbii*, and many others. These glorious trees form forests by themselves, surpassing in grandeur those of any other temperate land; and every one of these grows freely and rapidly with us (which they do not in Eastern America), and, if grown under natural conditions, would probably attain nearly as great a size as in their native country. Their extreme beauty has, however, caused them to be almost always grown singly as specimens, and even thus the rapidity of their growth is often amazing. The Wellingtonia will reach twenty feet in ten years; the Douglas fir grows even more rapidly when young, and a specimen at Dropmore, fifty years old, is now more than a hundred feet high, while its branches, spreading on the ground, cover a space sixty-six feet in diameter. The beautiful grass-green *Pinus insignis* at the same place reached sixty-eight feet high in thirty-four years; and were these trees planted in masses, so as to draw each other upward, and cause the lower branches to drop off as in their native forests, they would almost certainly grow even more rapidly, and the younger members of the present generation might live to walk amid forests of these noble trees not much inferior to those which excite so much admiration on the mountains of California and Oregon.

Here, again, there is no question of success. The experiment has been made already for us hundreds of times over, and we have only to profit by it. These trees succeed well in every part of England without exception, and they would certainly not fail at Epping. An expanse of a hundred or two hundred acres covered with the coniferous trees of Western America, planted in masses, groups, or belts, and with winding paths, broad glades, and occasional shrub-planted openings admitting of free access to every part of it, would probably be even more attractive than the forest of Eastern America. For many of these trees are exquisitely beautiful objects in their young state, the varying colours of the under and upper surfaces of their foliage and the delicate tints of the new growth in summer, being especially remarkable. Their different rates of growth would soon cause some species to tower above others, and thus produce that charm of variety which is wanting where large areas are planted with trees which all grow at about the same rate.

The next forest type of which we should have an example, is that of Eastern Europe and Western Asia, containing all those interesting trees of the European forest region which are not natives of our own country. Here we should grow the various European pines and firs, including the symmetrical pinsapo of Spain, the well-known silver fir of the Alps, and the allied but more beautiful Nordman's fir of Russia. Here, too, we should have the nettle-tree, the Judas-tree, the flowering ash, the wild olive, the hop-hornbeam, the almost evergreen Neapolitan alder, and our old favourites the plane, the walnut, the laburnum, and the Portugal laurel. Along with these we should plant the many beautiful and often sweet-scented shrubs of the same districts—laurestinus, myrtles, Spanish broom, coronillas, cistuses, Mediterranean heaths, the favourite lilac, and the luscious Philadelphus, or syringa. A smaller space would serve to exhibit these trees and shrubs in forest growth, as they are less numerous and generally not of large size; but as they comprise so many of our garden favourites, the forest of Eastern Europe would certainly be very attractive.

We now come to the most remarkable of all the forest regions of the temperate zone—that of Eastern Asia and Japan. This forest is even richer than that of Eastern America in deciduous trees, and at the same time richer than that of Western America in conifers;[1] and, as it is only partially explored, while the others are well known, its comparative richness will certainly increase as future discoveries are made. We find here a number of the deciduous trees of Eastern America represented by closely allied species, and, in addition, a number of altogether peculiar types. Among these are the well-known ailanthus,

---

[1] Deciduous trees, 123 species; conifers, 45 species.

on the leaves of which silkworms are fed, and which grows
with extreme rapidity; the beautiful paulownia, with
flowers like those of a foxglove; the handsome *Sophora
japonica;* and of smaller trees and shrubs, the winter-
flowering chimonanthus, the crimson-flowered japonica
which adorns our walls in early spring, the favourite
weigelia, the yellow-flowered forsythia, the red-berried
aucuba, and last, but not least important for our purpose,
the camellia. This glorious evergreen is really as hardy as
the common laurel, and will grow out of doors in perfect
health and vigour. Its beautiful flowers will, indeed, be
often destroyed by the wet and frosts of our springs, but
if a sunny bank in the midst of the protecting forest were
covered with these shrubs, they would blossom abundantly
whenever we had a mild spring, and would then, indeed,
be worth a journey to see; while at all times their splendid
glossy green foliage would be a delightful spectacle.

Even more varied and more beautiful than the conifers
of California are those of Japan and China, of which there
are no less than forty-five species belonging to nineteen
generic groups, many of which are altogether peculiar to
this region. Here are the elegant cryptomeria and retino-
sporas, the remarkable salisburia, or gingko-tree—a pine
with foliage like that of a gigantic maiden-hair fern, and
the hardly less curious sciadopitys, or umbrella-pine. To
these we may add the fine cunninghamia, the funereal
cypress, and some interesting species of arbor-vitæ.

The space required for this Asiatic forest would not at
first be large, as only the most distinct and interesting
species need be made use of, while many are not yet to be
obtained in this country. Some of the Japanese trees
grow slowly, but it is not improbable that when planted
in greater quantities they might make more rapid progress. Anyhow, the plants themselves are usually so
peculiar and generally so beautiful, that in every stage of
their growth they would be sure to prove attractive to
the public.

We might, however, increase the extent of our Asiatic
forest by adding to it another small piece of land in
order to cultivate several beautiful trees which characterize

the temperate regions of the higher Himalayas, among which are the favourite deodara, some beautiful maples, birches, and oaks, the elegant leycesteria, some fine berberries, rhododendrons, and other interesting plants.

There remain the temperate forests of the Southern Hemisphere, chiefly represented in Chili and Patagonia, in Australia, and in New Zealand, and comprising a number of very interesting species, many of which will grow in this country. From Chili there is a peculiar pine, libocedrus, and the well-known araucaria, which when grown in avenues or masses produces a very grand effect. Many of our favourite shrubs come from this region, as the golden-balled buddlea, the lovely flowering evergreens, escallonia and berberis, and the pretty cross-leaved veronica. These would form exquisite flowering-thickets to set off the stiff forms of the araucarias. From Australia and New Zealand more variety may be obtained, though comparatively few of the trees of these countries have yet been proved to be perfectly hardy. The common *Eucalyptus globulus*, celebrated as a remover of miasma, suffers much from frost when young, but may possibly become hardier as it grows older. Other species of eucalyptus are much more hardy and more ornamental. One raised from seed by myself has, in an exposed situation, reached a height of twenty feet in five years, though once cut down by frost. Another mountain species, raised at the same time, is only five feet high, but is perfectly hardy, the leaves being quite uninjured by frost, and it will probably grow into a lofty tree.[1] Some of the acacias are also probably hardy, as they grow well and flower beautifully out of doors; but the most elegant of these southern trees are the pittosporums of New Zealand, which in five years have formed splendid bushes nearly six feet high, and as much in diameter, with delicate foliage of a pale green colour which does not appear to suffer the least from any ordinary winter's frost. These will grow into small flowering-trees fifteen or twenty feet high, having an appearance quite distinct from anything

---

[1] This tree, *Eucalyptus Gunnii*, is now (1899) over 30 feet high, the stem being nearly 3½ feet in circumference.

at present in cultivation. The celebrated huon pine of Tasmania is another fine tree of this region ; and one of the proteaceæ (*Lomatia longifolia*) has lived more than twenty years in a garden near London. These, with such shrubs as the white-flowered leptospermum and the purple veronicas, will form a group of plants well illustrating the beautiful evergreen woods of the Southern Hemisphere.

There remain still the climbing plants, which form a conspicuous ornament of all these forests, and many of which are quite as hardy as the trees they decorate. We might adorn our North American forests with festoons of the Virginia creeper and wild vines, while the red trumpet-creeper and the passion-flower of the Southern States would form beautiful objects, climbing over the bushes and among the branches of trees, and displaying their showy blossoms, which are hardly surpassed by the denizens of our hothouses. The Asiatic forest would in like manner be ornamented with lilac-flowered clematises, the Japan honeysuckle, the evergreen banksian rose, the winter-flowering yellow jasmine, and the glorious wistaria, the very queen of climbing plants. It is the opinion of some eminent horticulturists, that even the superb Chilian *Lapageria rosea* would grow freely out of doors in a suitable soil and situation, and it might well be tried in association with the trees and shrubs of the same country.

Treatment of the Native Forest.

Quitting now that portion of Epping Forest which requires to be replanted, we find extensive tracts still more or less covered with wood, and which require, comparatively speaking, little to be done to them ; but that little should be well considered and carefully executed. The preservation of "the natural aspect of the forest," as specially mentioned in the Act of Parliament, should always be kept prominently in view, and this principle should influence the character of such foot-bridges, dams, banks, or other building or engineering works as may be found absolutely necessary. Every such work should be carefully studied, so as to be at once in harmony with the surroundings, permanent, and picturesque. Unpainted wood and stone, both as bold and substantial as possible, should alone be employed, brick being, whenever possible, avoided as both commonplace and unsightly. Wherever possible, earthwork or natural masses of rock should be used, so as to blend imperceptibly with the surrounding forest scenery. Among the works absolutely needed for the enjoyment of the forest are numerous footpaths; and these should be systematically laid out in connection with broader "rides" traversing the larger wooded tracts between well-marked points on either border, thus serving as a means of extricating any unfortunate tourist who may have lost his way. Grassy or shrubby openings might also be occasionally formed in the most densely wooded portions, such clear spaces being very pleasing, admitting air and sunshine, and forming agreeable contrasts. Trees which are any way remarkable for their age, size, or picturesque beauty should be cleared of surrounding thicket, so that they may be properly seen and admired; and this comprises nearly all that need be done here, beyond the ordinary forester's duty of keeping up a sufficient stock of healthy young trees to supply the place of those which die or are accidentally destroyed.

Among the powers conferred upon the conservators is that of draining where needed, and as very great misconception prevails on this subject a few remarks here may not be out of place. People have been so accustomed to hear "draining" spoken of as one of the greatest and most necessary of improvements, that they may not unnaturally think it equally necessary in a forest as in a farm or private estate. It is true that where some particular timber is to be grown for profit, draining may be necessary, but when you only require trees growing naturally, so as to produce beauty and picturesque effects, then every variety of soil and every degree of moisture are beneficial. Forests as a rule grow better in damp than in dry soils, and there is no ground so wet that some kinds of trees will not flourish in it. It is only necessary, therefore, to plant the right kinds of trees, and the wet places may be covered with wood even more quickly than the dry.

It must be remembered, too, that a proportion of bog and swamp and damp hollows, are essential parts of the "natural aspect" of every great forest tract. It is in and around such places that many trees and shrubs grow most luxuriantly; it is such spots that will be haunted by interesting birds and rare insects; and there alone many of the gems of our native flora may still be found. Every naturalist searches for such spots as his best hunting-grounds. Every lover of nature finds them interesting and enjoyable. Here the wanderer from the great city may perchance find such lovely flowers as the fringed buck-bean, the delicate bog pimpernell and creeping campanula, the insect-catching sundew, and the pretty spotted orchises.[1] These and many other choice plants would be exterminated if, by too severe drainage, all such wet places were made dry; the marsh birds and rare insects which haunted them would disappear, and thus a chief source of recreation and enjoyment to that numerous and yearly-increasing class who delight in wild flowers, and birds, and insects, would be seriously interfered with.

There is also a wider and more general point of view from which it may be important to survey this question of drainage. Epping Forest lies within the area of scanty rainfall, which extends over much of the eastern part of England, and as its surface consists largely of gravel, the rain-water rapidly passes away, and thus tends to create an aridity not favourable to luxuriant vegetation. Now, every marsh and bog and swampy flat acts as a natural reservoir, retaining a part of the rainfall, and permanently moistening both the atmosphere and the surrounding soil. In order to improve the climate and foster the vegetation of the forest, it should be the object of its conservators to retain as much as possible of the rainfall-water within the area under their jurisdiction. The forest streams might be dammed up at intervals, so as

[1] Besides those above mentioned, the following rare or interesting marsh or bog plants inhabit Epping Forest: marsh St. John's wort (*Hypericum Elodes*), opposite-leaved golden saxifrage (*Chrysosplenium oppositifolium*) red cranberry (*Vaccinium oxycoccos*), bladderwort (*Utricularia vulgaris*), water-violet (*Hottonia palustris*), and the royal fern (*Osmunda regalis*), but this last is, perhaps, extinct.

to form permanent ponds or lakes, by which means, combined with the natural reservoirs already alluded to, and aided by the check to evaporation which additional planting will produce, the forest itself and even the surrounding country will be permanently benefited. By extensive draining, on the other hand, water is carried away rapidly from the district, and with it much fertilising matter; the climate is made drier, and the growth of herbage as well as of trees and shrubs is rendered less luxuriant.

Differences of Temperate Forests, and their Causes.

Coming back now to the general question of forest distribution in the Northern Hemisphere, many of my readers must have been struck by the singular inequality and remarkable contrasts of the four great temperate forests of which we have proposed that illustrations should be grown at Epping. In a lecture recently delivered before the Harvard University Natural History Society, Professor Asa Gray has given an explanation of these contrasts, which will commend itself to all naturalists who know how important has been the agency of the glacial period in bringing about the existing relations between Alpine and Arctic plants.

Let us first consider the remarkable difference between the forest vegetation of Eastern America and that of Europe and Western Asia. The latter area is the more extensive and more varied of the two, yet its trees, both deciduous and coniferous, are scarcely half as numerous or half as diversified. Why, we naturally ask, is America so rich? Professor Asa Gray answers, it is not America that is exceptionally rich, but Europe that is exceptionally poor. This is shown in two ways. Firstly, because America, rich as it is, is surpassed by Eastern Asia; and, secondly, because Europe itself was formerly at least as rich as America is now. During the later Miocene or Pliocene periods, Europe possessed most of the generic groups of trees now confined to North America and East Asia, and was wonderfully rich in different kinds. The later Tertiary deposits of Switzerland alone have yielded, according to Professor Heer, 291 species of trees and 242 shrubs, or far more than the present rich flora of Eastern Asia added to the poorer one of Europe. It is true that this number includes the species of several distinct deposits of somewhat different ages. But in the beds of one single locality and period, at Œninghen, the remains of nearly two hundred species of trees have been found; and it is in the highest degree improbable that all which lived there have been preserved, while it is certain that the flora of Œninghen was not so rich as that of Switzerland, and was, *à fortiori*, very much poorer than that of Europe. Making, therefore, all necessary deductions for imperfect determinations of species, it is impossible to doubt that the kinds of trees inhabiting Europe in late Tertiary times were far more numerous and varied than they are now even in Eastern Asia, which, as we have seen, is the richest part of the north temperate zone. Since the period of these deposits the climate of all these regions has greatly deteriorated, culminating in a Glacial epoch which has only recently passed away; and to this is naturally imputed the wonderful change from riches to poverty which has come over the woody plants of Europe. But we have still to ask, Why did not Eastern America and Eastern Asia become equally poor? And Professor Asa Gray has now answered that question for us in a very satisfactory manner.

We must first call attention to the fact that when Europe enjoyed a milder climate, with a rich and varied flora, there was also an abundant vegetation, very similar in character to that which now clothes our north temperate latitudes, extending northward to the Arctic circle and far beyond it. In Arctic America, in Greenland, and even in Spitzbergen, there have been found well-preserved remains of maples, poplars, birches, and limes, like those of Europe; of magnolias, hickories, sassafras, and Wellingtonias, like those of America; as well as of gingko-trees and several other kinds now peculiar to Japan. The period when these Arctic woods flourished was no doubt earlier than that of the forests of Œninghen (though both are usually termed Miocene), the northern plants having migrated southward owing to the lowering of the mean temperature. As the severer cold of the Glacial epoch came on, the same species could only live by migrating still farther south; and then, when the cold period had passed away, they moved back again, and many of them now occupy the same countries as they did before the Glacial epoch.

And now we arrive at the explanation of the exceptional poverty of Europe. If we look at a good map or large globe, we shall see that in North America the Alleghany Mountains run north and south, and the lowlands east and west of them extend uninterruptedly to Florida, to Texas, and to the Gulf of Mexico. There was, therefore nothing to prevent the southward migration of the flora, when the mountains were covered with snow and ice, and its return afterwards. But in Europe the geographical conditions are very different. There is a great chain of mountains, the Alps and Pyrenees, running in an east and west direction, and farther south a great sea, the Mediterranean, also running east and west. As the Glacial epoch came on, the icy mantle crept southward from the Arctic Ocean and downward from the mountain heights, thus preventing the plants of Central Europe from migrating southward, and destroying all that were not capable of enduring a very severe climate, or which did not also exist south of the Alps. But here, too, the Mediterranean prevented any southern migration; and being crowded into a diminished area between the mountains and the sea, many species must have perished. When the cold passed away, the survivors spread northwards and rapidly covered the whole country, but their greatly diminished numbers and the prevalence of a few hardy species over very wide areas, sufficiently attest the severe ordeal they have passed through.

The correctness of this explanation can hardly be doubted, more especially as it equally serves to explain the superior riches of Eastern Asia. For here we find a far greater extent of northern land from which the existing forest-trees originally came, and also a greater extent of southern lowlands extending uninterruptedly into the tropics, for them to retreat to during the period of cold. All the conditions were here favourable, first for the production and next for the preservation of a rich flora.

The poverty of Western America in deciduous trees and its richness in conifers, Professor Asa Gray considers to be a more difficult and at present an insoluble problem. But here, too, a consideration of the physical character of the country suggests an intelligible explanation. Conifers are more especially mountain plants, while deciduous trees abound most in the lowlands. Now in North-west America there is a vast stretch of mountains from the extreme north to the far south, and no extensive lowlands—exactly the reverse of what obtains in Eastern America, where the lowlands are vastly more extensive than the mountains. Conifers, therefore, most likely always abounded most on the western side of the continent, and during their enforced southern migrations always found suitable mountain habitats. The deciduous trees, on the other hand (always, probably, few in number), were many of them exterminated in their migrations first southward and again northward, for want of suitable places of growth, or were overpowered by the greater vigour of the competing coniferous trees.

Turning again to Eastern Asia we find a combination of both these conditions. Ample mountain ranges traverse every part of it from the Arctic circle to the tropics, but these are everywhere interrupted by great river-valleys and extensive plateaus of moderate elevation, thus offering equally favourable conditions for the preservation of both kinds of trees; and here we accordingly still find the richest and most perfectly balanced woody vegetation of the north temperate zone.

The marvellous history that we have here sketched in the merest outline, teaches us that our own country has been denuded of its proper share of wild trees and shrubs by a great natural catastrophe—the Glacial epoch—which destroyed them just as a hurricane or a conflagration might have destroyed them, only more gradually, and at the same time more thoroughly. In replanting the same or similar trees as those which inhabited Europe before the Glacial period, we may be said to be only bringing back our own, and again clothing our land with those forest denizens which at no very distant epoch it actually possessed.

CHAPTER V

WHITE MEN IN THE TROPICS

CAN the tropics be permanently colonized by Europeans, and particularly by men of the Anglo-Saxon race? This is the question that now occupies much attention in view of the mad struggle among the chief European Governments for a share of all those parts of tropical Africa and Asia still held by inferior races. And the general opinion seems to be that there is something in the tropical climate inimical to Europeans, who cannot live and work there as the natives can, and who must, therefore, be content with a few years' residence, occupying the country solely as rulers, and as exploiters of native labour. Again and again the statement is made in the public press, and by writers of some authority, that "white men cannot live and work in the tropics;" and this dogma is made the foundation of theories as to our conduct toward the natives, and is often held to justify us in inducing or compelling them to work for us by methods which do not very much differ in their results from modified slavery. It therefore becomes important to ascertain whether this dogma is true or false; and on this question, having myself lived and worked for twelve years within ten degrees of the equator, in the Amazon valley and in the Malay Archipelago, I have formed a very definite opinion.

A few preliminary remarks are needed to avoid misconception. In the first place, we must clearly distinguish between the *climate* and the *diseases* of the tropics. Most people form their opinions from the effects of those tropical diseases which prevail in the cities and towns where Europeans most congregate, or of the climate in the very worst portions of the tropical regions. The great trading centres of tropical America, from Havana and Vera Cruz to Rio de Janeiro, owe their extreme unhealthiness to two main causes—the absence of all effective sanitary arrangements among the native population, and the fact that they were for several centuries emporiums of the slave trade. It is to this latter cause that Dr. C. Creighton, one of the greatest authorities on the history of epidemic diseases, traces the origin and persistence of the fatal yellow fever, which is only endemic in the slave trade area on the two sides of the Atlantic. The slave ships reached their destination in a state of indescribable filth, which year after year was poured out into the shallow water of the harbours, and soon formed a permanent constituent of the soil between high and low water marks. In the East there were no such slave ships and there is no yellow fever; but the overcrowding in all centres of population, and the neglect of sanitation, both by the natives and by their English rulers in India, who, knowing better, are most to blame, produces and propagates plague and other zymotic diseases. But these are in no way due to the tropical climate, since three centuries ago plague was as prevalent in the cities of England as it is now in those of India.

Still more commonly associated with the tropics are the various forms of malarial fevers, but these also are in no sense due to the climate, but simply to ignorant dealing with the soil. My own experience has shown me that swamps and marshes near the equator are perfectly healthy so long as they are left nearly in a state of nature —that is, covered with a dense forest or other vegetation. It is when extensive marshy areas are cleared for cultivation, and for half the year are dried up by the tropical sun, that they become deadly. I have lived for months together in or close to tropical swamps, both in the Amazon valley, in Borneo and in the Moluccas, without a day's illness; but when living in open cultivated marshy districts I almost invariably had malarial fever, though I believe the worst types of these fevers are due to unwholesome food. But here again, malaria was equally prevalent in England less than two centuries ago.

If we take the great belt, about two thousand miles wide, extending from twelve to fifteen degrees north and south of the equator, we have an enormous area, by far the larger part of which is not only well adapted for European colonization in the true sense, that is, for permanent occupation by white men, but is also with proper sanitary precautions the most healthy and enjoyable part of the world, and that in which the labourer can obtain the maximum return with the minimum of toil. I formed this opinion in 1851 when returning down the Rio Negro and Amazon after four years' residence there, and my subsequent eight years' experience in the East has only confirmed it. I then wrote as follows:

"It is a vulgar error, copied and repeated from one book to another, that in the tropics the luxuriance of the vegetation overpowers the efforts of man. Just the reverse is the case: Nature and climate are nowhere so favourable to the labourer, and I fearlessly assert that here (on the Rio Negro) the primeval forest can be converted into rich pasture or into cultivated fields, gardens and orchards, containing every variety of produce, with half the labour, and, what is of more importance, in less than half the time that would be required at home."

Then, after giving some details as to the various crops that may be grown and the varieties of fruits, vegetables and animal food that can be easily had, I conclude thus:

"Now I unhesitatingly affirm that two or three families, each containing half a dozen working and industrious men and boys, and being able to bring a capital in goods of £50 ($250), might in three years find themselves in possession of all I have mentioned. Supposing them to become used to the mandiocca and maize bread, they would, with the exception of clothing, have no one necessary or luxury to purchase; they would be abundantly supplied with pork, beef and mutton, poultry, eggs, butter, milk and cheese, coffee and cocoa, molasses and sugar. Delicious fish, turtles and turtles' eggs, and a great variety of game would furnish their table with constant variety, while vegetables would not be wanting, with fruits, both cultivated and wild, in superfluous abundance and of a quality that we at home rarely obtain. Oranges and lemons, figs and grapes, melons and watermelons, jack-fruit, custard-apples, cashews, pineapples, etc., are among the commonest, while numerous palm and other forest fruits furnish delicious drinks and delicacies which every one soon gets very fond of. Both animal and vegetable oils can be procured for light and cooking. And then, having provided for the body, what lovely gardens and shady walks might be made! How easy to form natural orchid bowers and ferneries! What elegant avenues of palms might be planted! What lovely climbers abound to train over arbours or up the walls of the house!"

But, it is objected, this cannot be done without hard work, and we *know* that "white men cannot live and work in the tropics." But I maintain that we know nothing of the kind. It is *not* the fact that white men cannot permanently live and work in the tropics. Work of some sort, there as here, is a condition of healthy life. But with a reasonable amount of work—and such is the beneficence of nature that little is needed—man can not only live permanently but most healthily and enjoyably in those portions of the tropics I am referring to, and probably, with special precautions, in every part. I will now give some of the facts bearing upon this question.

My own experience assures me that I owe my long life and comparatively good health to my twelve years' residence in the uniform climate and pure air of the equatorial forests, although I suffered frequently from fevers, and on one occasion was brought to the very point of death. I was a very delicate child, with weak lungs, and at the age of sixteen or seventeen suffered from serious ulceration of the lungs, and was only saved by the application of Dr. Ramage's common-sense air-treatment, somewhat analogous to that now being introduced for consumption. When I came home in 1862, although much weakened by other illnesses, my lungs were quite sound; and I distinctly trace my recovery to an open-air life in an equable, warm, pure atmosphere. My work as a collector of natural history specimens led to my being out of doors for six or seven hours during the heat of the day, and I found that I could take as much exercise without fatigue as I could at home.

At Para, in 1848, I saw a striking case of how a white man *can* work in the tropics. A tall, gentlemanly young Scotchman, finding no suitable occupation, and seeing that good milk was scarce in the city, determined to turn milkman. He hired a hut and some sheds about half a mile away, surrounded by second-growth forest and coarse grassy fields, obtained three or four cows, and when I made his acquaintance had got his business in full swing; and his work was certainly rather heavy. He lived absolutely alone; all the fodder for his cows when in milk had to be cut with a scythe and carried to the sheds where they were kept; water had also to be brought to them and the sheds kept clean. Early in the morning the cows were milked, filling two large cans, when he immediately started for the city, carrying them from a yoke across the shoulders in the orthodox manner, and making his rounds to all the houses he served. Returning, he had to get his own breakfast. Then for several hours there was grass-cutting and attending to the cows, and getting his own dinner. Yet often in the early evening he was dressed and made calls, often at the very houses he had served with milk in the morning. Notwithstanding this hard work, with the thermometer from 80 to 90 degrees or upward every day, he was the picture of health, and appeared to enjoy his life.

It is a well-known fact that in Ceylon and India the men who have the best health are the enthusiastic sportsmen who seize every opportunity of getting away from civilization, and who often submit to much privation and fatigue with benefit rather than injury to their health. Our soldiers, again, even in the unhealthy climate of India, most of which is really outside the tropics, have to do a good deal of work, and when marching against an enemy undergo much fatigue, and we do not hear that they are unequal to it on account of the heat. The same is even more clearly the case with our sailors, who do their regular work when stationed in the tropics, and do not suffer injury either from the climate or the work, if not exposed to infectious disease while on shore. The editor of the *Ceylon Observer*, commenting on my letter on this subject in the *Daily Chronicle*, adduced case after case of officers, planters, doctors, &c., who had lived from twenty-five up to fifty-eight years in Ceylon and have retained almost continuous good health. He also refers to Dutch families descended from settlers who came out from 150 to 200 years ago, and who have maintained average good health even in the hot country of the plains. In the Moluccas there are even more striking examples, many of the Dutch families having been continuously on the islands for 300 years, and they have still the fair complexions and robustness of form characteristic of their kinsfolk in Holland. The Government physician at Amboyna, a German, assured me also that the race is quite as prolific as in Europe, families of ten or a dozen children being not uncommon. The Dutch, however, live sensibly in the tropics, doing all their official work between the hours of 7 and 12 a.m., resting in the afternoon, and going out in the evening.

But perhaps the most conclusive example is that of Queensland, the climate of which is completely tropical; yet white men work in every part of it. Whether as gold miners, sheep shearers, sugar workers or railway builders, there has never been any complaint that white men *cannot* work; while almost all the heavy mechanical work of the country, engineering of every kind, carpentering and all the various building trades, and the scores of varied industries of a civilized community are carried on by white workmen without any difficulty and with no special effect on their general health. In an article on "Industrial Expansion in Queensland" (*Westminster Review*, March, 1897), Mr. T. M. Donovan tells us that many of the large estates have now been broken up into small farms of about eighty acres each, and sold to white farmers, and he adds:

"Where a few years ago there was a large plantation worked by gangs of South Sea Islanders, there are now twenty or thirty comfortable homesteads. And the contention that white European labour could not stand the field work is blown into thin air by the practical experience of thousands of white labourers all along the coast. The black labour question is settling itself; it is only a matter of time until the sugar industry can entirely do away with Kanaka labour."

This experiment in Queensland really settles the question.

The fact is that white men *can* live and work anywhere in the tropics, *if they are obliged*, and unless they are obliged they will not, as a rule, work even in the most temperate regions. Hence, wherever there are inferior races, the white men get these to work for them, and the kinds of work performed by these inferiors become *infra dig.* for the white man. This is the real reason why the myth, as to white men not being able to work in the tropics, has been spread abroad. It applies in most cases to agricultural work only, because natives can usually be got to do this kind of work, while that of the skilled mechanics has usually to be done by white men. And another reason is that it is only by getting cheap labour in quantity that fortunes can be made in most tropical countries. But when people come to recognize that the fortune-makers, whether by gold mining, speculating or any of the various forms of thinly-veiled slavery, are not by any means the happiest, the healthiest or the wisest men, whereas those who really *work*, under the best conditions, so as to receive the whole produce of their labour, may be both healthy and happy, will usually live longer and enjoy life more, and by working in association may obtain all the necessaries and comforts of existence —then the enormous advantage of living in the best parts of the tropics will become evident. For not only is nature so much more productive that equal amounts of produce may be obtained with half or perhaps a quarter of the labour required in northern lands, but the essentials of a happy and an easy life are so much fewer in number. Houses may be slighter and far less costly; clothing may be reduced to less than half what is required here; fuel is only wanted for cooking; while the enjoyability of the early morning hours is so great that everybody rises before the sun, and thus comparatively little artificial light is required. When all this is fully realized we may hope to see co-operative colonies established in many tropical lands, where families of the same grade of education and refinement may so live as really to enjoy the best that life can give them. Thus only, in my opinion, can the best use be made of the tropics.

CHAPTER VI

HOW TO CIVILIZE SAVAGES

Do our missionaries really produce on savages an effect proportionate to the time, money, and energy expended? Are the dogmas of our Church adapted to people in every degree of barbarism, and in all stages of mental development? Does the fact of a particular form of religion taking root, and maintaining itself among a people, depend in any way upon race—upon those deep-seated mental and moral peculiarities which distinguish the European or Aryan races from the negro or the Australian savage? Can the savage be mentally, morally, and physically improved, without the inculcation of the tenets of a dogmatic theology? These are a few of the interesting questions that were discussed, however imperfectly, at a meeting of the Anthropological Society in 1865, when the Bishop of Natal read his paper, "On the Efforts of Missionaries among Savages;" and on some of these questions we propose to make a few observations.

If the history of mankind teaches us one thing more clearly than another, it is this—that all true civilizations and all great religions are alike the slow growth of ages, and both are inextricably connected with the struggles and development of the human mind. They have ever in their infancy been watered with tears and blood—they have had to suffer the rude prunings of wars and persecutions—they have withstood the wintry blasts of anarchy, of despotism, and of neglect—they have been able to survive all the vicissitudes of human affairs, and have proved their suitability to their age and country by successfully resisting every attack, and by flourishing under the most unfavourable conditions.

A form of religion which is to maintain itself and to be useful to a people, must be especially adapted to their mental constitution, and must respond in an intelligible manner to the better sentiments and the higher capacities of their nature. It would, therefore, almost appear self-evident that those special forms of faith and doctrine which have been slowly elaborated by eighteen centuries of struggle and of mental growth, and by the action and reaction of the varied nationalities of Europe on each other, cannot be exactly adapted to the wants and capacities of every savage race alike. Our form of Christianity, wherever it has maintained itself, has done so by being in harmony with the spirit of the age, and by its adaptability to the mental and moral wants of the people among whom it has taken root. As Macaulay justly observed in the first chapter of his history:

"It is a most significant circumstance that no large society of which the tongue is not Teutonic has ever turned Protestant, and that, wherever a language derived from that of ancient Rome is spoken, the religion of modern Rome to this day prevails."

In the early Christian Church, the many uncanonical gospels that were written, and the countless heresies that arose, were but the necessary results of the process of adaptation of the Christian religion to the wants and capacities of many and various peoples. This was an essential feature in the growth of Christianity. This shows that it took root in the hearts and feelings of men, and became a part of their very nature. Thenceforth it grew with their growth, and became the expression of their deepest feelings and of their highest aspirations; and required no external aid from a superior race to keep it from dying out. It was remarked by one of the speakers at the Anthropological Society's meeting, that the absence of this modifying and assimilating power among modern converts—of this absorption of the new religion into their own nature—of this colouring given by the national mind—is a bad sign for the ultimate success of our form of Christianity among savages. When once a mission has been established, a fair number of converts made, and the first generation of children educated, the missionary's work should properly have ceased. A native church, with native teachers, should by that time have been established, and should be left to work out its own national form of Christianity. In many places we have now had missions for more than the period of one generation. Have any self-supporting, free, and national Christian churches arisen among savages? If not—if the new religion can only be kept alive by fresh relays of priests sent from a far distant land—priests educated and paid by foreigners, and who are, and ever must be, widely separated from their flocks in mind and character—is it not the strongest proof of the failure of the missionary scheme? Are these new Christians to be for ever kept in tutelage, and to be for ever taught the peculiar doctrines which have, perhaps, just become fashionable among us? Are they never to become men, and to form their own opinions, and develop their own minds, under national and local influences? If, as we hold, Christianity is good for all races and for all nations alike, it is thus alone that its goodness can be tested; and they who fear the results of such a test can have but small confidence in the doctrines they preach.

The views here expressed are now, after more than thirty years, receiving unexpected support, if we may rely on a well-written and thoughtful article by Mr. E. M. Green in the *Nineteenth Century* of November, 1899. It appears that in our Colonies in South Africa the educated Kaffirs are beginning a movement for a church of their own with native ministers and native organisation. There is said to be ample education, talent, and religious enthusiasm to support such a church; but instead of being welcomed and fostered by encouragement and assistance, it seems to be viewed with suspicion and dislike by the official representatives of the local churches. The *South African Congregational Magazine*, for example, writing on this movement, remarks:

"The ground of their revolt appears to have been a sense of resentment against the social barriers in the way of their advancement to the chief seats of official authority in their ecclesiastical system. Conceiving they had a grievance on the ground of such suppression of their self-importance, the dream of a formation of a native Church, dissociated from all European influence and control, began to impress itself on their imaginations."

The writer goes on to say that as there was no hope of financial aid from any section of the colonial constituencies, a new idea struck the "curly pow" of the Rev. Mr. Dwaine, which was to get the negroes of America to take up the movement. Then the writer tells us that this "Rev. Mr. Dwaine" is an accomplished linguist (although a Kaffir), "speaks English as to the manner born," as well as Dutch and his own native tongue, and has a record of unsullied reputation and honourable Christian service; that he went to America, and "was enthusiastically received into the fellowship of the Methodist Episcopal Church, blessed by its bishops, and sent back with the assurance that the new cause would be taken up and backed by the available resources of the denomination in America."

Mr. Green visited this Mr. Dwaine, and tells us that he was dressed as a clergyman, and that his English was excellent. He said:

"The missionaries cannot understand how we feel about our old customs, and we think that if all the ministers for natives were natives themselves it would be better. You tell us that we are all the same in God's sight, but your people will not worship in the same church with our people."

Mr. Green adds, that as Dwaine's position is national rather than doctrinal, it is probable that he will influence his people in large numbers; and I told him that I had never attended a missionary meeting in London about Africa without hearing that a native ministry was the end to keep in view. His reply was: "They say that in London, but they do not say it here."

Nothing more strikingly illustrates the way these educated natives are treated in the Colonies than the fact that when Dwaine visited England to get funds in order to found a South African College for natives, he wished much to see St. Paul's Cathedral, but was afraid of being turned out. But some one told him to walk in, and he did so, and finding he was not turned out, he went again, and also went several times to Westminster Abbey to hear noted preachers, and he was surprised at the toleration of the white man—in London. Here we have the skin-deep Christianity that preaches brotherhood and equality, but acts the very opposite; while the colonial dislike of the idea of a native church is evidently due to another form of that love of place and power which, notwithstanding fine promises and theories, still refuses all self-government or political rights to the countless millions in British India, as well as to these educated Kaffirs who are still subjected to the most irritating and degrading subjection to petty officialdom, as strikingly illustrated by cases which Mr. Green gives us.

Yet these people are quite as intelligent and as capable of benefiting by a good education as are average Europeans. This is well shown by a letter to the *Queenstown Free Press*, from a Basuto named Pelem, which is given in Mr. Green's article. This letter is not only very good sense, but is written in clearer and better English than are the average letters that appear in our own local newspapers, showing to what a marvellous extent education has spread among these people, and how high are their natural capacities.

But we are told to look at the results of missions. We are told that the converted savages are wiser, better, and happier than they were before—that they have improved in morality and advanced in civilization—and that such results can only be shown where missionaries have been at work. No doubt, a great deal of this is true; but certain laymen and philosophers believe that a considerable portion of this effect is due to the example and precept of civilized and educated men—the example of decency, cleanliness, and comfort set by them—their teaching of the arts and customs of civilization, and the natural influence of the superiority of race. And it may fairly be doubted whether most of these advantages might not be given to savages without the accompanying inculcation of particular religious tenets. True, the experiment has not been fairly tried, and the missionaries have almost all the facts to appeal to on their own side; for it is undoubtedly the case that the wide sympathy and self-denying charity which gives up so much to benefit the savage, is almost always accompanied and often strengthened by strong religious convictions. Yet there are not wanting facts to show that much may be done without the influence of religion. It cannot be doubted, for example, that the Roman occupation laid the foundation of civilization in Britain, and produced a considerable amelioration in the condition and habits of the people, which was not in any way due to religious teaching. The Turkish and Egyptian Governments have been, in modern times, much improved, and the condition of their people ameliorated, by the influence of Western civilization, unaccompanied by any change in the national religion. In Java, where the natives are Mohammedans, and scarcely a Christian convert exists, the good order established by the Dutch Government and their pure administration of justice, together with the example of civilized Europeans widely scattered over the country, have greatly improved the physical and moral condition of the people. In all these cases, however, the personal influence of kindly, moral, and intelligent men, devoted wholly to the work of civilization, has been wanting; and this form of influence, in the case of missionaries, is very great. A missionary who is really earnest, and has the art (and the heart) to gain the affections of his flock, may do much in eradicating barbarous customs, and in raising the standard of morality and happiness. But he may do all this quite independently of any form of sectarian theological teaching, and it is a mistake too often made to impute all to the particular doctrines inculcated, and little or nothing to the other influences we have mentioned. We believe that the purest morality, the most perfect justice, the highest civilization, and the qualities that tend to render men good, and wise, and happy, may be inculcated quite independently of fixed forms or dogmas, and perhaps even better for the want of them. The savage may be certainly made amenable to the influence of the affections, and will probably submit the more readily to the teaching of one who does not, at the very outset, attack his rude superstitions. These will assuredly die out of themselves, when knowledge and morality and civilization have gained some influence over him; and he will then be in a condition to receive and assimilate whatever there is of goodness and truth in the religion of his teacher.

Unfortunately, the practices of European settlers are too often so diametrically opposed to the precepts of Christianity, and so deficient in humanity, justice, and charity, that the poor savage must be sorely puzzled to understand why this new faith, which is to do him so much good, should have had so little effect on his teacher's own countrymen. The white men in our Colonies are too frequently the true savages, and require to be taught and Christianized quite as much as the natives. We have heard, on good authority, that in Australia a man has been known to prove the goodness of a rifle he wanted to sell, by shooting a child from the back of a native woman who was passing at some distance; while another, when the policy of shooting all natives who came near a station was discussed, advocated his own plan of putting poisoned food in their way, as much less troublesome and more effectual. Incredible though such things seem, we can believe that they not unfrequently occur wherever the European comes in contact with the savage man, for human nature changes little with times and places; and I have myself heard a Brazilian friar boast, with much complacency, of having saved the Government the expense of a war with a hostile tribe of Indians, by the simple expedient of placing in their way clothing infected with the smallpox, which disease soon nearly exterminated them. Facts, perhaps less horrible, but equally indicative of lawlessness and inhumanity, may be heard of in all our Colonies; and recent events in Japan and in New Zealand show a determination to pursue our own ends, with very little regard for the rights, or desire for the improvement, of the natives. The savage may well wonder at our inconsistency in pressing upon him a religion which has so signally failed to improve our own moral character, as he too acutely feels in the treatment he receives from Christians. It seems desirable, therefore, that our Missionary Societies should endeavour to exhibit to their proposed converts some more favourable specimens of the effect of their teaching. It might be well to devote a portion of the funds of such societies to the establishment of model communities, adapted to show the benefits of the civilization we wish to introduce, and to serve as a visible illustration of the effects of Christianity on its professors. The general practice of Christian virtues by the Europeans around them would, we feel assured, be a most powerful instrument for the general improvement of savage races, and is, perhaps, the only mode of teaching that would produce a real and lasting effect.

CHAPTER VII

THE EXPRESSIVENESS OF SPEECH; OR, MOUTH-GESTURE AS A FACTOR IN THE ORIGIN OF LANGUAGE

THE science of language, as treated by its modern students and professors, is so largely devoted to tracing the affinities and the laws of growth and modification of existing and recently extinct languages, that some of the essential characteristics of human speech have been obscured, and the features that contribute largely to its inherent intelligibility overlooked. Philologists have discovered, as the result of long and laborious research, what they hold to be the roots or fundamental units of each of the great families of language; but these roots themselves are supposed to be for the most part conventional, or, if they had in the very beginning of language any natural meaning, this is held to have been so obscured by successive changes of form and structure as to be now usually undiscoverable. As regards a considerable number of the words which occur under various forms in a variety of languages, and which seem to have a common root, this latter statement may be true, but it is by no means always, and perhaps not even generally, true. In our own language, and probably in all others, a considerable number of the most familiar words are so constructed as to proclaim their meaning more or less distinctly, sometimes by means of imitative sounds, but also, in a large number of cases, by the shape or the movements of the various parts of the mouth used in pronouncing them, and by peculiarities in breathing or in vocalisation, which may express a meaning quite independent of mere sound-imitation.

These naturally expressive words are very often represented by closely allied forms in some of the Teutonic, Celtic, or other Aryan languages, and they have thus every appearance of constituting a remnant of that original imitative or expressive speech, the essential features of which have undergone little change, although the exact form of the words may have been continually modified. But even when it can be shown that a word which is now strikingly suggestive of its meaning has been derived from some other words which are less, or not at all, suggestive of the same idea, or which even refer to some totally different idea, the obvious conclusion will be that, even in the present day, there is so powerful a tendency to bring sound and sense into unison, as to render it in the highest degree probable that we have here a fundamental principle which has always been at work, both in the origin and in the successive modifications of human speech.

Many writers have discussed the interjectional and imitative origin of language—especially, in this country, Dean Farrar and Mr. Hensleigh Wedgwood—but neither in their volumes, nor in any other English work with which I am acquainted, is the subject elaborated with any approach to completeness, while many of its most important features appear to have been overlooked. One of the most celebrated philological scholars and writers has treated it with extreme contempt, and has christened it the " Bow-wow and Pooh-pooh theory; " and, perhaps in consequence of this contempt, its advocates often adopt an apologetic tone, and, while urging the correctness of the principle, are prepared to admit that its application is very limited, and that it can only be used to explain a very small portion of any language. This is, no doubt, true, if we go no further than the ordinary classes of interjectional and imitative words—the Oh! of astonishment, the Ah! and Ugh! of pain, the infantile Ba, Pa, and Ma, as the origin of father and mother terms, and the direct imitation of animal or human sounds, as in cuckoo, mew, whinny, sneeze, snore, and many others, together with the various words that may be derived from them. But this is merely the beginning and rudiment of a much wider subject, and gives us no adequate conception of the range and interest of the great principle of speech-expression, as exhibited both in the varied forms of indirect imitation, but more especially by what may be termed speech or mouth-gesture. During my long residence among many savage or barbarous people I first observed some of these mouth-gestures, and have been thereby led to detect a mode of natural expression by words which is, I believe, to a large extent new, and which opens up a much wider range of expressiveness in speech than has hitherto been possible, giving us a clue to the natural meaning of whole classes of words which are usually supposed to be purely conventional.

Mouth-gestures.

My attention was first directed to this subject by noticing that, when Malays were talking together, they often indicated direction by pouting out their lips. They would do this either silently, referring to something already spoken or understood, but more frequently when saying *disána* (there) or *itu* (that), thus avoiding any further explanation of what was meant. At the time, I did not see the important bearing of this gesture; but many years afterwards, when paying some attention to the imitative origin of language, it occurred to me that while pronouncing the words in question, impressively, the mouth would be opened and the lips naturally protruded, while the same thing would occur with our corresponding English words *there* and *that*; and when I saw further that the French *là* and *cela*, and the German *da* and *das*, had a similar open-mouthed pronunciation, it seemed probable that an important principle was involved.[1]

The next step was made on meeting with the statement, that there was no apparent reason why the word *go* should

[1] The botanical explorer, Martius, describes lip-pointing as used by certain Brazilian tribes, but he does not seem to have connected it with the character of the word accompanying the gesture, or to have drawn any conclusions from it.

not have signified the idea of coming, and the word *come* the idea of going; the implication being that these, like the great bulk of the words of every language, were pure conventions and essentially meaningless: or that if they once had a natural meaning it was now wholly lost and undecipherable. But, with the cases of *there* and *that* in my mind, it seemed to me clear that there was a similar open-mouthed sound in *go*, with the corresponding meaning of motion away from the person speaking; and this view was rendered more probable on considering the word with an opposite meaning, *come*, where we find that the mouth has to be closed and the lips pressed together, or drawn inwards, implying motion towards the speaker. The expressiveness of these two words is so real and intelligible that a deaf person would be able to interpret the mouth-gestures with great facility. The fact that words of similar meaning in several other European languages are equally expressive, lends strong support to this view. Thus for *go*, we have the French *va*, the Italian *vai*, the German *geh*, and the Anglo-Saxon *gân*, all having similar open-mouthed sounds; while the corresponding words for *come*—*venez, vieni, komm*, and *kuman*—are all pronounced with but slight movements of the mouth and lips, or even with the lips closed.

If, now, we assume that the word-gestures here described afford us indications of the primitive and fundamental expressiveness of what may be termed natural, as opposed to mere conventional speech, we shall be prepared to find that the same principle has been at work in the formation of many other simple words, though in some cases its application may be less obvious. We must, however, always bear in mind that, though to us words are for the most part mere conventions, they were not so to primitive man. He had, as it were, to struggle hard to make himself understood, and would, therefore, make use of every possible indication of meaning afforded by the positions and motions of mouth, lips, or breath, in pronouncing each word: and he would lay stress upon and exaggerate these indications, not slur them over as we do. The various examples of these natural forms of speech which will now be adduced will be almost wholly confined to the English language, since I have no sufficient knowof foreign tongues. I also think that the importance and reality of the principle will be better shown by illustrations drawn from one language only, while such a method will certainly be both more intelligible and more interesting to general readers.

First, then, we have a considerable number of pairs of words which are pronounced with mouth-gestures very similar to those of *go* and *come*. Thus we have *to* and *from*, *out* and *in*, *down* and *up*, *fall* and *rise*, *far* and *near*, *that* and *this*; in all of which we have, in the first series, the broad vowels *a* or *o*, pronounced, expressively, with rather widely-open mouth, while in the second series we have the thin vowels *e*, *i*, or *u*, or the terminal consonants *m*, *n*, or *p*, which are pronounced either within the mouth or with closed lips; and in each special case the action will be found to be expressive of the meaning. Thus, in *to* the lips are protruded almost as much as in *go* (always supposing we are speaking impressively and with energy), while *from* requires only a slight motion of the lips ending with their complete closure; in *out* we have an energetic expiration and outward motion of the lips, while *in* is pronounced wholly inside the mouth, and does not require the lips to be moved at all after the mouth is opened; in *down* we have a quick downward movement of the lower jaw, which is very characteristic, since the word cannot be spoken without it; while in *up* the quick movement is upward, after having opened the mouth as slowly as we please; in *fall* we require a downward motion of the jaw as in *down*, but slower, and the word is completed with the mouth open, indicating, perhaps, that *fall* is a more decided and permanent thing than *down*, which implies position rather than motion, while in *rise* we have a slight parting of the lips, and the meaning would probably be made clearer by the gesture of raising the head, which is natural during inspiration. In repeating the lines—

> "On the swell
> The silver lily heaved and fell,"

we *feel* the motion in our heaving and falling chest, and we may be sure that with early man, such motions, when they helped the meaning of the words, were always fully emphasized.

Of the same general character as the words just considered, are the personal pronouns—*thou, you, he, they*—all of which are pronounced with outward breathing, and more or less outward motion of the lips, as compared with *I, me, we, us*, which require only slightly parted lips, thus clearly marking the difference between inward and outward, self and not-self. In like manner, *there* is spoken open-mouthed, and with strong outward breathing, while *here* requires but a slightly open mouth, and is only slightly aspirated.

Mr. E. B. Tylor has called attention to "the device of conveying different ideas of distance by the use of a graduated scale of vowels," as being one of great philological interest, on account of "the suggestive hint it gives of the proceedings of the language-makers in most distant regions of the world, working out in various ways a similar ingenious contrivance of expression by sound." He then gives a list of the words for this and that, here and there, I, thou, and he, in twenty-three languages of savage or barbarous tribes in both hemispheres, in all of which the ideas of nearness and distance, or self and not-self, are conveyed by the "similar ingenious contrivance" of different vowel-sounds.[1] But he does not appear to have observed that there is a method in the use of vowels, and that they are not therefore merely "ingenious contrivances," or contrivances at all in the true sense of the word, but are natural expressions of the difference of meaning in the way here pointed out. This is decidedly the case in eighteen out of the twenty-three languages given by Mr. Tylor, the broad, open-mouthed sounds *ah, o,* and *u,* being used to express outwardness or distance, while the contrasted vowels *e* and *i,* occur whenever self-hood or nearness is implied. In the other five languages the vowels are apparently reversed, which may be due either to a mistake of the compiler of the vocabulary—not at all

---

[1] *Primitive Culture*, vol. i., p. 199.

an uncommon thing when vocabularies are obtained through interpreters—or, possibly, to a real change of the letter used, owing to some of the numerous causes which bring about modifications of language, and even reversals of the original meaning of words. The tendency to preserve or add to the expressiveness of speech evidently varies much among different peoples, and we must not, therefore, be surprised at finding some incongruities in the use of even the most simple and natural sounds.

We now come to a series of words in which the action of breathing is the expressive part, the motion of the lips being very slight or altogether imperceptible; such are *air*, which is merely a modulated breathing; *wind*, in which more movement of the lips is required, with a slight indication of the characteristic murmuring sound; while in *blow* we almost exactly imitate the action of blowing. The words *breath* and *life* are related, inasmuch as the life-giving action of breathing is the fundamental part of both, modified by a different slight action of the lips and tongue, and it is suggestive that in many languages breath is used for spirit or life. *High* and *low* are also breath- or throat-words, the former being pronounced with open mouth, and, probably, with the accompanying gesture of raising the head, the latter with the tongue and palate only, the lips being but slightly parted. Small modifications of the former word would lead to *sky*, and perhaps also to *fly*, in both of which the idea of height is prominent.

We next have a group of words of which the essential character seems to be that the mouth remains open when they are spoken, as in the word *mouth* itself, in which the lips, teeth, and tongue are all employed; and in *all*, in which the mouth is still more widely opened. This is especially the case in words denoting round objects, such as *moon, ball, ring, wheel, round*, in all of which, as well as in many of the corresponding words in other languages, the chief feature is that the lips are held apart, and the mouth more or less rounded in pronouncing them. *Sun* may well belong to the same group, if it is not the chief of them, since it is the only object in nature that is always perfectly round, a feature that would be more easily represented in primitive speech than the light or heat which to us seems its most important characters. The root SU, and the various forms of sun in other Aryan languages, have all the same character of open-mouthed pronunciation, and the term for *south*, or sunward, is clearly derived from it. In Mr. Kavanah's work on *Myths traced to their Primary Source in Language*, the symbol O, representing the sun, is held to have been the first word and symbol used by primitive man, and a vast wealth of illustration from various sources is brought together to support the somewhat fantastic idea.

Other characteristic mouth-words are *mum* (silence), a mere parting and closing of the lips, whence comes *mumble* and perhaps *dumb*. *Spit* also is a labial imitative word, but it imitates the action of spitting as well as the sound. *Sleep* may also be considered a mouth-word, and in pronouncing it we gradually close the mouth in a very suggestive manner, while in *wake*, *awake*, we abruptly open it.

We now pass on to words for *nose* and whatever appertains to it, which, in a considerable proportion of known languages, are formed by nasal sounds, such as are represented by our letters *m*, *n*, *ng*, with the sibilants *s* or *z*. Thus we have *snout*, *nozzle*, *nostril*, *snore*, *snort*, *sneeze*, *sneer*, *sniff*, *snivel*, all things or actions immediately connected with the nose, while *smell*, *stink*, *stench*, and *nasty*, are also expressive nasal words.

A distinct set of words, appertaining to the teeth, tongue, or palate, are characterized by *t, d, s*, and *n* sounds, and are pronounced wholly within the mouth without any definite action of the lips. Thus, besides *tooth* and *tongue*, we have *tusk*, *eat*, *gnaw*, *gnash*, and *taste*; while perhaps *knee*, *knot*, *knob*, *knoll*, *knuckle*, and some other words of doubtful derivation, may get their characteristic type from the analogy of a tooth-like projection. It is to be noted that nasal and dental sounds characterize words of similar meaning, not only in European languages, but more or less all over the world.

Continuous or abrupt Sounds and their Meanings.

Before passing on to consider the various modes in which sounds, actions, and even qualities, are expressively represented in speech, attention must be called to the way in which certain groups of consonants are utilized to indicate differences in the general character of sounds and motions. When either of the following letters—*f, l, m, n, ny, r, v, s*, or *z*—occur at the end of a word, either with or without a final vowel, we can dwell upon them and thus give them a continuous sound; and the more important of these have been termed *liquids*, because they seem to flow together and form one continuous sound. But the letters *b, d, g, k, p,* and *t*, have a very different character, and when any of them comes at the end of a word, and are not silent, the sound ends abruptly, and we find ourselves altogether unable to dwell upon and lengthen out the sounds of these letters as we can those of the first group; neither does the addition of a final *e* help us to dwell upon them. Compare, for instance, the words *ball* or *bear* with *bat* or *dog*. In the former the sound of the final letters can be continued indefinitely, while in the case of the latter we come to a dead stop, and by no effort can continue the sound.

Now, the various sounds which occur in nature may be broadly divided into two classes, the continuous and the abrupt; and it is a most suggestive fact that these two classes of sounds are almost always represented in our language by words which, owing to their terminal letter, are of a corresponding character. Thus, among continuous sounds we have *roar, snore, hiss, sing, hum, scream, wail, purr,* and *buzz,* all of which end in letters of the first series, enabling us to dwell upon the word as long as we please. But when we name abrupt sounds, such as *rap, clap, crack, tick, pop, thud, grunt,* and many others, we find that the word ends as abruptly as does the sound it represents, and that the final letter does not in any case admit of being dwelt upon and drawn out as in the case of words of the first series.

But even more curious is the fact that the same law of expression applies in the case of motions. These, too, are either continuous or abrupt; and these are also represented by words whose terminal letters either can or cannot be dwelt upon. Of the former kind are—*fly, run, swim, swing, move, crawl, turn, whirl*, and *slide;* and these words all indicate the continuity of the various kinds of motion by their terminal sounds being indefinitely continuous. But motions whose chief characteristic is their abrupt termination, such as *step, hop, jump, leap, halt, stop, drop, bump, wink*, or actions which imply such motion as *strike, hit, knock, pat, slap, stamp, stab, kick*, all have a corresponding ending in non-continuous letter-sounds.

This remarkable series of correspondences is highly suggestive of a law of primitive word-formation. At a very early stage in the growth of speech, it would be observed that some vocal sounds were capable of being drawn out, while others necessarily had an abrupt termination; and, as natural sounds and motions had also these contrasted features of abruptness or continuity, it was the most natural thing in the world to make the names of these sounds, motions, or actions, agree in this respect with the things named. Most of these words are very similar in other Teutonic languages, and however much they may have changed in the course of ages, they have, as we see, retained this particular form of expressiveness in a very remarkable degree. In all this we have no mere convention or ingenious contrivance, but a natural imitative expressiveness, arising out of the very nature and limitations of articulate speech.

Words imitating Sounds.

We will now proceed to a brief discussion of the various classes of words which are more directly sound-imitations; and though many of these are among the most familiar examples adduced by the exponents of the imitative origin of language, yet their great range, the variety in their modes of imitation, and their marvellous power of indicating not only sounds, but even motions, actions, and physical qualities, have hardly received sufficient attention.

Human cries have already been referred to when noticing the difference between abrupt and continuous sounds, but there are a few points of detail that may be noted here. In the word *whistle* we have the nearest representation a word can give to the action of whistling; in *babble* we have the *ba ba* of infancy; in *whisper* we have a word which is a mere articulate breathing or aspirate; in *hush!* we have a gentle aspirate alone; in *cough, wheeze,* and *spit,* we have not merely the sounds but the actions closely represented in words; in pronouncing *yawn* we open the mouth and produce a throat sound as in yawning; in *scream, screech, squall,* and *yell,* we have a fair imitation of loud and energetic cries due to sudden pain or anger; while in *moan, groan, wail, sigh,* and *sob,* we hear the more subdued indications of grief or continuous pain. *Stutter* and *stammer* almost exactly reproduce the acts indicated.

In naming the sounds or voices of animals we use words which are almost universally imitative, and are so well known that they need not be here given; but we may note how well *chirp* and *warble* represent the voices of the less and more musical of our small birds, as do the *cawing* of the rook, and the *cooing* of the dove, those of larger species.

It is when we come to the varied sounds of inanimate nature that we begin to realize the wonderful expressiveness and picturesqueness of our every-day speech, and how far superior it is to any purely conventional language as a means of conveying to another person a description of the varied scenes, actions, and passions of life.

And first, how well the word *murmur* serves to represent the low, modulated sound of a gentle wind among trees, or of the distant waves; while *breeze* indicates the distant rustle of leaves shaken by a stronger wind; and from these sounds and motions the word *trees* and *tremble* have not improbably arisen, as they occur with but slight modifications in all the Teutonic languages. Then, again, how well the minute differences of quality between various common sounds are represented in their names—the light and moderately sharp *tap*, the much sharper *snap*, the fuller and broader *clap*, with the less abrupt *flap*, the duller *flop*, and the softer and still duller *thud*.

Sounds which have an element of vibration in them are represented by words containing *r* or *cr* when harsh, as in *creak* and *crack*; but when the vibration is of a more pronounced or musical character we have *clang*, *ring*, and *sing*; and when vibratory objects strike together we have *clink* and *clash*. How well the sound of boiling liquids is represented by *bubble*; the confused sound of various hard objects striking together by *clatter* or *rattle*; while *hiss*, *whizz*, and *fizz* well represent the effects of rapidly escaping air or gases.

Words imitating the sounds of various kinds of breaking objects are highly characteristic. Beginning with *squash*, which applies best to soft fruits, we find *crush*, in which the *cr* represents the somewhat harsh sound of the initial break, as in *crack*; and *crunch*, in which we seem to hear the final crushing up of the hard pieces into which the first *crack* reduced the object. In *grind* we have this final breaking up into dust alone represented; while in *crumble* we have the disintegration of a much softer substance under moderate pressure. *Split* represents the sudden, sharp sound of splitting wood; *tear*, the violent pulling asunder of a woven fabric; and *rip*, the still harsher sound when a seam is cut or torn apart. In *scratch*, we have the sound first represented, followed by the interjectional *ach* of pain which is the result of the action. In the word *saw* we have an imperfect imitation of the sound produced by sawing, though in Sanscrit, and in many of the languages of semi civilized peoples, it is more exactly imitative.[1]

The sounds produced by liquids in motion are often indicated by *sh*, as in *wash*, *splash*, and *dash*; a quantity of liquid falling to the ground causes a *slop* which repre-

---

[1] See Tylor's *Primitive Culture*, vol. i., p. 191, where a rather full account is given of imitative words in the languages of all parts of the world.

sents the sound it makes, as does *drop* when caused by a small globular portion; while *quench* well represents the noise produced by water used in sufficient quantities to extinguish a fire.

Many natural objects appear to have been named from their characteristic sound. *Brass* and *glass*, from their resonance; *tin*, from its more delicate, tinkling sound; *iron*, perhaps from its peculiar harsh vibration when struck; *lead* and *wood*, from the dull sound, or *thud*, which they produce. In *ice* we have probably the indications of the *sh* of "shiver" caused by touching it, and its transparency may have led to the use of the somewhat similar term for *glass*. In pronouncing the word *fire* we seem to imitate with the lips and breath the wavy flickering motion of flame, and the name for the *fir* tree, almost identical in many of the Scandinavian and Celtic languages, is doubtless in reference to the upward-growing, pointing form, like that characteristic of fire. *Glow* seems to represent the steady light of embers as contrasted with the incessant motion of *fire*, for while the latter word requires a double motion of the lips, the former is pronounced wholly inside the mouth by means of the tongue and palate, the lips remaining motionless. In the words *step*, *stamp*, and *stop*, we have a very close representation of the sound of the bare foot upon the ground in walking, and it seems quite probable that the root *sta*, from which they are said to be derived, had this origin.[1]

Sounds which represent Motions.

We now pass on from mere sounds to the various kinds of motions to be observed in nature; and we shall find that these also are represented by curiously expressive combinations of vocal utterances, often requiring imitative motions in the organs of speech. The modes of indicating the difference between continuous and abrupt motions have already been referred to, but each particular kind of motion has also its characteristic combination of

---

[1] A considerable number of these directly imitative words are given in Dean Farrar's *Essay on the Origin of Language*, chap. iv.

letters. The word *slow*, to be spoken distinctly and impressively, must be pronounced slowly, while *quick* and *swift*, on the contrary, must be spoken rapidly. *Move* takes time to pronounce it distinctly, and implies slow and smooth motion, as *fly* implies swifter motion. In *crawl*, the harsh sounds at the beginning and end of the word imply slow and difficult motion, and the still harsher sound in *drag* recalls the noise of a heavy object forcibly drawn over an irregular surface. In *flutter* and *flicker* we have complex motions of the lips, tongue, and palate, corresponding to those they indicate; in *hurry* and *flurry* we seem to hear the rapid breathing of a tired or excited person; while in *wobble* and *hobble*, the clumsy movements are reproduced in the mouth of the speaker. How perfectly is smoothness of motion imitated while we say *slide* or *glide*; while the slow down and up motion of the lips in pronouncing *wave* is highly suggestive of wave-motion. The more rapid wave-movement we term vibration is indicated by the *br* in *vibrate*; while in *tremble* we have a more irregular shaking denoted by the *tr* at the beginning, and the *bl* at the end of the word. When we say *twist* or *screw*, there is a tendency to twist the mouth; while *shiver* represents a trembling motion accompanied by the *sh* of cold. In *stream* and *flow* the liquid consonants well represent the smoothness and continuity of liquid motion; and in *glow* we have, as already stated, a corresponding word to imply the smooth and steady light of incandescent matter, so different from the unsteady flicker which is characteristic of flame. A similar use of liquid sounds in *blush* and *flush* serves to indicate a gradual and steady increase of colour.

Qualities represented by Sounds.

We have now to take another step—and a most important one—in the development of language, and to show how the various qualities or properties of inanimate objects, and even the powers and faculties of men and animals, are clearly indicated by characteristic combinations of vocal sounds, affording us many striking examples of the expressiveness of speech.

Just as certain motions were seen to be distinguished by the use of harsh or liquid sounds, so are the qualities of objects on which these varied kinds of motion often depend equally well characterized. Compare, for example, the words *smooth, even, polished*, with *rough, rugged, gritty*, and we at once see that these are not merely conventional terms, but that they are as truly and naturally expressive as are the most direct imitations of human or animal cries. Corresponding to these, we have the names of many smooth substances—as *oil, soap, slime, varnish*, characterized by smooth or liquid sounds; and, on the other hand, such objects as *rock, gravel, grit, grouts, ground*, all containing the harsh sounds implying roughness. When we pronounce the words *sticky*, or *clammy*, we seem to feel the tongue and palate stick together, and have to pull them apart; and the same peculiarity applies to the words *cling* and *glue*.

There are in all languages words allied to *foul, putrid, pus*, &c., which are usually traced to the interjectional expressions of disgust, *puh! fie!* Similar expressions are shown by Mr. Tylor to be used among the most widely separated races in all parts of the world, and the reason of this identity is to be found in the natural and almost involuntary action of blowing away, through both mouth and nostrils, the emanations from putrid matter—as when we draw back the head and say *puh!*—an action more or less common to all mankind.

The words *hard* and *soft* are also expressive, though it is more difficult to define why. The former word, however, is pronounced with a strong aspirate, and the terminal *rd* requires more effort to pronounce than the gentle sibilant and terminal *ft* of soft. But when we consider the various terms designating contrasts of size, we have no such difficulty. The words *great, grand, huge, vast, immense, monstrous, gigantic*, are all pronounced with well-opened mouth and with some sense of effort, and the more stress we lay upon the word, the more distinctly we show our meaning by the wide opening of the mouth.

In the correlative words *small, little, wee, tiny, pigmy*, on the contrary, we use no effort, and hardly need to open the mouth at all, the pronunciation being effected almost wholly by the tongue and teeth. Even when new words are invented they follow the same rule, as in Swift's "Brobdingnag" and "Lilliput;" while the languages of uncivilized peoples are usually, as regards these words, equally characteristic. Though usually limiting my illustrations to our own language, I will here give the words for great and small in several of the languages of the Malay Archipelago; thus—*busar, bagut, baké, lamu, ilahe, maina*, all with broad open-mouthed vowel-sounds, mean great or large; while *kichil, chili, kidi, koi, roit, kemi, anan, fek, didiki*, all meaning small in the same languages, are in every case pronounced inside the mouth, and with but slightly parted lips.

Even more expressive are the words by which we indicate power or effort, such as *might, strive, strenuous, struggle, laborious, strong, strength*—this last being one of the most remarkably expressive in the language, consisting, as it does, of no less than seven consonants and only one vowel, all the consonants being fully and distinctly sounded. To pronounce this word clearly and emphatically requires a considerable effort, and we thus seem to be exerting the very quality it is used to express. How different are the words of opposite meaning, such as *weak, weary, languish, faint*, which can all be spoken with the minimum of effort, and with a hardly perceptible motion of the lips; and the same contrast is found in the common adjectives, *difficult* and *easy*.

How much the natural expressiveness of words adds to the beauty of descriptive poetry may be seen everywhere. In Pope's well-known lines—

"When Ajax strives some rock's huge weight to throw,
The line too labours and the words move slow,"

the very nature of the words which are of necessity employed, produces that effect of appropriateness which we are apt to think is due wholly to the skill of the poet. In another couplet from the same poem—

"A needless alexandrine ends the song,
And like a wounded snake drags its slow length along,

the natural expressiveness of the words, *drags, slow*, and *length*, is what conveys such a sense of appropriateness to the simile. Tennyson also is full of such naturally descriptive passages. The lines—

"The myriad shriek of wheeling ocean-fowl,
The league-long roller thundering on the reef,"

owe much of their force and beauty to the natural expressiveness of our common words; and the same is the case in the still more beautiful lines—

"Myriads of rivulets hurrying through the lawn,
The moan of doves in immemorial elms,
And murmuring of innumerable bees."

A few examples of words that are especially expressive may now be given, in order to illustrate some of the varied ways in which the principle has acted, and how largely it has influenced the formation of language. The word *growth* is expressive of the gradual extension of a young plant owing to the circumstance that we begin its pronunciation far back in the mouth, and that it seems to move outwards till the tongue touches the teeth or even the protruded lips. If we watch carefully we shall see how curiously, when we say "growth," we imitate with our vocal organs the very process which the word implies. From this foundation the name of the colour *green* has been derived, as that of growing things, and probably also *grass, graze*, and even *ground*. This last word is usually supposed to be allied to *grind*, as implying that the ground is dust, earth, or rock ground up. But this is surely a very unlikely idea to have occurred to primitive man, since the natural ground is usually firm and covered with some kind of vegetation or "growth," whence its name would be naturally derived.

When pronouncing the work *suck*, we are evidently imitating both the sound and the action of sucking, by drawing back the tongue during an inspiration; and in *taste* we are equally imitating the act of tasting, by moving the tongue twice within the mouth into contact with the palate, as we do when using it to move about and taste a savory morsel. So, in the word *sweet*, we seem to draw in and taste an agreeable substance; while in *sour* we open the mouth and the tongue remains free from either teeth or palate, as if we desired to get rid of a too biting flavour. Now *sweet*, with various modifications of form and meaning, occurs in all the Teutonic and Latin languages, but its whole significance as a naturally expressive word is lost when we are referred for its origin to the Aryan root *swad*, to please.[1] In Sanscrit, *swad* is to taste, and *svádu* sweet; and the more probable inference would be that the abstract root *swad*, to please, was derived from the more primitive and naturally formed terms for *taste* and *sweetness*.

Even moral qualities may be indicated by words which are naturally expressive, as in *right* and *wrong*. The former is, in most languages, connected with *straight* and *stretch*, the latter word being imitative of the sound produced when stretching a cord, the only straight line accessible to primitive man; while *wrong* is undoubtedly the same word essentially as *wrung*, from *wring*, *wry*, *wrench*, *wrest*, and other words meaning *twisted*, in pronouncing which and giving its full sound to the initial *w*, we seem naturally to give a twist to the mouth. When we speak of "rectitude," of an "upright" man, of "crooked" dealings, of a "perverted" disposition, we show how easy it is to describe moral characteristics by means of words applicable to mechanical or physical qualities only.[2]

---

[1] Skeat's *Etymological Dictionary of the English Language*, under "Sweet."

[2] As examples of this transference of meaning from the physical to the mental or moral, Dean Farrar gives, "imagination," the summoning up of an image before the inward eye; "comprehension," a grasping; "disgust," an unpleasant taste; "insinuation," getting into the bosom of a thing or person; "austerity," dryness; "humility," related to the ground; "virtue," that which becomes a man; "courtesy," from a court or palace; "aversion" and "inclination," a turning away from, and a bending towards anything; "error," a wandering; "envy," "invidious," a looking at, with bad intent; "influence," a flowing in; "emotion," a motion from within, or of the soul. (See *Origin of Language*, p. 122.) To which we may add, "evident," to be seen clearly; and from the same Latin words—*videre*, to see, and *visus*, sight—a whole series of English words are derived,

Summary of the Argument.

I have now briefly sketched and illustrated the varied ways in which many of the most familiar words of our language are truly expressive of the meaning attached to them, and have shown how far these carry us beyond the range of interjectional and imitative speech, as usually understood. Besides the more or less direct imitation of the varied sounds of nature, animate and inanimate, we have *form*, indicated by the shape of the mouth; *direction*, by the motion of the lips; such ideas as those of *coming* and *going*, of *inward* and *outward*, of *self* and *others*, of *up* and *down*, expressed by various breathings or by lip and tongue-motions; we find the distinct classes of abrupt or continuous *sounds*, as well as the corresponding contrasted *motions*, clearly indicated by the use of expressive terminal letters; *motion* of almost every kind, whether human, animal, or inorganic, we find to be naturally expressed by corresponding motions of the organs of speech; the physical *qualities* of various kinds of matter are similarly indicated; while even some of the mental and moral qualities of man, as well as many of his actions and sensations, are more or less clearly expressed by means of the various forms of speech-gesture.

If we consider the enormous changes every language has undergone; that words have often taken on new meanings, or have been displaced by foreign words quite distinct in derivation and original signification; that inflections have been altered or altogether dropped; and that, in various other ways, words have been undergoing a continual process of growth and modification, the wonder is that so much of the natural foundations of our language can still be detected. Philologists give us innumerable

among which are "advice," according to a person's judgment or seeing; "provide" and "prudent," to act with foresight; "visit," to go to see a person; "visage," the face or seeing part; "view," that which is seen; and many others. Hence it is easy to perceive that, once given terms for the physical characteristics and qualities of objects, and the whole range of language which refers to mental and moral qualities can be easily developed.

examples of how words come to be used in ways quite remote from their original meanings, and how several quite distinct words grow out of a common root—as when *cannon*, a great gun, and *canon*, meaning either a dignitary of the church or a body of ecclesiastical or other laws, are alike derived from *canna*, a cane or reed, used either as a tube, or as a ruler, while from *canistrum*, a reed basket, we get *canister*, now used chiefly for metal cases of a particular form.

The late Mr. Hyde Clarke has shown how very widely the primitive terms for mouth, tooth, tongue, &c., are applied to other things of like form or motions, or having a supposed or real analogy to them; thus, languages can be found in which the words for head, face, eye, ear, sun, moon, egg, ring, blood, and mother, are derived from *mouth*, for reasons which we can, in most cases, perceive or guess at. It follows that, whenever people use any form of written symbol for words or things, the growth of language goes on more rapidly, because symbols, which were at first actual representations of the object, for convenience become conventionalised, and then other objects which resemble the modified symbol are given either the same or an allied name. With us, *door* is named after the opening used for entering a house, and is allied to *through*; but, according to Mr. Hyde Clarke, in some languages the door-way or opening is derived from mouth—as we ourselves say the mouth of a pit or cavern—while the door itself is formed from a word meaning *tooth*.[1] The words *hill* and *mountain* have no light thrown upon their real origin by a reference of the former to the Latin *collis*, and the latter to *mons* and *montanus;* but, on the principles here set forth, both of them, as well as the German *berg*, owe their characteristic form of open-mouthed aspirated words to the natural panting ejaculations of those who ascend them. Yet in many languages they have been named from their form-resemblances, as in the well-known terms *dent, sierra, peak, pap, ben,* &c.

[1] See letter in *Nature*, vol. xxvi., p. 419, and vol. xxiv., p. 380.

How Speech originated.

Some of the correspondences which have been here pointed out between words and their meanings, will doubtless be held by many to be mere fantastic imaginings But if we try to picture to ourselves the condition of mankind when first acquiring and developing spoken language, and struggling in every possible way to produce articulate sounds which should carry in themselves, both to the speaker and the hearer, some expression of the things, motions, or actions represented, it will seem quite natural that they should utilize everything connected with the act of speaking which could in any way further that object. We are apt to forget that, though speech is now acquired by children solely by imitation, and must be to them almost wholly conventional, this was not its original character. Speech was formed and evolved, not by children, but by men and women who felt the need of a mode of communication other than by gesture only.[1] Gesture-language and word-language doubtless arose together, and for a long time were used in conjunction and supplemented each other. It is admitted that gesture-language is *never* purely conventional, but is based either on direct imitation or on some kind of analogy or suggestion; and it is therefore almost certain that word-language, arising at the same time, would be developed in the same way, and would never originate in purely conventional terms. Gesture would at first be exclusively

[1] One of the critics of this article ignored this very obvious fact, and argued that the only way to gain correct notions of the origin of language was, through close observation of the speech of children,— that this was the *scientific* method, and mine altogether unscientific. I venture to maintain the contrary. Men and women would *need* language, while to children it would be quite unnecessary. And at its *first* origin it could not have been conventional, since there would have been no means of explaining the conventions or coming to any agreement as to their use. Both gesture and speech must have originated in actions or sounds which were felt by the hearers to be expressive. Every one would be seeking after such modes of expression, and amidst these various efforts the fittest would survive. New words are even now formed and adopted in this way, and hardly any other mode is conceivable.

used to describe motion, action, and passion; speech to represent the infinite variety of sounds in nature, and, with some modification, the creatures or objects that produced the sounds. But there are many disadvantages in the use of gesture as compared with speech. It requires always a considerable muscular effort; the hands and limbs must be free; an erect, or partially erect, posture is needed; there must be sufficient light; and, lastly, the communicators must be in such a position as to see each other. As articulate speech is free from all these disadvantages, there would be a constant endeavour to render it capable of replacing gesture; and the most obvious way of doing this would be to transfer gesture from the limbs to the mouth itself, and to utilize so much of the corresponding motions as were possible to the lips, tongue and breath. These mouth-gestures, as we have seen, necessarily lead to distinct classes of sounds; and thus there arose from the very beginnings of articulate speech, the use of characteristic sounds to express certain groups of motions, actions, and sensations which we are still able to detect even in our highly-developed language, and the more important of which I have here attempted to define and illustrate.

It may be well to give an example of how definite words may have arisen by such a process. Each of the words—air, wind, breeze, blow, blast, breathe—has to us a definite meaning, and a form which seems often to have nothing in common with the rest. Yet they possess the common character that the essential part of each is a breathing, more or less pronounced and modulated; and at first they were probably all alike expressed by a strong and audible breathing or blowing. For convenience and to save exertion, this would soon be modified into an articulate sound or word which would enable the act of blowing to be easily recognized. Then, as time went on and the need arose, some one or other of the different ideas comprised in the word would be separated, and this would be most effectually done by the use of different consonants with the same fundamental form of *breathing* or *blowing*, and the distinction caused by the *r* and *l* in these two words well illustrates the principle. Thus, every such class of expressive words would have a natural basis, while the detailed modifications to differentiate the various ideas included in it might be to a considerable extent conventional.

In conclusion, I venture to submit the facts and arguments here set forth as a contribution to the fascinating subject of the origin of language. Of their novelty and value I must leave Anthropologists and Philologists to judge.[1]

[1] The fundamental idea of mouth-gesture was stated by the present writer in a review of Mr. E. B. Tylor's "Anthropology," in *Nature*, vol. xxiv. p. 242 (1881).

CHAPTER VIII

COAL A NATIONAL TRUST [1]

It has now become an axiom with all liberal thinkers that complete freedom of exchange between nations and countries of the various products each has in superabundance and can best spare, for others which it requires, is for the benefit of both parties; and this principle is thought to be so universally applicable, that, even when it produces positive injury to ourselves and is certain to injure our descendants, hardly any public writer who professes liberal views ventures to propose a limitation of it. It seems clear, however, that there are limitations to its wholesome application, and that there are certain commodities which we have no right to exchange away without restriction, for others of more immediate use to the individuals or communities who happen to be in possession of them. These commodities may be briefly defined as those natural products which are practically limited in quantity, and which cannot be reproduced.

What is meant may perhaps be best explained by taking what may be considered a very extreme case as an illustration. Let us suppose, for instance, a country in which the springs or wells of water were strictly limited in number, but sufficiently copious to supply all the actual needs of the people, who had always had the use of them on making a nominal payment to the owners of the land on which they were situated. Acting on the princi-

[1] This article appeared in the *Daily News* of September 16th, 1873. It is even more applicable to-day.

ples of unrestricted free trade, and anxious to increase their wealth, one after another of the landowners sold their springs to manufacturers, who used up all the water except that required to supply the wants of their own workpeople, thus rendering the remainder of the country almost uninhabitable. A still more extreme case, but one rather more to the point, would be that of a country possessing a surface soil of very moderate depth, but of extreme fertility, and supporting a dense population on its vegetable products. The landowners might find it very profitable to them to sell this surface soil to the wealthy horticulturists of other countries; and if the principle of free trade is unlimited, they would be justified in doing so, although they would permanently impoverish the land, and render it capable of supporting a less numerous and less healthy population in long future ages.

Most persons will admit that in both these cases the exercise of the unrestricted right of free trade becomes a wrong to mankind, and should on no account be permitted; and it will perhaps be said that such cases could never occur in a civilized community, as public opinion would not allow the landowners to act in the manner indicated even were they disposed to do so. I believe, however, it may be shown that, under circumstances far worse than those here supposed, the landowners in the most civilized community on the globe do act in a very analogous manner, and, moreover, are not yet condemned by public opinion for doing so. Let us first, however, deduce from such supposed cases as those above given a general principle determining what articles of merchandise are and what are not the proper subjects of free trade. A little consideration will convince us that most animal or vegetable products or manufactured articles, the production and increase of which are almost unlimited in comparatively short periods, are those whose free exchange is an unmixed benefit to mankind; the reason being that such exchange enriches both parties without impoverishing either, and by leading to improved modes of cultivation and an increased power of production, adds continually to the sustaining power of the earth, and benefits future generations as much as it does ourselves.

On the other hand, all those articles of consumption which are in any way essential to the comfort and well-being of the community, and which are, either absolutely or practically, limited in quantity and incapable of being reproduced in any period of time commensurate with the length of human life, are in a totally different category. They must be considered to be held by us in trust for the community, and for succeeding generations. They should be jealously guarded from all waste or unnecessary expenditure, and it should be considered (as it will certainly come to be regarded) as a positive crime against posterity to expend them lavishly for the sole purpose of increasing our own wealth, luxury, or commercial importance. Under this head we must class all mineral products which are extensively used in domestic economy, the arts or manufactures, and which are in any way essential to the health or well-being of the community, and more especially those which from their bulk, weight, and extensive use could not be imported from distant regions without a very serious addition to their cost, such as is pre-eminently the case with coal and iron.

Now, it will be seen that we have here to deal with a case quite as extreme in reality as those supposititious cases with which we commenced this inquiry. For coal and iron are almost as much necessaries of life to the large population of this country as are abundance of water and a fertile soil; but there is this difference, that the water might be restored to its legitimate use, and the soil might be renewed by a sufficient period of vegetable growth; whereas coal burned, and iron oxydized, are absolutely lost to mankind, and we have no knowledge of any restorative processes except after the lapse of periods so vast that they cannot enter into our calculations. It may be replied, that the quantity existing on the globe is vast enough for the necessities of mankind for any periods we need calculate on; but even if this be so (of which we are by no means certain), it may none the less be shown that numerous and wide-spread evils result from our present mode of recklessly expending the stores in certain countries, while the same products remain totally unused in many of the countries they are exported to. For a number of years we have been increasing our production of coal and iron at an enormous rate, and sending vast quantities of both to all parts of the world, civilized and uncivilized, and have thereby produced, so far as I can see, only evil results in various forms some of which have hitherto received little attention.

Briefly to state these:—In the first place, we have seriously, and perhaps permanently, increased the cost of one of the chief necessaries of life in so changeable a climate as ours—fuel. This is in itself so great and positive an evil that no considerations of mere convenience to remote nations, such as the construction of railways in New Zealand or in Honduras, ought even to be mentioned as an excuse for it. Coal in winter is a question of comfort or misery, even of life or death, to millions of the people whose happiness it is our first duty to secure; and shall we coolly tell them that the Antipodes must have railroads, and that landowners, coalowners, and contractors must make fortunes, although the necessary consequence is the yearly increasing scarcity of one of their first necessaries and greatest comforts?

In the second place, by destroying for ever a considerable and ever-increasing proportion of the mineral wealth of our country, we have rendered it absolutely less habitable and less enjoyable for our descendants, and we have not done this by any fair and justifiable use for our own necessities or enjoyments, but by the abuse of increasing to the utmost of our power the quantity we send out of the country, never mind for what purpose, so that it adds to the wealth of our landowners, capitalists, and manufacturers.

In the third place, we have brought into existence a large population wholly dependent on this excessive production and export of minerals, and therefore not capable, under existing conditions of society, of being permanently maintained on their native soil. In proportion as other nations make use of their own mineral productions, and as our own minerals, from the ever increasing difficulty of procuring them, become necessarily more costly, so must our excessive exports diminish, and with it must diminish our power of maintaining our present abnormal mining population. A period of adversity will then probably set in for us, only faintly foreshadowed in intensity and duration by those arising from mere temporary fluctuations in the demand for minerals and their manufactured products.

Fourthly, we not only injure ourselves and our successors by thus striving to get rid of our mineral treasures as fast as possible, but we probably do more harm than good to the nations to whom we export them; for we prevent them from deriving the various social and intellectual benefits which would undoubtedly arise from their being compelled to utilize for their own purposes the mineral products of their own lands. The working of mines and the establishment of manufactures bring into action such a variety of the mental faculties, and so well vary and supplement the labours and the profits of agriculture or trade, that a people who wholly neglect these branches of industry can hardly be said to live a complete and healthy national life. By considering our rich stores of coal and iron as held in trust by us for the use of the present and future populations of these islands, we should probably stimulate and advance a healthy civilization in many countries which the most lavish expenditure of our own minerals, aided by our capital and engineering skill, fail to benefit.

Lastly, I would call attention to the way in which the lavish production of minerals disfigures the country, diminishes vegetable and animal life, and destroys the fertility (for perhaps hundreds of generations) of large tracts of valuable land. It would be interesting to have a survey made of the number of acres of land covered by slag-heaps and cinder-tips at our iron and copper works, and by the waste and refuse mounds at our various mines and slate quarries, together with the land destroyed or seriously injured by smoke and deleterious gases in those "black countries" which it pains the lover of nature to travel through. The extent of once fertile land thus rendered more or less permanently barren would, I believe, astonish and affright us. How strikingly contrasted, both in their motive and results, are those noble works of planting or of irrigation which permanently increase both the beauty and productiveness of a country, and carry down their blessings to succeeding generations!

This brief sketch of some of the more salient features of the subject of mineral export will serve to show how many and various are the evil results which flow from allowing these invaluable treasures to be wasted at the dictates of mad speculation and the eager race for wealth. These considerations have a very practical bearing at the present time. The recent great rise in the price of coal has brought up the question of the advisability of an export duty upon it. The press, almost without exception, has opposed this as being "contrary to the principles of free trade;" and it has further been argued that such a duty would have little or no effect, because the real cause of the high price of coal is that so much is used in the excessive manufacture of iron. But it is evident, from the considerations here set forth, that the export both of coal and iron requires to be regulated or forbidden, and for the same reasons; and if the "principles of free trade" are opposed to this, so much the worse for those "principles," since they will be opposed not only to the true economy of human progress, but also to the clearest principles of social and national morality. Many persons will now ask whether those can be true principles which lead to the exhaustion of our coal-fields for the purpose of lighting South American cities with gas or for building railways in every insolvent South American Republic, while our own hard-working population has to suffer the pangs of cold in winter, in consequence of the high price of coal which such reckless projects tend to cause. And the fact that all parties concerned—landowners, colliery proprietors, speculators, and legislators—are so far from seeing anything wrong in what they are doing that their one aim at all times is to secure a larger annual output, and an increased export, will be to many an additional argument for taking the property in land altogether out of private hands. Waiving that question, however, for the present, I maintain that it is a wrong to our own population, and a still greater wrong to the next generation, to permit the unlimited export of those mineral products which are absolute necessaries of life, but which once destroyed we can never reproduce. To do so is to sell and alienate for ever a portion of our land itself, and should no more be permitted to private individuals than the selling of the land surface to a foreign State.

Whether or not the period of the total exhaustion of our coal-fields can be approximately estimated, it is clear that the present vast and increasing rate of consumption must be stopped. The numerous evils of the present system I have briefly indicated—where are the benefits which counterbalance them ? And the benefits, if they exist, must be large and clear and positive indeed to justify us in recklessly scattering over the whole world the mineral products of our land. It is to their possession that we attribute much of our wealth and power and national prosperity, yet we are doing our best to deprive future generations of any of the advantages we have derived from them.

It appears, then, to be clearly our duty to check the further exhaustion of our coal supplies by at once putting export duties on coal and iron in every form, very small duties at first, so as not to produce too sudden a check on the employment of labour, but gradually increasing them till, by stimulating an increased production in other countries, they may no longer be required. If other nations should see the wisdom and justice of following our example, each may in future develop and enjoy its own mineral products, may help to supply what is necessary to the welfare of those countries which do not possess these natural gifts, and may still leave an ample supply to their descendants.

CHAPTER IX

PAPER MONEY AS A STANDARD OF VALUE

THE proposition embodied in this heading will seem to most persons to be an absurdity; but I hope to be able to show from the statements and admissions of orthodox authorities that paper money, under proper regulations, would be the most permanent, and therefore the best, possible standard of value. I presume that the late Prof. W. Stanley Jevons was a trustworthy authority on the subject; and in his volume on *Money and the Mechanism of Exchange* he gives some important facts and principles bearing upon this question, and these I shall take as the basis of my argument.

1. He shows that gold has undergone great changes of value during the last hundred years, as determined from the average prices of fifty or a hundred of the chief necessaries of life. The difference amounted to a fall of 46 per cent. from 1789 to 1809; while from 1809 to 1849 it rose 145 per cent. Since 1849 it fell about 20 or 25 per cent.; while in the last twenty or thirty years all the authorities declare that it has risen considerably.

2. Having thus shown that gold does not even approximate to a permanent standard of value—though I believe the alleged fluctuations are enormously exaggerated, for reasons which it would take too long to give here—he goes on to explain the various proposals which have been made to obviate the evils of such fluctuations by means of a "Tabular Standard of Value." A Government official—who might be called the Registrar of Prices—would collect

the market prices of the list of commodities fixed upon to determine the value of money, and would publish the result monthly or quarterly, and the value of money so determined would be used to regulate all payments of debts, salaries, &c. "Thus suppose a debt of £100 was incurred on July 1, 1875, and was to be paid July 1, 1878, and the Registrar's table showed that in that interval gold had fallen in value six per cent., then the creditor would claim to be paid an increase of six per cent., while, if there had been a rise in the value of gold then the debtor would have a right to pay proportionally less than the amount nominally due."

He says there are only two difficulties—the determination of the commodities chosen to fix the standard value, and the complexity introduced into the relations of debtors and creditors. The latter is, no doubt, a real objection, but it does not arise (as I shall presently show) when paper money alone is used. Neither is there any real difficulty in the former. What is needed is to take a representative selection of all the *necessaries* of life. These may be roughly classed as food, clothing, houses, fuel, and literature. For the first we might take meat, bread, potatoes, sugar, tea, butter, and beer; for houses timber, bricks, iron, glass, lime, cement, slates, and building land—and so on under the other headings. But the most important consideration is, that each item be taken in the proportion in which it is consumed in the country. The need of this was seen by the original proposer of the method—Joseph Lowe, in 1822—but has been neglected by some modern writers. It would, therefore, be necessary first to estimate the total quantities of each item consumed in the kingdom in a year, which could be done without much difficulty by experts, and then, representing the smallest quantity of the whole series by one or ten, to give all the others their due proportions. The prices of these several commodities being ascertained on the average of a number of years to be fixed upon, a table would be formed, giving the money-value of the due proportion of each of the commodities. Then, by adding up these values, we should have a sum total which would represent with considerable accuracy the average cost of all the chief necessaries of life in the proportions in which they are consumed by the whole community.

But in order that money may retain the same purchasing power, and thus constitute a real standard of value, this same amount of money must always purchase the same amounts on the average of all these commodities. This can never be the case with gold or silver money, or with the two combined, but I will now show that paper-money may be so regulated as to have always the same purchasing-power.

Prof. Jevons states the chief objections to inconvertible paper-money as follows:

1. The great temptations which it offers to over-issue and consequent depreciation.

2. The impossibility of varying its amount in accordance with the requirements of trade.

The first of these objections does not arise when the whole purpose of adopting a paper-currency is to secure a permanent standard of value, and definite arrangements are made to preserve that constancy. The second objection must have been stated without due consideration, since nothing is more simple than to produce this " variation of amount;" and when the variation is such as to keep average prices steady, that steadiness will exist *because* the quantity issued *is* in accordance with the requirements of trade. This objection, which is stated at length under the heading, " Want of Elasticity of Paper Money" (p. 237), is really completely answered by the method of the tabular " Standard of Value " (p. 329), but the two things are not brought together as parts of one system.

In order to show how Prof. Jevons's "impossibility" may be easily overcome, let us suppose the transition period to have been passed over: all gold coin having been called in or having ceased to be a legal tender, and paper-currency issued to the same amount. The Registrar of Prices, having determined that during the preceding year the purchasing power of this money is two or three per cent. greater than that of the standard as determined by his table of average values, and having had experience of the effect produced by a given increase or diminution of the currency, instructs the Mint to issue fresh money at a given rate per week. This money is sent to the Treasury and is at once brought into circulation by being paid away in salaries, wages, purchase of materials, &c., in the various Government departments. There is thus no difficulty whatever in increasing the amount of the currency and thus diminishing its purchasing power. The Registrar of Prices carefully watches the effect upon the markets week by week, and month by month, and when he sees that the standard is very nearly attained he instructs the Mint to stop further issues.

On the other hand, when prices are rising, owing to there being rather more money in circulation than is necessary, instructions are sent to the Treasury to cancel a certain amount of the money paid in for taxes, stamps, &c., till the balance is restored. But this will very seldom, perhaps never, be necessary. The continuous increase of the population requires a constant increase in the currency, while another constant renewal is required to make good the losses by fire, water, and other accidents. And as the amount required to keep average prices steady would be so carefully watched, the mere stoppage of the normal issues would in most cases suffice to bring back average prices when they showed any tendency to rise above the standard rate.

The total gain to the country of such a currency would be very great. All the additions required to keep up with increase of population and to make up for accidental losses would be clear gain, and would probably amount to a considerable annual revenue; while during the transition from gold to paper an enormous amount of coin would be accumulated by the Treasury which might be kept as a reserve against foreign war expenses or might be supplied to merchants as bullion of guaranteed quality for foreign payments. Silver and bronze coins for payment of wages and small transactions might be continued in use, as they are both customary and convenient, but their actual value in metal might be reduced, thus giving a larger profit to the Government on their issue than there is now.

A convenient form for £1 and £5 notes would probably be very thin tough cards of the size of railway tickets, and of different colours. They would thus be very portable and easily distinguishable. They would constitute the legal tender of the country, and would always purchase, on the average, the same quantities of the chief necessaries of life. They would thus constitute a permanent standard of value—the ideal perfection of money; and would have the additional advantage of being a steady source of revenue to the country.

CHAPTER X

LIMITATION OF STATE FUNCTIONS IN THE ADMINISTRATION OF JUSTICE [1]

AMID the endless discussions that have taken place as to the sphere and duties of Government, all parties are agreed that there are two great and primary functions which every efficient Government must perform if it deserve the name: it must guard the country against attack by foreign enemies; and it must make such arrangements for the administration of the laws, that every man may obtain justice—as far as possible free and speedy justice—against wilful evil-doers.

The fact that there is an absolute unanimity as to these two important functions of a good Government, while almost everything else that Governments do, or attempt to do, has been denounced by great thinkers as beyond their proper sphere of action, renders it probable that these are, at all events, the primary and most important functions of the State. It may not, perhaps, be easy to determine which of these two is of the greatest importance; for even admitting that conquest by a foreign foe is an evil incalculably greater than any wrong which individuals may suffer, yet the one is of so much more frequent occurrence—every member of society being daily exposed to it, while attempts at conquest occur only at distant and uncertain intervals—that repetition in the one case may make up for magnitude in the other. We are therefore pretty safe

---

[1] This article appeared in the *Contemporary Review* of December, 1873. It is now considerably extended.

in assuming that they are of equal importance, and in affirming that it is as much the duty of Government to protect its individual subjects from wrong to person or property committed by their fellows, as to protect the entire community from foreign enemies.

But if we look around us to see how these primary duties are performed, it becomes evident, either that existing Governments do not consider these duties as equally imperative upon them (even if they are not of absolutely equal importance), or that the former duty is a very much more difficult one than the latter. In every country we find an enormous organization for the purpose of national defence, which occupies a large portion of the wealth, the skill, and the labour of the community. No cost is too great, no preparations are too tedious, in order to deter an enemy from venturing to attack us, or to secure us the victory should he be so bold as to do so. For this end we keep thousands of young and healthy men in a state of unproductive activity, or idleness; for this we pile up mountains of debt, which continue to burthen the country for successive generations. New ships, new weapons, every invention that art or science can produce, are at once taken advantage of, while the less perfect appliances of a few years ago are thrown aside with hardly a thought of the vast sums which they represent.

If we now turn to see how the other paramount duty of the State is performed, we find a very different condition of things. Here everything is antiquated, cumbrous, and inefficient. The laws are an almost unintelligible mass of patchwork which the professional study of a life is unable to master; and the mode of procedure, handed down from the dark ages, is often circuitous and ineffective, notwithstanding a number of modern improvements. It may be admitted that in criminal cases tolerably sure, if not very speedy, punishment falls on the aggressor; but the sufferer receives, in most cases, no compensation, and often incurs great expense and much trouble in the prosecution. He gets revenge, not justice. That relic of barbarism, the fixed money fine, the same for the beggar and the millionaire, though almost universally admitted to be unjust, is not yet abolished. It is, however, in cases of civil wrong that individuals find the greatest difficulty (often amounting to an absolute impossibility) of obtaining justice. This arises, not only from the enormously voluminous and intricate mass of enactments and precedents, and the tedious mode of procedure, involving grievous delay and expense to every applicant for justice, but also to the vast accumulation of cases which are allowed to come before the courts, many of which are of such a complex nature as to some extent to justify the strict forms of procedure which bear so hardly on those who seek relief in much simpler cases. The result is, that it is often better for a man to put up with a palpable wrong than to endeavour to obtain redress; and the assertion that in our happy country there is "not one law for the rich and another for the poor," though literally true, is practically the very opposite of truth, since in a large number of cases the wealthy alone can afford to pay for the means of obtaining justice.

How to Simplify the Law.

Our system of law is, in great part, the product of times when the security of property was held to be of more importance than protection to the person. The legislators being almost always the great landowners, a large part of the law was adapted to secure them the power of dealing with the land (the most important of all property) in any imaginable way; and in their bungling attempts to do this, they have produced a system of law of real estate of almost unimaginable intricacy. To interpret and carry out this and other branches of the law of property, occupies a large and influential portion of the legal profession. Lawyers exist upon the complexity of the law. It is not to their interest that we should be able to obtain cheap and speedy justice; nor is it their interest to reduce the number of suitors at the courts. We cannot reasonably expect them to do either of these things, which are yet of vital importance to us who are not lawyers. They may, indeed, so modify, and to some extent simplify, procedure as to take away a portion of the terrors of "going to law" in the estimation of aggrieved parties, and so induce a larger number than before to seek their aid against oppression and wrong; but they will never make any radical reform, or attempt to do what every intelligent suitor knows might be done. Our interests are directly opposed to theirs, and it is mere madness to expect any thorough simplification of the law from lawyers. Such a reform requires the common sense of minds untrammelled by legal technicalities or legal interests. The people must be shown that such a reform is possible—nay, easy—and they will then demand that this matter shall be taken altogether out of the hands of lawyers. It is in the hope of showing how one great branch of this much-needed reform may be made, that the present writer ventures to attack a problem generally considered far beyond the reach of laymen.

A first step, and a very important one, towards rendering cheap and speedy justice possible for every man is, so to simplify the law of property as to free the courts from a large proportion (perhaps one-half, perhaps much more than one-half) of the cases which now occupy them. This would not only render it far easier to dispose promptly of the much simpler cases—which, however, are those which are often of more real importance to the parties affected—but it would allow of the whole method of procedure being altered to suit those simpler cases which would then form the bulk of the business of the courts. Now, this great diminution of cases can be effected without denying redress for any grievance, or a remedy for any wrong, by simply putting out of court a host of matters which ought never to have been taken cognizance of by the law. Here, as in so many other instances, it will be found that reform must begin by a " limitation of State functions;" and that it is because Governments have undertaken to do much that is unnecessary and even injurious, that they are not able to fulfil one of their first and plainest duties—that of giving free, speedy, and substantial justice to the weakest and most indigent, as well as to the most powerful and most wealthy, of their subjects.

Trusts.

The first, and perhaps the largest, group of cases which ought to be taken out of the cognizance of our courts of law are those which may be comprised under the general term of "trusts." At present any one may place property in the hands of another, either during his own life or to take effect after his death, for certain specified purposes, and if these purposes are neither illegal nor positively immoral, the law will compel the trustee to carry out these purposes to the very letter. They may be trivial or absurd, or even injurious, but the man who once gets a trustee to accept a trust (and even this is not necessary when it is created by a will) becomes thereby an absolute potentate, who has at his command the whole power of a great State employed to see that his most minute directions are carried out. The number of cases of this kind is enormous, including all those which involve the interpretation and carrying into effect of the provisions of trust-deeds, settlements, and wills; so that a considerable portion of our machinery for administering justice is devoted to ascertaining and giving effect to the whims of individuals for years, and often for scores of years, after they are dead. Under the same general head may be included the power of determining by deed or will the contingent succession to property, and of creating any number and kinds of disqualifications with regard to it. The supposed necessity for providing for every imaginable exercise of this power has led to such endless complications in the law relating to the transfer of land in all its forms and modes, that years of study are required to comprehend them. They furnish the materials for perhaps the majority of the cases that come before our civil courts, and give occupation to a very large section of the legal profession.

But in the whole group of cases here referred to there is no question of administering justice. For a Government not to carry out a man's wishes after his death is not a wrong, but quite the reverse, since it may with much reason be maintained that, for any Government to occupy itself with carrying out the whims of every man (whether he be sage or fool) who may wish to make his relations or successors subject to his orders in the application of property no longer his, is a positive wrong to the community, inasmuch as it is incompatible with the performance of duties of a paramount nature. What the law may do, and all that it should do, is, to recognize and enforce gifts or transfers of property of all kinds, to living individuals, absolutely. It should utterly refuse to recognize any desires, whims, or fancies of individuals as to the applications of the property, or any limitation to the future owner's absolute possession of it. It should not even recognize any alternative applications of the property in the case of the death of the legatee before that of the testator, who could in that case have altered his will, and if he has not done so the legacy should pass to the legal representatives of the legatee. Property should always be considered by the law to be in the possession of some person absolutely, who can transfer it to another person absolutely, but cannot enforce any stipulations whatever as to the use of it on the next owner. Life interests in landed and other property, with all their attendant evils, would thus never exist.

The wishes of the donor or testator of property, although not a proper subject for the interference of the law, could be in many cases carried out by means of what may be termed a voluntary and amicable trust. The trustee (who would be really the legatee) would be chosen on account of friendship, integrity, and sympathy with the objects and desires of the testator, and he would give just so much effect to those desires as his reason and his conscience impelled him to give. The law would consider him only as the owner of the property, and would in no way interfere with the manner in which he thought proper to interpret the wishes of his friend. To provide for children and minors, property might be either left absolutely to their nearest relative or friend to stand to them *in loco parentis*, or it might be left to themselves, in which case an officer of the court would be their official trustee, and would prevent any misappropriation of their property by relations or guardians till they came of age. We should in this way greatly simplify wills, and almost abolish will-cases, while the courts would be relieved from that great mass of causes of the most tedious kind, in which trust-deeds, settlements, legal estates, shifting uses, entails, and trustees bear a prominent part.

It has been so long and so universally the practice in civilized countries for the law to recognize and enforce the wishes of individuals as to applications of their property other than the simple transfer of it to others, that to many, perhaps to most persons, it will at first seem to be a positive injustice to take away from them the power to do so. Yet the law itself recognizes that the practice is beset with evils, and from a very early period legislative restrictions have been applied to it. Hence the laws of mortmain, and the long series of amendments, relaxations, or restrictions of those laws; as well as the limitation of the power of entailing estates for any longer period than a life in being and twenty-one years afterwards. These restrictions prove that the unlimited power of disposition of property has been held to be a law-given custom, not an inherent right; for if the latter, every restriction of its exercise must be a wrong to the parties restricted, which it has never been held to be. The whole question is, however, so very important, and has so many and such wide applications, that it deserves a somewhat fuller discussion.

The establishment of the Endowed Schools Commission has struck the first real blow at the system of a perpetual and blind submission to the wills of dead men; but the new principle, even in its limited application to endowed schools and charities, often excites much opposition. Many liberal and intelligent men still look upon the "intentions" of those who in past ages endowed churches, schools, hospitals, almshouses, and other institutions, as something sacred, which it is almost impious to ignore, and which it is our plainest duty to carry out with only such slight modifications as the changed conditions of society absolutely necessitate. But it is here contended that this notion is not founded on any true conception, either of what is just or what is politic, but that it is, on the contrary, altogether erroneous in principle and mischievous in practice; whence it follows that the sooner it can be got rid of the better for society.

Let us, then, seriously ask, what sufficient reason can be adduced why the State should interfere to carry into effect the desires, whims, or superstitious fancies of any man, for generations, or perhaps for centuries, after his death? Why should the more enlightened future be bound by the behests of the less enlightened past? Why should we allow, and even encourage, men to hold and administer property after they are dead? For it really comes to that. A man may, justly and usefully, be allowed full liberty (within the bounds of law and order) to use his property as he pleases *during his life;* but why should we go out of our way and make complex arrangements enabling him to continue to do the same after he is dead? During a man's lifetime he can *give* property to whom he thinks fit, or he can apply it to any purpose that he has at heart, without the State's interference; but he absolutely requires the State's assistance in order that his property may continue to be applied precisely in accordance with his ideas of what is best, after his death. The question is, *why* the State should take any cognizance of the matter? It is here contended that this is one of those things quite beyond the proper functions of a Government, and that it has produced, as such excess of authority always does produce, a vast amount of evil. When a man dies he generally has what may be termed natural heirs, that is, children or relatives dependent upon him for a provision in life. For these he is morally bound first to provide, and any surplus beyond their needs, and beyond what the law may give to the State, he may rightly claim the power of bequeathing to any living individuals; and the State on its part is bound to exercise that minimum of interference necessary to secure the property to the respective persons indicated by him. But on what grounds can the testator claim the interference of the State for the purpose of compelling the recipients of the property to do with it what *he* pleases ? claim—that is—that *he* shall still be considered to be the real owner of the property *after he is dead?* The thing is so intrinsically absurd, and perhaps even immoral, that nothing but long and universal custom could blind us to the absurdity of it.

What a man may do, and ought to be enabled to do, either during his life or at his death, is to *give* property, and *recommend* (not command) what use he wishes to be made of it. If his morals and his intellect are both good and his judgment sound, his chosen legatees will, at their discretion, carry out his wishes. But to compel them to do so absolutely is monstrous. It implies that the *right* to property continues after death, and that when a man can no longer use it himself he *ought* to be enabled to restrict the freedom of others in the use of it. It implies also that a man with much property to leave is necessarily wise, so wise as to know what will be best for people years after his death. A living agent can modify or supplement his plans as occasions arise or as circumstances require, and he generally does see reason to modify them after a few years' experience. Even acts of parliament, the concentrated essence of the nation's wisdom and foresight, one year, often require alteration in the next. But that every man who chooses to do so should be encouraged to make his little "act" before he dies, minutely directing what shall be done with his property for years after his death, and that this "act" should be held to be a fixed law, against which there can be no appeal, all changes of circumstances notwithstanding, and should be enforced by the whole power and authority of the State, is a circumstance which will one day be looked back upon as an amazing anachronism, since it would seem only fitted to exist in a country where the established religion was the worship of ancestors.

We English are wisely jealous of too much government interference in the details of our social life; yet our rulers are living men, imbued with all the ideas and habits and feelings and passions of the age, and are often men of high intellectual attainments, and far in advance of the average of the community. Such a government interferes, at all events, with full command of the most recent knowledge, and with open eyes; yet we will not submit to such interference. But, strange to say, we do submit, and almost pride ourselves in submitting, to have various important social matters determined for us by self-chosen dead men, who are therefore necessarily *behind the age*, and who were sometimes too ignorant, conceited, or superstitious to be up to the intellectual level even of the age in which they lived. It is by such blind guides that we to this day submit to be, in great part, governed in the all-important matters of religion, education, and the administration of charity; and in submission to the immutable laws of these dead rulers we have allowed vast wealth to be misemployed or wasted in the hands of irresponsible and antiquated corporations, which, well bestowed, might have enlightened our people or beautified our land. Who can doubt that the nation would have greatly benefited had our churches and colleges, our schools and charities, our guilds and companies, been free to develop, from age to age, in accordance with the wants and feelings of the living, untrammelled by any slavish adherence to the expressed or implied wishes of the dead?

From the considerations now adduced, it will be evident that the cessation of State interference in the way here objected to, would produce other beneficial results besides that of facilitating the administration of justice. These may be briefly summarized as follows :—

It would take away one of the existing inducements to a life-long devotion to the pursuit of wealth, for if a man could neither make use of it himself nor enjoy the sense of power felt in directing absolutely how it should be employed by others, he would pause in his career of accumulation, and perhaps endeavour to do something useful with it during his own lifetime, rather than run the risk of having it all go entirely beyond his control.

It would have the effect of inducing many who now leave their wealth for charitable and philanthropic purposes at their death, to found such institutions as they wished to have established, during their own life-time, in order to see the working of them, and so adapt them to the fulfilment of an admitted good end as to ensure that they would be preserved by future generations. This active charity or philanthropy would have a most beneficial effect on character, and would undoubtedly lead to more good results than the mere passive bequeathing of money to be employed in some fixed, but often ill-considered and comparatively inefficient manner.

It would prevent the establishment of institutions not adapted to the requirements of the age, and would thus abolish a great bar to mental and moral progress. For the notion of "sacredness" attached to the wishes and commands of the founders of religious, educational, and charitable institutions has done a vast amount of evil, in confusing our notions of what is right and what is useful, and in keeping up the obsolete ideas and practices of a bygone age, long after they have become out of harmony with a more advanced state of society.

There is no fear, as some may imagine, that under the modification of the law here suggested, such institutions would want stability and would be subject to constant fundamental changes in accordance with the ideas of each successive body of governors, for the conservative tendencies of mankind in general, and especially of all governing bodies, are very strong, and customs or practices, even when pernicious or absurd, seldom get changed till long after their hurtfulness or foolishness are universally acknowledged. In proof of this we may adduce the case of our own representative government, which attaches no idea of sacredness to old laws, and is subject to the powerful influence of public opinion; yet we do not find any dangerous instability in our legislation, but rather a slow, many think far too slow, march onward in a tolerably well-defined course of reform.

The change here advocated would also be beneficial, by helping to rid us of the notion that a man can infallibly prescribe what is good for his successors, or that even if he could, he ought to be allowed so to prescribe; for the next generation will be quite as well able to attend to its own affairs as the last was, and will certainly not be benefited by being debarred from the freest action. Once this notion is abolished, our truest philanthropists would be more willing than heretofore to devote their wealth to public purposes, because they would feel confident of its being permanently useful. They would know that each succeeding generation would watch its application critically, and insist that no obsolete customs or erroneous teachings should be perpetuated by means of it,—that it should never become a drag on the wheels of progress, as has been the case with many such institutions, but rather resemble a powerful engine capable of helping on the necessarily slow march of society towards a higher civilization.

If the main principle here advocated—namely, that it is intrinsically absurd and morally wrong that a dead man's will or intention should have power to determine the mode of application of property no longer his—be a sound one, it will have a most important bearing on a question that is now much discussed, as to how far endowments of the National Church by private individuals may be properly claimed by the State. Even writers of very liberal views see in this a stumbling-block to the complete disendowment of the Church of England, because they cannot get rid of the notion that it is something like a robbery to take property given for one purpose and apply it to any other purpose. It is, therefore, a maxim with them, that when any change in the application of such a fund is demanded by public policy, it should still be kept as near as possible to the intentions of the original donor. It is, however, to be remarked, that when the property in question has already been forcibly applied to other uses than those originally intended, the most scrupulous do not propose that it should be brought back to its ancient use; and this seems to imply a doubt of the soundness of their principle. A large part of the existing endowments of the Church of England, for example, were certainly intended to maintain the teaching and services of the Roman Catholic religion. If the donor's intentions are "sacred," these should be given back to the Roman Catholic Church. If it be said that the intention was to maintain the religion of the country, whatever that might be, then the revenues should be fairly divided among all existing sects for the time being,—but that is "concurrent endowment," and is almost universally repudiated. The only consistent, and it is maintained the only true, view, is, that dead men should have no influence (beyond their personal influence on their friends) other than what is due to the intrinsic value of their opinions; and that property cannot be left in trust to carry out dead men's wishes, on the common-sense ground that the living know better what is good for themselves than the dead can do, and that the latter have no just or reasonable claim to coerce a society to which they no longer belong. To hold the contrary view is, practically, to allow men to continue to be the possessors of property after they are dead, and to give more weight to the injunctions of those who had no possible means of knowing what is best for us now, than we give to the deliberate convictions of men who still live among us and who have made our welfare their life-long study.

The dead are not truly honoured by sacrificing the interests of the living to their old-world schemes; and if, as we may reasonably suppose, the future state is one of progress, at least as rapid as that which obtains on earth, it may be that they are afflicted with unavailing regrets at our blindness in insisting on being guided by the feeble and uncertain light which they once had the presumption to imagine would for ever be sufficient to illuminate the world.

Debtor and Creditor.

Another group of property cases which occupy the time of the courts even more largely than wills and trusts, are those connected with the recovery of debts of various kinds, culminating in the proceedings of the Bankruptcy Courts. Here again, in by far the larger portion of the cases, there is no necessity whatever for the law to intervene, while it is not improbable that its total abstention would in various ways be beneficial to the whole community. Let us consider a few of the more familiar cases which come before the courts.

First we have the gigantic system of trading on credit, continually offering temptations to fraud and leading in so many cases to bankruptcy. Repeated efforts have been made to improve the law of bankruptcy, but with small results, since, although there has been of late years some diminution in the number of bankruptcies and amount of bankrupts' liabilities, this is far more than compensated by the enormous increase in the liquidations of Companies under the Limited Liability Act, so that both the losses of the public and the temptations to fraud have continually increased. The cost of the Bankruptcy department paid for by the public is about £150,000 a year, besides the legal costs paid by the estates of the bankrupts and by their creditors. Now this vast system of credit, and all the temptation to speculation and fraud that arises out of it, is the creation of the law, and the interference of Government is in no way necessary for the protection of individuals. Instead of attempting to amend the law of bankruptcy it should be abolished altogether, and no claim for the value of goods sold on credit should be recognized by the courts. The effect of this simple and common-sense principle would be, that no credit would be given to any individual until he had proved that he was worthy of credit; and even then it would be at the creditor's risk. There cannot be the least doubt, that this would be for the benefit of everybody concerned; since speculative trading would cease except by those who chose to risk their own capital instead of that of other persons. If a young man was known to be honest, sober, industrious, and with some business capacity, he would be sure to have friends who would advance money to start him in life, on his personal security. If he did not possess or had not exhibited these qualities it is better for himself and his friends that he should not be able to speculate with other people's property until he has acquired experience and given proofs of integrity. There is absolutely no reason whatever for the Government to keep up a costly organization for the purpose of protecting people who choose, with their eyes open, to lend money without security. Let all business be on a cash basis, and if persons who have confidence in a friend's honesty give credit, let it be at the giver's own risk and responsibility.

In the case of shopkeepers it may be said that they would be liable to loss by persons having goods sent home and then refusing to pay for them. But this would be very easily remedied by either having cash with the order or declining to leave any goods at the house of an unknown or imperfectly known customer, till paid for. The whole thing would right itself in a few months or even weeks, if the Government and the law did not undertake a supposed duty which is wholly unnecessary, and which inevitably tends to endless developments of speculation and fraud.

So far I have touched only on debts incurred without security, but there are also many kinds of security which the law should not recognize. Such are all those which involve injury to others, and especially the loss or deterioration of the family home. Bills of sale on furniture and mortgages of the dwelling or homestead should therefore be alike illegal and valueless. They are contrary to public policy as well as to individual well-being, and there is no more necessity for them than there is to allow a man to pledge his wife or children for a debt. The home should be as sacred from such profanation as the family itself.

For analogous reasons nothing in the nature of *post obits*, or any agreement to pay money out of a future expected inheritance, should have the slightest legal value. By enforcing such agreements the law has been the direct cause of an overwhelming flood of demoralization, and has produced more vice and suffering than have been caused by the most irrational customs of the lowest savages. And there is not the slightest need for such interference. Why should the Government of any country calling itself civilized use its vast organized power to tempt young men to borrow money on their future expectations, almost always for purposes of the wildest extravagance, debauchery and vice? Let the law simply ignore all such debts and claims, and no one will lend money to persons in such a position, unless they are so well-known as to render it certain that they will make a proper use of it. In such a case some relative or personal friend will advance the money at his own risk, and the recipient will consider it a debt of honour, to be repaid at the first opportunity.

There would thus be only two modes of borrowing money or goods—either from some personal friend who could trust to the character of the borrower to repay him; or by giving in pledge some portable article of value to be returned when the money was repaid, and in neither of these cases would the law be called upon to interfere. In all departments of genuine business between persons of known integrity, bills and promissory notes would pass as at present, the only difference being that they would represent debts of honour only with which the law would have nothing whatever to do.

So far as I can see, the interference of the law in all money relations of individuals with each other is wholly unnecessary and produces nothing but evil. In an endless variety of ways it offers a premium on dishonesty, while through all the ramifications of business it is the honest buyers who pay ready money or punctually discharge their obligations, that really have to pay the loss caused by the great army of defaulters and swindlers, which the law itself, under pretence of enforcing the execution of contracts and the payment of debts, has really brought into existence. Our law of property and of debtor and creditor is the actual cause of almost all fraud and dishonesty, a very small portion of the evils due to which it can ever succeed in remedying; while by affording endless scope for the rogues to satisfy their wants at the expense of the honest men it is perhaps the greatest corrupter of morals that now exists.

I am quite aware that in raising my voice against the abuses of governmental action here touched upon, I am one of a very small minority. Yet I feel sure that, short of a complete reorganization of society either on socialistic lines or on those of true individualism as indicated in the last chapter of this volume, no more beneficial reforms could be effected than the strict limitation of the functions of the State in the two directions I have here indicated. Not only would it be an enormous saving of public and private expenditure that is altogether useless, but by making all success in business absolutely dependent on consistent integrity it would become a great power in the moral elevation of society.

CHAPTER XI

RECIPROCITY THE ESSENCE OF FREE TRADE

It is usually said that the English are a practical people; that they prefer experience to theory, and will seldom follow out admitted principles to their full logical results. But this hardly represents them fairly, and many facts in their history might lead an outside observer to give them credit for exactly opposite qualities. He might even say that the English race are more guided by principles than any other, because, though it takes them a long time to become satisfied of the truth of any new principle, when they have once adopted it they follow it out almost blindly, regardless of the contempt of their neighbours, or of loss and injury to themselves. As one example, he might point to the English race in America, who, having at length seen that slavery was incompatible with the principles of their own declaration of independence, not only made all the slaves free, but at once raised the whole body of those slaves, degraded and ignorant as they were, to perfect political equality with themselves, allowing them not only to vote at parliamentary elections but to sit as legislators, and to hold any office under Government. In England itself he would point to our action in the matter of education and free trade. Till quite recently, public feeling was overwhelmingly in favour of leaving education to private and local enterprise; it was maintained that to educate children was a personal not a public duty, and that you should not attempt to make people learned and wise by

Act of Parliament. At length a change came. Public opinion and the legislature alike agreed that to educate the people was a national duty; and so thoroughly was this idea carried out that, till a few years ago, food for the mind was looked upon as of more importance than food or clothing for the body, and parents who could not earn sufficient to keep their children in health were fined, or at all events made to lose time, which was to them often the cost of a meal, because they did not send their children to school, and either themselves pay the school fees or become paupers.

An even more remarkable instance of devotion to a principle might be adduced in our action with regard to free trade. Till a generation ago we put heavy import duties on food of all kinds, as well as on many other raw products and manufactured articles. On this question of the free import of food for the people, the battle of free trade was fought, and, after a severe struggle, was won. The result was that the principle of universal free trade gradually became a fixed idea, as something supremely good and constantly to be sought after for its own sake. Its benefits were, theoretically, so clear and indisputable to us, that we thought we had only to set the example to other nations less wise than ourselves, who would be sure to adopt it before long, and thus bring about a kind of commercial millennium. We did set the example. We threw open our ports, not only to food for our people, but to the manufactured goods of all other nations, though those goods often competed with our own productions, and their unrestricted import sometimes produced immediate misery and starvation among our manufacturing classes. But, firm to a great principle, we continued our course, and notwithstanding that after fifty years' trial other nations have *not* followed our example, we still admit their manufactures free, while they shut out ours by protective duties.

These various instances do not support the view that we are especially practical in our politics, but rather that we are essentially conservative. We possess as a nation an enormous *vis inertiæ*. A tremendous motive force is required to set us going in any new direction, but when once in motion an equally great force is requisite in order to stop or even to turn us. After spending so much mental effort and so much national agitation in deciding to adopt a new principle, we hate to have to review our decision, to think we have done wrong, or even that any limitations or conditions are to be taken into account in the application of it. This rigid conservatism is well shown in the treatment of the demand of many of our manufacturers and some of our politicians for a fresh investigation of the subject of free trade by the light of the experience of the last fifty years. They put forward "reciprocity" as the principle on which we should act, and they are simply treated with derision or contempt. They are spoken of as weak, or foolish, or ignorant people, wanting in self-reliance, and seeking to bolster up home productions by a return to protection; and this is the tone adopted by the press generally, and by all the chief politicians, both Liberal and Conservative. Little argument is attempted; the facts of increased imports and diminished exports, and of widespread commercial distress, are explained away, as all facts in such a complex question can be, and the names of Adam Smith and Cobden are quoted as having settled the question once and for ever.

Now this mode of treating an important subject which affects the well being of the nation is not satisfactory. No one believes more completely than myself in the benefits of free trade, and the impolicy of restricting free intercourse between nation and nation any more than between individual and individual; but, like most other principles, it must be subject in its application to the conditions imposed upon us by the state of civilization and the mutual relations of the independent countries with whom we have dealings. Nobody advocates free trade in poisons, or explosives, or even in alcoholic drinks; and few believe that we are bound to allow Zulus or Chinese to become armed with breech-loaders and rifled cannons if we can prevent it; and the mere fact that restriction in these cases is admitted to be necessary should make us see that no commercial principle, however good in itself, can be of universal application in an imperfect human society.

The essence of free trade is its mutuality. Its especial value depends upon the almost self-evident proposition, that, *if each country freely produces that which it can produce best and cheapest, and exchanges its surplus for the similarly produced products of other countries, all will derive benefit.* As an argument against the old policy of bounties and monopolies and prohibitory import duties, and the idea that it was best for a country to produce everything for itself and be independent of all its neighbours, this was irresistible, and it did good work in its day. But people were so impressed with its self-evident commonsense (which it yet took them so many years of hard struggle to force upon a reluctant and conservative population) that having once got it, they set it up on high and worshipped it, as if it were a moral truth, instead of a mere maxim of expediency calculated to produce certain economical effects if properly carried out. They have thus been led to overlook two important aspects of the question, which must be carefully studied and acted upon if we are to obtain the full benefits to be derived from free trade. The one is, that, even if universally adopted—that is, if no artificial restrictions were imposed by any nation on the trade of any other nation with it—there are yet many conceivable cases in which its full application would produce injurious results, morally, physically, and intellectually, which might so overbalance the mere commercial advantages it would bestow, as to justify a people in voluntarily declining to act up to the principles it enunciates. The other is, that even the commercial advantages depend on the *whole programme* of free trade being carried out, and that if the first half of it is neglected—that is, if each country does not *freely produce* that which it can *naturally* produce best and cheapest—then it may be demonstrated that one entire section of the benefits derivable from free trade, and perhaps the most important section for the real well-being of a nation, is destroyed. These two points are of such importance that they deserve to be carefully considered.

Some Limitations of Free Trade.

Admitting that free trade will necessarily benefit a country materially, it does not follow that it will be best for that country to adopt it. Man has an intellectual, a moral, and an æsthetic nature; and the exercise and gratification of these various faculties is thought by some people to be of as much importance as cheap cotton, cheap silk, or cheap claret. We will suppose a small country to be but moderately fertile, yet very beautiful, with abundance of green fields, pleasant woodlands, picturesque hills, and sparkling streams. The inhabitants live by agriculture and by a few small manufactures, and obtain some foreign necessaries and luxuries by means of their surplus products. They have also abundance of coal and of every kind of metallic ore, which pervade their whole country, but which they have hitherto worked only on a small scale for the supply of their own wants. They are a happy and a healthy people; their towns and cities are comparatively small; their whole population enjoy pure air and beautiful scenery, and a large proportion of them are engaged in healthy outdoor occupations. But now the doctrines of free trade are spread among them. They are told that they are wasting their opportunities; that other nations can supply them with various articles of food and clothing far cheaper than they can supply themselves; while they, on the other hand, can supply half the world with coal and iron, lead and copper, if they will but do their duty as members of the great comity of nations, and develop those resources which nature has so bountifully given them. Visions of wealth and power float before them; they listen to the voice of the charmer; they devote themselves to the development of their natural resources; their hills and valleys become full of furnaces and steam-engines; their green meadows are buried beneath heaps of mine-refuse or are destroyed by the fumes from copper-works; their waving woods are cut down for timber to supply their mines and collieries; their towns and cities increase in size, in dirt, and in gloom; the fish are killed in their rivers by mineral solutions, and entire hill-sides are devastated by noxious vapours; their population is increased from ten millions to twenty millions, but most of them live in "black countries", or in huge smoky towns, and, in default of more innocent pleasures, take to drink; the country as a whole is more wealthy, but, owing to the large proportion of the population depending upon the fluctuating demands of foreign trade, there are periodically recurring epochs of distress far beyond what was ever known in their former condition.

With this example of the natural effects of carrying out the essential principles of free trade, another people in almost exactly similar circumstances determine that they prefer less wealth and less population, rather than destroy the natural beauty of their country and give up the simple, healthful, and natural pleasures they now enjoy. They accordingly, by the free choice of the people in Parliament assembled, forbid by high duties the exportation of any minerals, and even regulate the number of mines that shall be worked, in order that their country shall not be changed into a huge congeries of manufactories. A balance is thus kept up between different industries, all of which are allowed absolutely free development so long as they do not interfere with the public enjoyment, or cause any permanent deterioration to the water, the soil, or the vegetation of the country. They are in fact protectionists, for the purpose of preserving the beauty and enjoyability of their native land for themselves and for their posterity. Free trade under the guidance of capitalism would destroy these, and give them instead cheaper wine and silk, stale eggs instead of fresh, and butter ingeniously manufactured from various refuse fats. They prefer nature to luxury. They prefer intellectual and æsthetic pleasures, with fresh air and pure water, to an endless variety of cheap manufactures. Are they morally or intellectually wrong in doing so?

Again: there may be, and probably are, countries which produce nothing that some other country could not supply them with at a cheaper rate. But as populations must work to live, they have to contravene the essential principle of free trade and produce the necessaries of life dearly for themselves. Such people could hardly export anything. They must necessarily be poor, and their surplus population must emigrate; but these very conditions might be highly favourable to social and moral advancement and a not inconsiderable share of happiness. Theoretically, such a people ought not to exist, since they only produce what can be produced with much less labour elsewhere; yet conditions approaching to these have led to the development of one of the freest and most enviable people of Europe—the Swiss.

It is indeed fortunate that most countries are so varied as they are, and that none are so peculiar as to be adapted for the economical practice of one industry only. For if they were, the principles of free trade would in time lead to the whole population being similarly employed; they would become parts of a great machine for the growth of one product or the manufacture of one article. It surely will be admitted that such a state of things would not be desirable for any country; and it thus seems as if nature herself had taught us that the principle of each country limiting its energies to the one or two kinds of industry it can practise best and cheapest, though commercially sound, cannot always be carried out without injury, and must always be subordinated to considerations of social, moral, and intellectual advantage. These arguments, which certainly go to the very root of the matter, have so far as I know, never been answered, never even been considered by the advocates of absolute free trade.

The whole Programme.

We will now come to the other essential point – that the whole programme of free trade must be carried out if its advantages are not to be overbalanced by disadvantages. That programme is, that each country shall *freely* produce that which it can *naturally* produce best, and that all shall *freely* exchange their surplus products. But after fifty years' example on our part, no other country approaches to this state of things. By means of protective duties they all artificially foster certain industries, which could not long survive under that open competition which is the essence of free trade. In all the popular articles and discussions on this subject which I have seen, the extreme free traders, without exception, maintain that this makes no difference, and that because the competition of such artificially supported industries keeps down prices here, therefore it benefits us and injures only the protectionist peoples. But this argument entirely ignores the element of *stability* and *healthy growth*, perhaps the most vital essential to the prosperity of all industrial pursuits, and of every manufacturing or trading community. When a country is developing its natural resources without the artificial stimulus of bounties or protective duties, its progress may not be very rapid, but it will be sure, and for long periods permanent. It will depend upon the attraction of capital to the industries in question, the training up of skilled workmen, the making its way in foreign markets, and other similar causes; and under a system of general free trade, these will not be subject to extreme fluctuations, and the industry in question will be stable as well as prosperous. No one can doubt that such stability in the various industries of a country is the very essence of true prosperity, leading to a steady rate of wages and an assured return both to labour and capital; whereas the contrary condition of instability and fluctuation is the most disastrous and disheartening. But such instability is the necessary result of the trade of one country being subject to the ever-changing influences of the protectionist legislation of other countries. When, after acquiring a natural supremacy in any industry, we are suddenly shut out of a market by prohibitive duties, and subjected to the competition which those duties bring upon us, disturbance, loss, and suffering are sure to be caused both to capitalist and workman. Here then we are deprived of what is really the most important advantage of free trade, by the action of other countries. Is there either reason or justice in passively submitting to this deprivation, and is there any mode of action by which we can gain for ourselves the benefits of that system of freedom which we have so long magnanimously offered to all the world? I venture to say that there is, and that by a consistent and clearly marked course of action we can, to a considerable extent, prevent other nations from injuring us by their various phases of protectionist policy, while we retain whatever benefits free trade can give us; and further, that while thus ourselves carrying out the essential spirit of a free-trade policy, we shall be in possession of the most powerful conceivable engine to convert others to its adoption.

Before proceeding to explain my plan, let us see what other schemes have been put forth by the advocates of reciprocity. As far as I can make out, they are two only: the one to put a small uniform *ad valorem* import duty on all foreign-manufactured articles; the other to arrange, by treaties of commerce or otherwise, a scheme of reciprocal import duties which shall be adjusted so as to benefit both parties to the arrangement. The objection to the first is, that it is giving up the whole principle of free trade, and neither public opinion nor the legislature would sanction it; while the second is vague, and involves innumerable questions of detail, and equally gives up the principle of free trade with the first. To these objections I add one of my own, that by neither plan could we secure that stability and unchecked development of our resources which is the most valuable of all the results of complete free trade. We should not thereby prevent other nations from influencing our industries prejudicially when with changes of government come changes of policy. Bounties might still be given or increased, import duties might be raised or lowered, and the capital invested in some of our industries in order to supply both a home and foreign demand now, might be greatly depreciated or even rendered worthless by the unexpected action of some foreign protectionist minister a few years hence.

Hoping to get some further light on this subject, I turned to Professor Fawcett's volume on *Free Trade and Protection*, feeling sure that I should there find the question fairly stated and the reasons against "reciprocity" fully set forth. To my great astonishment, however, I find that Mr. Fawcett's arguments are entirely directed, not against "reciprocity" of import duties, as I understand

the term, but against two totally distinct things—
"retaliation" towards such foreign countries as tax our
products, and renewed "protection" of our domestic
industries—both of which are clearly proved by him, and
are freely admitted by me, to be useless or injurious to
ourselves. Thus at p. 63 he says: "If we desire to
retaliate with effect upon America for the injury which, by
her tariff, she inflicts on our commerce;"—and on the
same page, "If, therefore, we desire to make the American
people *suffer* some of the same loss and inconvenience
which they inflict on our commerce;"—and again, at p.
162, he speaks of the objection "against imposing a duty
on some article of French manufacture, with the view
of *punishing* the French for refusing to renew the Commercial Treaty." Surely such expressions as these which
I have italicised, are unworthy of an argumentative work
on political economy and of Professor Fawcett's high
reputation. The desire of our manufacturers and workmen to enjoy the legitimate benefits of free trade, and to
be guarded against the injury admitted to be done to
them by the arbitrary and uncertain departures from its
principle by other nations, is a very different thing from
" retaliation " or a revengeful wish to make others suffer.

The late Professor Fawcett further argued, as it appears
to me very unsoundly, that because the import of goods
which compete with our manufactures is often *comparatively* small, therefore the injury done or the distress
caused is proportionately of small amount. Surely he
must know that there is often a very narrow margin
between profit and loss in manufactures, and that the
importation of a *comparatively* small quantity may determine the price at which a much larger quantity must be
sold. It is a well known fact that the increased economy
in working to the full power of a factory is such that the
surplus so produced may be advantageously sold abroad at
less than the actual cost, owing to the increased profit on
the bulk of the goods manufactured at a lower average
cost. Foreign manufacturers, protected by import duties
against competition by us, enjoy practically a monopoly in
their own countries, and can secure such a profit on the

bulk of their goods sold at home that they can afford to undersell us with their surplus stocks. These vary of course with variations of trade, and thus our manufacturers are at any time liable to great fluctuations of prices owing to such importations. It is a weak and miserable answer to say that the people benefit by the low prices thus caused; for the great mass of our people are wage-earners and producers as well as consumers, and almost every article we either produce or manufacture is subject to the injurious effects of the influx of surplus stocks from protected countries, by which wages are lowered and numbers of men and women thrown out of work. There is no comparison between the great loss and suffering thus caused, and the small advantage to the consumer in an almost infinitesimal and often temporary lowering of the retail price of goods the majority of which are *not* prime necessaries of life.

What Reciprocity means.

But there is a very simple mode by which we *can* obtain that stability which general free trade would give us, and which, as I have endeavoured to show, is its greatest recommendation. It is to reply to protectionist countries by putting the *very same import duty* on *the very same articles* that they do, changing our duties as they change theirs.

This will restore the balance, and, so far as we are concerned, be almost equivalent to general free trade. It may, perhaps, even be better for us, in some respects, for we shall get some revenue from these duties; but the great thing is, that we shall obtain *stability*. Our capitalists and workmen will alike feel that foreign protectionist governments can no longer play upon our industries as they please, for their own benefit. They will know that they will be always free from unfair competition, while neither asking nor receiving a shred of protection from that fair competition of naturally developed industries which is alone compatible with the principles of free trade. There will then be every incentive to exertion in order to bring our manufactures up to the highest standard, so that they may compete with the best productions of other nations, without any fear that when they have achieved an honourable success they may be deprived of their reward by an additional weight of protective duties against them.

It is urged against the advocates of reciprocity that they are vague in their suggested remedies, and, when asked to specify their proposals, " escape in a cloud of generalities." No one can make this charge against my proposal. It is sufficiently clear and sufficiently definite. Neither are Professor Fawcett's objections—that " a policy of reciprocity is impracticable," and that, once embarked on it, trade after trade would claim protection—at all more to the point. Every trade and industry would be treated alike. All would have a free field and no favour. And as regards foreign countries we should strictly do as we are done by and as we would be done by, and no more. We should make no attempt to injure them or retaliate on them, but should simply and exactly neutralise their interference with free trade as between ourselves and them.

As I am here discussing an important question of principle, to which, if it can be clearly established, our practice should conform, I am spared the necessity of adducing that array of statistics which is generally made use of in arguments on this subject. It is well, however, to give one or two illustrative cases. Professor Fawcett has clearly proved, that the effect of the French sugar bounties is, that sugar is sold in England under its cost price in France, and that the only people who benefit by it are the proprietors on whose land beet-root is grown, and the people of this country who get sugar somewhat cheaper. He admits, however, that "considerable injury is, no doubt, inflicted on English sugar-refiners by the French being bribed by their Government to sell sugar in the English market at a price which, without a State subvention, would not prove remunerative;" but, he adds, " if we embark on the policy of protecting a special trade against the harm done to it by the unwise fiscal policy of other countries, we shall become involved in a labyrinth of commercial restrictions," &c. Surely this is a very vague and unsatisfactory reason why our home and colonial sugar manufacture should be left at the mercy of a foreign Power. For if the French Government at any time and for any reason still further increase the sugar bounties, they might completely ruin many of our manufacturers. Then, some future Ministry might abolish these bounties altogether, and later, when fresh capital had been drawn to the manufacture in England, it might be again ruined. Are we to submit to this, on account of the shibboleth of what is miscalled " free trade," when the imposition of an import duty of *the same amount as the bounty* would prevent all such fluctuations ? By this course we should leave to France the full benefit of her *natural sugar-producing capacity*, only taking away from her the power to cause commercial distress in our country and our colonies by a course of action which is liable to unforeseen changes at the whim of a Minister or a political party. Exactly the same arguments apply to our paper-manufacture, which is injured in the same way by foreign export duties on the raw material and import duties on the manufactured article; and, on the true principles of free trade, it is entitled to have those duties *neutralised*, until the countries which impose them think fit to abolish them altogether.

In almost every civilized country, including our own colonies, the people naturally wish to develop their own resources to the utmost; and we must all sympathize with this desire. But as they have in the first instance to struggle against old-established industries in other countries, the difficulties and risk are too great to attract the necessary capital, and they therefore endeavour to restore the balance in their favour by means of protective duties, professedly as a temporary resource till the new industry is well established. But Professor Fawcett assures us that, in the United States, in no single instance has a protective duty when once imposed been voluntarily relinquished, but, on the contrary, each case is made a ground for seeking, and often obtaining, further protection ; and for about a century American protective duties have been constantly increasing. The same thing applies more or less in the case of other civilized nations with whom we have commercial intercourse, and thus all security for the investment of capital in any manufacture is taken away from our people. Whether in our mineral products or our hardware, our cotton, paper, silk, or sugar, or any other of the thousand industries on which the prosperity of our producers and workers depends, all alike are subject to periodical floods of the surplus stocks of other countries, from whose markets we are shut out by protective and generally prohibitive duties.

The advantage to foreign manufacturers, on the other hand, of having an open market for their surplus goods, while they are themselves protected from competition at home, is so obvious and so great, that, instead of our example having any tendency to make them follow in our steps, it really becomes a premium to them to continue their system of exclusion. They obtain all the advantages of free trade, we all the disadvantages of protection. Internal competition keeps down prices in a protected country to a fair standard, and thus the consumers do not greatly suffer: while the free market we offer for surplus stocks gives to the manufacturers the great advantage of utilising their plant and machinery to its full extent, and thus working with a maximum of economy. Our boasted freedom of trade, on the other hand, consists in our being at a great disadvantage in half the markets of the world, and in being further handicapped by the irregular influx of surplus stocks which foreign manufacturers are (in the words of Professor Fawcett) "bribed to sell us under cost price!" How differently do we act when there is a suspicion of prison-manufactured goods competing with those of regular traders! The representations of those traders are always listened to with respect by our Government, and it is invariably admitted that they have a genuine case of grievance. They are never told that the people benefit, and therefore they must suffer; that prison mats and brooms can be sold at least a penny in the shilling lower than the usual prices, and that the public must not be deprived of this advantage, even though mat and broom makers starve.[1] Yet this is the very argument used (and almost the only argument) in favour of our present system. The public (or a section of it) get iron goods, and silk, and paper, and cotton, and sugar fractionally cheaper, owing to the influx of foreign-manufactured goods sold under cost price; therefore our manufacturers of all these things, and the large proportion of our population who are engaged directly or indirectly in such manufactures, must alike suffer. The weakness of this argument has already been exposed, while its inconsistency, cruelty, and selfishness are no less obvious.

I have now, as I believe, pointed out a mode of action which we may, as free traders, consistently adopt; which will satisfy all the just claims of our manufacturers and workmen; which will give stability to our industries, and inspire confidence in our capitalists; and which, by neutralising the effects of the protectionist policy of other countries, will place us as nearly as possible in the position we should occupy were they all to become free traders. I have shown, that as long as we continue our present course of action we really offer them the strongest inducements to continue, or even to extend, their present policy of protection; while it is evident that if we simply neutralize every step they take in this direction, they will have no motive, so far as regards us, for continuing such a system. Arguments in favour of free trade will then have fair play, since they will not be rendered nugatory by the bribe our policy now offers them to uphold protection.

Objections Answered.

The objections that I anticipate to my plan are: first, that it is too complex, as it would compel us to adopt as against each country its own tariff, however cumbersome; secondly, that it would not satisfy those who now ask for another kind of reciprocity in the shape of special protective duties; thirdly, that it would diminish our

---

[1] Of late years this ground *has* been taken against any restriction in the import of foreign prison-made goods.

commerce; and, fourthly, that it would be systematically evaded, and is therefore impracticable.

As to its complexity, I reply that it would really be the most simple of all tariffs, since it would be determined by a single self-adjusting principle. The fact that the various lists of duties imposed by foreign nations would be lengthy, is really of no importance whatever. When alphabetically arranged, it is not more difficult to find one item among a thousand than it is among five hundred. It may also be said that we could not ascertain in many cases what the foreign duties really are, owing to the complications introduced by bounties, drawbacks, and various kinds of imposts distinct from the nominal import duty. But if we could not precisely estimate the amount of protection afforded in every case, we certainly could do so approximately; and we might trust to our consuls and our customhouse officials to arrive at a sufficiently accurate estimate.

If my proposal should not at first satisfy the present demand of our manufacturers for reciprocity, I am sorry for it; but that does not in the least affect the proposal itself, which has to be judged by the rules of logic, common sense, and expediency. I put it forward as being strictly in accordance with the essential spirit of free trade; as a principle of action which has nothing in common with protection in any form, since its whole purport and effect is to neutralise all attempts at a protectionist policy by other countries. Argument and example have alike failed to influence them, but a checkmate of this kind may have a different result.

As to the third objection I maintain, that commerce exists, or ought to exist, for the good of the nation, not the nation for the good of commerce. If I have shown that the system of strict and detailed reciprocity here proposed would give us the most important of the benefits and blessings of free trade, and would thus be for the advantage of our entire industrial population, I need not concern myself to show that a section of the community which may have gained by the present false and one-sided policy will suffer no inconvenience should that policy be changed; for such arguments have always been put aside as irrelevant when free-trade principles have been at issue.

To the fourth objection, that our reciprocal duties would be evaded by passing goods through countries where they were allowed free entry, I reply, that the duty might be levied on each article *as being the product of a certain country*, from whatever port it was shipped to us. In most cases our custom-house experts would at once be able to say where the article was manufactured, and we might further protect ourselves by requiring satisfactory proof (such as a certificate from the manufacturer) that it was really the product of the country from whose port it was shipped, in order to be admitted duty-free. Even if we should be occasionally cheated, I cannot see that this is a valid objection against adopting a sound and beneficial course of action.

A FEW WORDS IN REPLY TO MR. LOWE.[1]

Although the subject of "Reciprocity" is not yet of sufficient popular interest to be the subject of another article in the *Nineteenth Century*, I beg to be allowed to say a few words in reply to Mr. Lowe's very forcible, not to say violent and contemptuous, article.

I have often been at once amused and disgusted at a common practice in the House of Commons, of flatly denying facts which a previous speaker had alleged as being undisputed, or had proved on good evidence; but I hardly expected that, in an article deliberately written and published, so eminent a politician as Mr. Lowe would condescend to similar tactics, and attempt to overthrow an adversary by the mere force of his weighty *ipse dixit*. Yet the most important part of his reply to me, that which he thinks—" so complete and absolute that I am convinced, had it occurred to Mr. Wallace, his article would never

[1] This reply is printed here because no other criticism than Mr. Lowe's has ever appeared, and because it well illustrates the methods so often adopted by the popular advocates of free trade.

have been written "—consists in the assertion that my proposal, even if carried out, would be quite inoperative, because, when foreign countries protect any class of manufactures, they thereby acknowledge that they cannot compete with us in our own or in any neutral markets, and that " by the conditions of the problem it is impossible " that they should do so.

But the fact that such protected goods *are* imported into this country, and *do* compete successfully with our own, must surely be known to Mr. Lowe; and I am afraid the most charitable view we can take is, that his article was written with some of that want of consideration which he so confidently alleges against myself. What does he say to the fact that the United States sent to this country in 1877 manufactured goods to the value of £3,559,521, including large quantities of cotton and iron goods, sugar, and linseed oil-cake, although every one of these manufactures is protected by almost prohibitive duties? Again, we have paper imported to the value of more than half a million a year, although the manufacture is heavily protected in every country but our own; and the competition of this protected foreign article, which, according to Mr. Lowe, *cannot* compete with ours, has yet ruined many of our paper manufacturers. So iron goods of all kinds are heavily protected in France, Belgium, America, and some other countries; yet iron and steel in various forms were imported in 1877 to the value of over £1,500,000. Our total imports of manufactured goods (including metals) in 1877 amounted to £64,635,418; and almost the whole of these goods are protected in the countries which export them. Most of them, in fact, are sent to us *because* they are protected, the manufacturers finding it to their advantage to work to the full power of their plant and capital, selling the larger portion of their output at a good profit in the home market, and, with the surplus, underselling us, which they are enabled to do *because all the fixed charges of the manufacture are already paid out of the profits of the domestic trade.*

Having thus disposed of Mr. Lowe's main attack, and shown that what he declares to be " impossible " nevertheless constantly occurs, I have only to notice his singular attempt to put me in the wrong by giving a new and unjustifiable meaning to one of the plainest words in the English language. He says that I am quite mistaken in considering " free trade " to be essentially mutual—to mean, in fact, what the component words mean—free commerce, free exchange, free buying and selling. On the contrary, says Mr. Lowe, it means free buying only, though selling may be ever so much restricted. But surely buying alone is not " trade," but only one half of " trade." Just as imports cannot exist without exports of equal value, so I have always considered that buying cannot long go on without selling, and that the two together constitute trade. Mr. Lowe, however, says I am historically wrong, but he does not give his authorities; and without very conclusive proof I cannot admit that the English language as well as the English commercial system, was revolutionised by the free-trade agitation.[1]

One of the most important of my arguments—that reciprocal import duties are just and politic, in order to secure " stability and healthy growth " to our manufactures —Mr. Lowe, with more ingenuity than ingenuousness, converts into a plea on my part for stagnation and freedom from competition ; and he maintains that the power of foreign governments to alter their import duties and bounties at pleasure, with the certainty that we shall take no active steps to neutralise their policy, is a healthy incentive to activity and enterprise !

The remainder of Mr. Lowe's arguments and sarcasms may pass for what they are worth ; but, while so many of our manufacturers, and that large proportion of our population who are dependent directly or indirectly on manufacturing industries, are suffering from the unfair competition brought upon them by foreign protection, the allegation that these form an insignificant *class*, and may be properly spoken of as " particular trades " whose prosperity is of little importance to the rest of the community in

---

[1] In Chambers's Dictionary *Trade* is defined as *Buying and Selling, commerce;* and this latter word as *interchange* of merchandise on a large scale.

comparison with that *summum bonum*—cheap goods—deserves a word of notice. I therefore beg leave to call attention to Richard Cobden's opinion of the supreme importance of these manufactures to England's welfare. He says:—

> "Upon the prosperity, then, of this interest [the manufacturing] hangs our foreign commerce; on which depends our external rank as a maritime state; our custom-duties, which are necessary to the payment of the national debt; and the supply of every foreign article of domestic consumption—every pound of tea, sugar, coffee, or rice,—and all the other commodities consumed by the entire population of these realms. In a word, our national existence is involved in the well-being of our manufacturers.
>
> "If we are asked, To what are we indebted for this commerce? we answer, in the name of every manufacturer and merchant in the kingdom, The *cheapness* alone of our manufactures. Are we asked, How is this trade protected, and by what means is it enlarged? the reply still is, By the *cheapness* of our manufactures. Is it inquired how this mighty industry, upon which depend the comfort and existence of the whole empire, can be torn from us? we rejoin, Only by the *greater cheapness* of the manufactures of another country."[1]

In another passage in the same volume he says:—

> "The French, whilst they are obliged to prohibit our fabrics from their own market, because their manufacturers cannot, they say, sustain a competition with us, even with a heavy protective duty, never will become our rivals in third markets where both will pay alike;"

from which it appears that he never contemplated the state of things that has actually come about, when by means of protective duties, and our open markets supplying all the world with cheap coal, iron, and machinery, other nations have been enabled to foster their manufactures till they have reached such a magnitude as not only to supply themselves, but, with their surplus goods, produced cheaply by means of protection, are actually able to undersell us at home. That time has, however, come; and I feel sure that if Cobden were now among us, his strong sense of justice and clear

[1] Cobden's *Political Writings*, vol. i. p. 227.

vision as to the true sources of our prosperity would lead him to advocate some such course of action as I have proposed, in order to bring about those benefits to the all-important manufacturing interests of our country, which the system of free imports—miscalled "free trade"—has not procured for it.

CHAPTER XII

THE DEPRESSION OF TRADE, ITS CAUSES AND ITS REMEDIES[1]

FOR more than half a century both our Government and our mercantile classes have acknowledged the importance of political economy, or the science of the production of wealth; and they have made it their guide in trade, in manufacture, in foreign commerce, and in legislation. During the same period we have had advantages that perhaps no nation in the world's history ever enjoyed before. It is during that period that steam has been applied to railways; during that time the great gold discoveries which added so much to our wealth, and gave such an enormous impetus to our trade, took place. We, especially, profited by these things, because we had as it were the start of other nations in possessing enormous stores of coal and iron, in the working of which we were pre-eminent. While the railway system was being developed all over the world, it was we who, to a large extent, supplied the coal and iron, and also the skill and labour, used in making these railways. During this same period, too, our colonies have increased with phenomenal rapidity, and have supplied us with customers for the commodities which we

[1] This article is the substance of a lecture delivered at Edinburgh in 1886, and published as one of a series of *Claims of Labour Lectures*. It is an abbreviation of my little book on *Bad Times*, and is reprinted here because I believe it contains views which are as true and as applicable now as they were then. I have not attempted to alter the colloquial style of the lecture, nor have I brought up the figures and statistics to the present time, because the argument it embodies is a general one and will continue to be applicable so long as our political financial, and social arrangements remain substantially unaltered.

produced, while they also afforded a magnificent outlet for our surplus population. With such advantages as these—advantages which we shall in vain search through history to find ever occurring before—it might be thought that we should have got on very well, and must have had a period of continuous prosperity even if we had had no infallible guide to teach us how to conduct our trade and commerce. Yet after fifty years of these unexampled advantages, after fifty years of following what was professed to be an infallible guide, we yet find ourselves at the present day (1886) in the terrible quagmire of commercial depression. All over the country trade is, and for many years now has been, dull; everywhere there are willing workers who cannot find employment. In all our great cities we have stagnation of business, poverty, and even starvation. Certainly, according to the doctrines of the political economy which we have followed none of these things ought to have happened; we ought to have had a continuous and enduring course of success.

Now the need of a thorough inquiry into what are really the causes of this commercial depression is very great, because until we clearly perceive what has produced it, we shall be virtually in the dark as to how to find a remedy for it. I consider, then, that a true conception of the various causes which have brought about this state of things, which, according to our professed teachers, ought never to have occurred, will enable us to lay down more surely what ought to be the radical programme of the future.

In 1885, when the matter became the subject of extensive discussion in the press and in Parliament, we had the most extraordinary chaos of opinion as to what was the real cause. I noted at the time at least eight different suggested causes. One great authority in Parliament stated that there was no accounting for it,—political economy did not explain it. Other great authorities agreed in this view, and the result was the appointment of a Parliamentary Commission of Inquiry. Another suggestion was that it was all a fallacy, and that there was really no depression at all. This was put forward by an eminent member of Parliament who was connected with money-making in the city. To this class of people no doubt there was no depression; money-making and speculation of all kinds went on as briskly as ever. Another suggestion—I am sorry to say the one adopted by the Conference of Trades Unions of England—was general over-production, an explanation which hardly needs refutation, since it has been refuted so often. Other suggestions were, that it was our free trade that caused it; and that it was due to the protection which still existed in foreign countries. Then, again, a very general view, and to some small extent a true one, is, that the continuous succession, for three or four years at all events, of bad harvests had something to do with it; but then there was another remarkable suggestion made, that the rather good harvest we had some few years ago was the cause of the more recent depression. That was seriously put forward in a pamphlet published under the authority of the Cobden Club, for it was stated that this good harvest rendered it unnecessary to import so much corn from America, and thus led to a depression in the shipping trade, and that affected all other trades. The last of this series of explanations was, that it was all due to the currency,—that it was due in fact to there having been an appreciation of gold and a depreciation of silver, one or both.

The Main Features of the Depression.

Now it appears to me that a little consideration of the true character, extent, and duration of the depression, will show us that none of these causes can possibly have been the real and fundamental one, nor even all of them combined. In the first place, the depression has lasted almost continuously for twelve years. It commenced suddenly at the end of the year 1874, and has extended not only throughout this country but, more or less, to every great commercial country in the world. I think, taking into account this long continuance, that no such depression is on record, at all events during the present century. Now the characteristic features of this depression are, as I have said, bad trade all over the country, both wholesale and retail, and in every department of industry, with a few exceptions which I shall point out presently. What is bad trade? Bad trade simply means that there is a deficiency of purchasers. Why is there a deficiency of purchasers? Simply because people who ought to be the purchasers have not got the money to purchase with. It is simply diminished consumption—universal diminished consumption—and the only direct cause of universal diminished consumption is poverty. Our purchasers, both in foreign countries and at home, have been less able to buy. There is not the slightest reason to believe that they have not been willing to buy, that they did not want the goods, but it was simply that they were not able to purchase them. This implies that whole communities are poorer than they were. The home trade suffering as well as the foreign trade show that the great body of our own people are poorer. I do not mean to say that the entire country is not more wealthy—I believe it is—but nevertheless the masses, who are always the chief support of our home trade and our staple manufactures, are poorer. The same thing is clear of our customers in the different countries of the world, the greater part of those that purchase from us are also poorer. Curiously enough, just in the very height of this depression, there appeared some authoritative pamphlets by Mr. Giffen, Mr. Mulhall, and Professor Leoni Levi, proving exactly the reverse, demonstrating, in their opinion, that the people were never so well off, and that they were far richer than they ever were before; and we were told to believe this when at the same time it was universally admitted that their purchasing power had diminished to such an extent as to cause this widespread diminution of trade!

This then, I say, is a statement of the immediate cause of the depression—universal impoverishment. Now we must endeavour to ascertain what is the cause of this universal impoverishment. To illustrate more clearly the period when the depression began, and what was its nature, I have drawn out a diagram giving our imports

192 STUDIES, SCIENTIFIC AND SOCIAL CHAP.

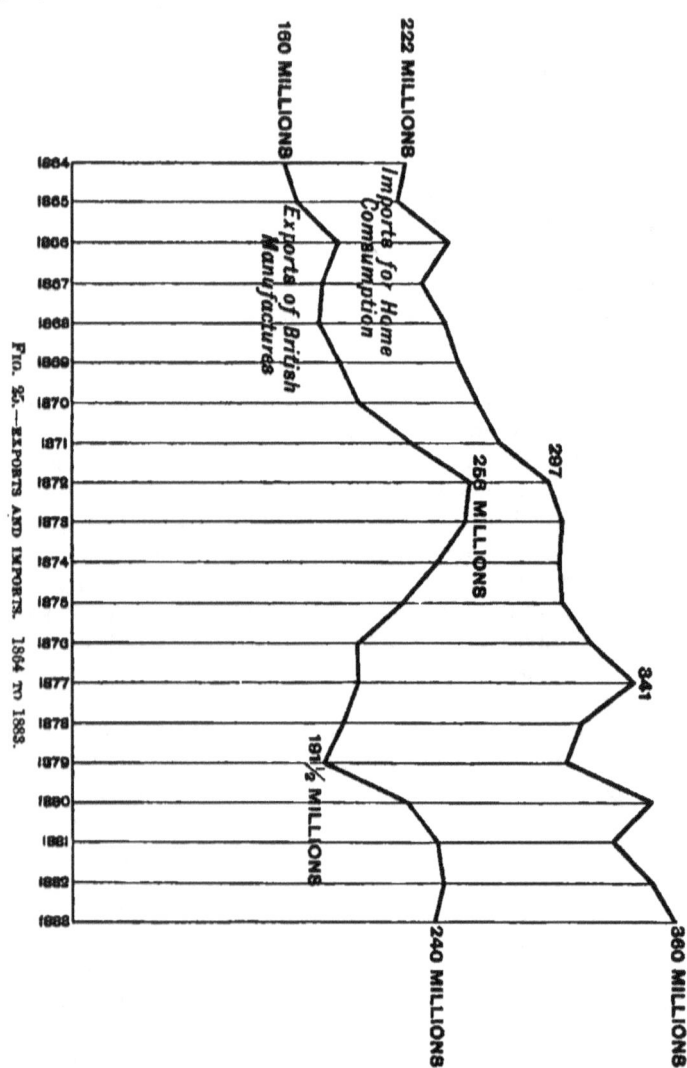

FIG. 25.—EXPORTS AND IMPORTS, 1864 TO 1883.

and exports—the upper line showing our imports, and the lower line our exports—from the year 1856 to 1884. If you look at this you will see that our imports, with the usual minor fluctuations, have gone on increasing steadily from the beginning of the period to the end, but our exports follow a totally different course. They went on increasing pretty steadily and regularly, and then rather suddenly, and especially suddenly from 1870 to 1872. The years 1872 and 1873 marked the culminating points of our commercial prosperity. Then there commenced—what I think cannot be found in all the records of our export trade—a rapid and remarkable decline, which continued right on, without any break, down to the year 1879. From that time it began to rise again, and has risen with fluctuations up to the present time; but even now (1885) it does not attain the culminating point it reached in 1872, thirteen years before. But owing to our increase of population and progressive increase of total wealth, we ought to have had a continuous *increase* of our exports much larger than that which has actually occurred.

Another indication of the course of the depression is afforded by the number of bankruptcies which took place during that same period. I will state briefly what are the facts. In the year 1870—that is, during the period of our prosperity—the annual bankruptcies were about 5,000, including bankruptcies and compositions with creditors. Shortly after the depression had commenced in 1875 they had reached 7,900. In the year 1879, when the depression had reached its height, they had amounted to no less than over 13,000. From that time they diminished in number to 9,000 in 1882 and 8,500 in 1883; and in 1884—almost all whose affairs were in a bad way having become bankrupt—they decreased to about 4,000. These numbers illustrate and enforce the diagram of exports, showing that the bankruptcies began to increase just after the culmination of our commercial prosperity, so that there is no doubt whatever that the real depression commenced about the year 1873 or 1874. This is important, because many writers insist upon leaving out of the question altogether this long continuance of the depression, and they treat it as a comparatively recent thing, which has entirely come on in the last two or three years; and, in fact, one of the two prize essays which have been recently published by Messrs. Pears never said a word about the depression having lasted ten or twelve years, but treated it as if it had commenced within the last three or four years.

True Causes of the Depression.

Now that we have got at what are, I think, the main facts, let us consider how we ought to set about to find what are the true causes. First, then, a cause to be worth anything must be a demonstrable cause of *poverty* in some large body of the people. Another essential point is, that it must have begun to act, or at all events must have acted with increased intensity, about the *period* when the depression commenced. Another point is, that it must have affected not ourselves alone, but several of the great manufacturing countries of the world. Now unless any alleged cause will answer to at least two out of these three tests, I do not consider that we ought to admit it to be a true cause; and you will find, I think, that none of those eight suggested causes which I summarised at the beginning of my lecture will at all answer to these conditions. After much consideration as to what are the real causes which answer to these conditions, and which are of sufficient importance and extent to account for the whole phenomenon, I have arrived at the conclusion that they are four in number. The first is, the excessive amount of foreign loans that were made during the period of prosperity; then there is the enormous increase of war expenditure by all the countries of Europe that also occurred about the same period; another cause is, the vast increase of late years (of which I shall give you proof) of speculation as a means of living, and the consequent increase of millionaires in this country; and the last, and one of the most important of all the causes, may be summarised in one of the results of our vicious land-system—the depopulation of the rural districts and the over-population of the towns.

Foreign Loans.

Now let us take these four causes in succession, and endeavour to see what was their extent, and how they acted. First, then, as to the foreign loans, to the effects of which very little attention has been paid. From the year 1862 to 1872 there was a positive mania in this country for foreign loans. The amount of these I endeavour to illustrate in this table by showing simply the new debts—the increase of former great national debts—created by Foreign powers between 1863 and 1875 :—

New Debts created 1863–1875.

France	£500,000,000
Italy	200,000,000
Russia	400,000,000
Turkey	200,000,000
Egypt	80,000,000
Tunis	7,000,000
Central and South America	73,000,000
	£1,460,000,000[1]

You will see that the total sum amounts to nearly £1,500,000,000 sterling. Now a very large portion of these loans was supplied by this country, and it is very important to consider what effect they had. First of all, you must remember what these loans were for, and what they were chiefly spent on. The greater part of them were spent in war or preparation for war, or to supply means for the reckless extravagance of foreign despots. Now, as I have pointed out, we at that time were the pre-eminent manufacturers in the world, and held the first place much more completely than we do now; so, as we supplied a large part of this money and had extensive commerce with all these countries, the natural result—at

---

[1] England probably lent half of this amount; and in five years only, 1870–75, we lent about £260,000,000 to foreign States, besides an enormous sum for railways and other foreign investments or speculations.

all events, the actual result—was, that a large part of this money was spent with us. Whether it was war material or new railways that were wanted, or jewellery or furniture or other luxuries required by the kings and despots who got the loans, a large part of it was spent with us. The consequence was that for a time everything seemed flourishing. Our trade went on increasing, as Mr. Gladstone said, "by leaps and bounds," and culminated in that wonderful period of apparent prosperity in the two years 1872 and 1873. About that time the money was nearly all spent. What happened then? Not only was there a sudden diminution in the demand—that was natural —but what was worse, there was a great diminution in the normal demand which had previously existed in those countries whose kings or despots had obtained these loans, for this reason, that up to that time the interest on the loans was paid out of capital, but when the money was all gone the interest had to be paid out of taxation; and from that moment, by the increasing taxation and oppression of these people whose governments had obtained these enormous loans, they were all impoverished to that extent, and therefore became worse customers to us and to every other country.

Now this is a real, an important, an inevitable cause. Perhaps some readers will understand it better, however, if I illustrate it by supposing a simpler case. Let us suppose, for instance, that there is a country town in which the people are fairly well off, and where trade is tolerably flourishing. There comes into this country town a body of money-lenders, and they offer everybody loans, on easy terms. Not only do traders and farmers and others get these loans, but all kinds of spendthrifts and idlers. Of course they spend the money they borrow, and during the few years they are spending there is an enormous amount of trade done in the place. Shopkeepers think there is a kind of millennium coming, and increase their stocks and expect to make fortunes. But after two or three years the lenders see that no more money can be safely lent, so they stop the supplies and immediately come down upon those who had the money for their interest.

We supposed that a very large portion of the community had borrowed money, and the consequence is they all suddenly become poorer by the amount of the interest they have to pay. Consequently not only do the shopkeepers lose their temporary increase of trade, but they do less trade than they did before that increase began. The last state of these men is in fact much worse than the first.

Increased War Expenditure.

We will now consider the next real cause of the depression, and that is the enormous increase of military and naval expenditure, which also began about the same time and has been continued almost up to the present day. It is a curious thing that up to the year 1874 our whole military expenditure had been for many years stationary. It was stationary at about £24,000,000— some years it was a little more, some years a little less. Then there commenced a sudden increase, corresponding with that of all the other nations of Europe, though not to so great an extent; and from that time—from 1874 to the present year (1885) it has increased rapidly till it is now £29,000,000 or £30,000,000. But that is nothing to the increase which has gone on with the other nations of Europe. They also had previously a tolerably fixed amount of war expenditure. But then two great events happened—one the Franco-German war, and the other the wonderful and continuous progress in the applications of science to warlike inventions. Not only did iron-clad ships rapidly increase in size, weight, and cost, but very soon steel began to be used, and cannons were made larger and larger in size. Every kind of projectile was improved till they have become works of art of the most costly description. The torpedo was invented, and in fact an amount of skill and science was devoted to this one destructive art perhaps greater than has been devoted to any other art in the world. The result was that owing to the dread of the increasing power of Germany, and the necessity of rivalling her in the application of science to destruction, the great military nations of Europe immediately commenced an enormous increase in war expenditure, and a few figures will show how great this increase was. I am speaking of the ten years 1874 to 1883. Austria increased her expenditure from £7,000,000 to £13,500,000; France, from £18,000,000 to £35,500,000, very nearly double; Germany herself, not so much, because she was in a very fine position before, from £17,000,000 to £20,000,000; Italy increased still more, from £9,000,000 a year to £19,000,000 a year; Russia, from £20,000,000 to £30,000,000 a year. The total of these shows that whereas up to 1874 these six great nations spent £96,000,000 a year on their warlike material and expenditure, in 1883 they spent £150,000,000. Here was an increase of £54,000,000 sterling, all newly added to the taxation of these countries, and, remember, the most utterly unproductive taxation that it is possible to conceive.

Evil Results of War Expenditure.

Now it is not generally considered how varied and extensive are the evil results of such expenditure. The losses involved by it may be summarised under three heads. We have, first, the large number of men employed unproductively; secondly, the increase of taxation; and, thirdly, the vast destruction and waste in war.

First, as to the unproductive men. I find that the European armies have increased since 1870 by 630,000 men—more than half a million. The present total is more that three and a-half millions of men, and this is what they call a peace establishment. Then it is not generally considered that this number of men by no means represents the number of men who are taken away altogether from productive work, for in addition to those who do nothing but drill and prepare for the purposes of destruction, you must have another army of men who are employed in supplying these with the materials for destruction; and I believe, if we could follow out all the war material to its source, and thus arrive at the total number of the men employed in preparing it and taken away from real production which adds to the wealth of the community, it would be found to constitute another army much larger than this vast army of 3,500,000 men. For you must remember that in one of our huge ironclads you do not merely have the men engaged in its construction, but you must go back to every ton of iron and coal used, to the men engaged in extracting the ore from the earth and in making the raw iron into its various forms, to the men engaged in making the elaborate machinery connected with it—the engines of war, and the wonderfully elaborate fittings so complicated that one of these great vessels is almost like a city—and if you follow all these back to their primary beginnings in all parts of the world, you will find that there must have been a large army of men employed in the construction of a single iron-clad. Add to that the wonderful machinery used in constructing our guns and torpedoes, the munition, clothes, food, everything that is used by these men; and if we further consider that armies waste perhaps more than they consume—taking all this into consideration, you will find that it cannot be less, but probably is much more, than another army of 3,500,000 men engaged in the service of the actual army. So that we have a total of 7,000,000 men at the present time entirely occupied in preparing for the work of destruction. If, as is admitted, the great armies of Europe have increased by 630,000 men, I think it more than probable that the increase of these armies which wait upon them have been proportionally much greater, because the appliances they require—the weapons, the ammunition, and the scientific paraphernalia of an army in the field—are so immensely more elaborate than they were forty or fifty years ago, so that it will be necessary to add near a million of men employed in this work, and we shall have an increase of about a million and a half of men whose labour is utterly wasted, besides those actually engaged in the destructive, wicked, and useless purposes of war.

We have a very striking indication, and to some extent a measure, of this enormous waste of human labour, in the increase of the total fiscal expenditure of these six great powers. Taking the different estimates of their annual expenditure for government purposes from 1870 to 1884, I find that these six great powers have increased their annual expenditure by £266,000,000 sterling. That is the increase of the six great powers of Europe, and that increase is almost wholly due to this terrible war expenditure which I have been trying to put before you. That £266,000,000 means, of course, £266,000,000 of additional taxation beyond what there was before. Surely this is a cause of the most terrible impoverishment, and sufficiently accounts for people not being so well able to buy as they were before. Then, again, we must remember that whenever this great engine is put to its destined use, there comes another loss in the actual destruction of property and life. In every country where war is carried on, as a necessary result towns and houses are battered down, vineyards and fields are rendered desolate, fruit trees are destroyed, and consequently we have an overwhelming amount of destruction of property whenever this war machine is put into motion; and here again is a cause of poverty, and therefore one of the most direct and immediate causes of the depression of trade.

Now this machine has been put into action almost continuously, either in greater wars or lesser wars, and as we supply goods to almost every nation in the world, it does not matter where the war is, one thing is certain, that a considerable number of our customers are killed and a much larger number are impoverished. Just consider; in 1872 we had the great Franco-German war; in 1875, the Ashantee war; in 1878, the terrible Russo-Turkish war; in 1879 and 1880, the Transvaal and Zulu wars; in 1881, the Afghan war; in 1883, the Egyptian war; in 1884-85, the Soudan war; and since then the French Tonquin war and then the Mahdi war. Now we have the Burmese war, and the Soudan war is still going on. Every one of these wars kills or impoverishes our customers; and consequently, not only by the cost of the huge armaments, but by the vast destruction of life and property they bring about, the war expenditure of Europe is the cause, to an unknown but enormous and incalculable extent, of the existing depression of our trade.

Now these two great causes—loans to foreign nations, at first inflating and then necessarily depressing our trade by the impoverishment of the people; and the increase of war-costs, which, as I have shown you, have been always enormous, and have been of late years ever increasing—these two may be considered to be the great external factors which have caused the depression of trade, by impoverishing our customers all over the known world. The effects of these two causes are clear as daylight; the result is an inevitable result; and the amount of the evil is so gigantic, that I think I am justified in placing them in the front as the most important and inevitable causes of the depression of trade. Yet, so far as I am aware, during the many months that the Royal Commission has sat not one word has been said about either of these causes; and I believe, when the final report of the Commission is issued, that you will probably not find one word about them.

The Increase of Millionaires as a Cause of Depression of Trade.

I now come to another branch of the subject, that which deals with our home trade—with the causes of the depression in our home trade in addition to that produced in our foreign commerce. I have given the increase of speculation and of huge fortunes made by speculation as one of the chief causes, and I will first adduce a few facts to prove that it is really the case that millionaires have been recently increasing.

The sums paid for probate duty have been published, and they show the amount of property on which probate duty is paid, but this only covers what is called the personal estate, it does not cover the landed estate; consequently, whatever the valuation is, it represents only a portion, and sometimes only a small portion, of the whole estate. To make it simple I have divided the results into two periods—the ten years previous to the commencement of the depression in 1874, and the ten years subsequent to it. Between 1862 and 1873 I find that 162 persons died with fortunes of over a quarter of a million. In the next ten years they had increased to 208 persons who had died with fortunes of over a quarter of a million. This is an increase of over 29 per cent. The detailed figures show still more remarkable results, because they show that the increase was still more rapid in very great fortunes, in fortunes over a million. In addition to that a very considerable number of great landowners have died who paid no probate duty, but whose capitalised fortunes have been from one to five millions sterling each. We have not the exact figures, but still we know that their fortunes have been of late increasing, owing to the increase of our large towns and the enormous increase of ground rents which have arisen in them. The main result is, that a few, that is comparatively few, have become much richer than they ever were before; and it appears to me that it is a demonstrable fact that, when those who are very rich suddenly become more numerous and still richer, without any increased power of wealth-creation independent of labour, then, as a necessary result, those who are poor become poorer.

This principle was laid down very clearly by Adam Smith, strange to say, in the very first sentence of his *Wealth of Nations*, but I do not know that much attention has been paid to it. The sentence is this. He says :

"The actual labour of every nation is the fund which originally supplies it with all the necessaries and conveniences of life which it actually consumes, and which consists always either in the immediate produce of that labour or in what is purchased with that produce from other nations."

This lays down a proposition perfectly clear, that there is no other source whatever of wealth in the country than the produce of the labour of its people. Hence it follows absolutely and indisputably, that if a larger proportion of that wealth goes to the few, a smaller proportion must remain with the many. As some people may not clearly see the bearing of this statement of Adam Smith, let me just illustrate it by a few particular cases. It is quite evident that all the wealth of the country is produced by labour, or by the use of labour and capital combined, and everybody who gets wealth must get a portion of this total amount. There is no other source from which he can get it. Whether he obtains it in the form of rent or from the taxes it comes exactly to the same thing, it can only come out of the produce of labour. In the same manner, whether he gets it in payment of wages or remuneration for professional services, those who pay it can only have got it, directly or indirectly, by labour. Consequently the fact is indisputable, that the produce of our labour measures the whole available wealth produced by us in the country, and that wealth has to be distributed by various methods among the whole community. Consequently, if it is clearly proved, as I think it is—to prove it in detail would require a much more complete examination of the statistics of the country, but I am sure it can be proved—that the large body of the very rich have been steadily growing richer, then it follows as a logical result that the remaining body, or at least a portion of the remaining body, must have been growing poorer in proportion.

A Proof of Increasing Poverty.

Of course this has been denied over and over again, but I have endeavoured to get some confirmation of it by examining the information given in the census returns.

The full census report, as you are probably aware, gives a great amount of detail as to the occupations of the people at different times, and I have looked up the facts as to the increase of the persons employed in particular trades and manufactures for the purpose of seeing what light it would throw upon this question, and I found that it supported in a remarkable manner the statement which I have laid down for your consideration, that is, that the great masses of the people have been growing poorer while the few have been growing richer. And it illustrates it in this manner:—whenever we have a manufacture which depends mainly on the consumption of the masses, we find that there has been either a decrease of those employed in it, or at all events that it has been stationary; on the other hand, where we have a special business or profession or trade which is supported wholly or mainly by the wealthy, we find an increase, and sometimes an enormous increase. When I use the word increase or decrease, I always mean an increase or decrease in proportion to the total population. Thus I find, taking the increase of population into account, between the two censuses of 1871 and 1881 (the last we had) the persons engaged in the cotton manufactures of this country diminished 20 per cent. in that period; persons employed in the linen and woollen trade diminished 15 per cent; metal workers remained stationary; and drapers diminished 7 per cent. Now these are all businesses and manufactures which certainly depend upon the consumption of the masses. Now we come to those which more especially depend upon the consumption of the wealthy. Milliners increased 4 per cent., more than the whole population increased; carpet makers increased 9 per cent; florists and gardeners increased 10 per cent.; musicians and musical instrument makers increased 23 per cent. These remarkable facts support my contention—and may almost be said to prove it—that the rich have grown richer and have been able to indulge in greater luxuries, while the poor have grown poorer and have been obliged to do with less of the bare necessaries of life.

The Increase of Speculation.

The census also gives some remarkable illustrations of the increase of speculation as a business,—and this is pre-eminently a non-productive business, and one that is impoverishing to all but the few winners.

In the same ten years I find that persons registered as bankers or bankers' clerks increased 21 per cent., and accountants 6 per cent.; and then there comes a most extraordinary item, which the census authorities note and say they are utterly unable to explain, and that is that persons who call themselves insurance agents or brokers have increased 300 per cent. I can only explain it by supposing that there are an immense number of people who live in the city by speculation who find it convenient to call themselves insurance agents or brokers. I think, as far as I can judge from advertisements in the newspapers, that this mania for speculation has been going on at an increasing rate; that is, that within the last few years it has increased more rapidly, and its effects therefore have been more injurious, than ever.

I now wish to point out to you another indication—another field as it were—in which this speculative mania has produced the most deplorable results, and has acted, in combination with other causes, so as to increase the poverty of one class and the wealth of another class, and has thus, as I shall show you, tended directly to produce depression of trade. Somewhat more than twenty years ago an Act was passed which was considered by the whole commercial world as one of the greatest boons ever given to it; this was the Limited Liability Act. This Act was universally approved of; was supported and praised by such a great and thoughtful writer and friend of the working classes as John Stuart Mill. But I do not think he could possibly have foreseen what would come out of it. About two years ago a short parliamentary paper was published giving a kind of summary of the results of this Act. It is a curious thing that this parliamentary report seems to be totally unknown, for I inquired of several friends in the city, particularly of one who is an accountant in the city, and whose business largely consists of winding up those companies, and he did not know of its existence. The report gives us some very startling facts. It covers a period of exactly twenty-one years, and is thus easily divisible into three periods of seven years each. In the first period I find that 4,782 companies were formed, being at the rate of about 700 per annum. In the next period the number increased to 6,900, and in the last seven years to 8,643. Out of this total of about 20,000 distinct companies formed in twenty-one years only 8,000 are now in existence, 12,000 having been wound up! It is also stated in this parliamentary report that the actual paid-up capital—not the nominal capital—of these 8,000 companies was £475,000,000; that is about £55,000 each on an average paid up, some of course very much more, and some very much less. Now, not to take an extreme estimate, suppose we reduce this average of £55,000 down to only £10,000, and consider that each of the wound-up companies involves a loss to its shareholders of £10,000, I think everybody who knows anything about them will think that absurdly low, and yet that would involve a loss of £120,000,000 sterling to the unfortunate shareholders.

Effect of Speculation in Depressing Trade.

Now let us think what is the effect of this continuous loss—and in many cases absolute ruin, of a large number of persons amounting to many hundreds of thousands—by the failure of these companies? I daresay there is not a person who reads this chapter but knows one, and most of you several, individuals who have been ruined by such things. A great number don't like to speak about the matter, and keep it secret, and therefore nothing is heard of it; but we have the absolute fact that thousands of individuals, mostly persons with small means, deluded by flattering prospectuses, were induced to invest their means in these companies—persons of the middle class, very often officers and widows and country clergymen, scattered over the country. These have lost, at the very least, £120,000,000, and much more likely three or four hundred millions sterling. Now just think what is the effect of the ever-increasing impoverishment of this large body of the middle classes, and we will take it in connection with the increasing mass of speculators who have become millionaires from the losses of these men. The one are counted by hundreds, and the other by tens of thousands. Some people will perhaps say, "What difference can it make to trade, if the money is there, and the money is spent?" But I want to show you that this is a most delusive idea, and that it really makes all the difference to trade. When you have a thousand families of the middle classes impoverished, it means that you reduce their outlay on all the staple manufactures of the country. In clothing, furniture, and everything in fact that makes life agreeable, they are obliged to economise, simply because it is more easy to economise in these than in absolute food. Therefore all over the country there is a diminution in the demand for the staple products of the country; but when this money is accumulated, and goes into the hands of a few speculators, it is spent on different things—on ornaments, entertainments, yachts, horse-racing, foreign travel, and hundreds of other ways, —it is spent on that which all economists tell us, and perfectly truly, is the most unproductive kind of expenditure. Consequently the loss to the manufactures and trade of the country by every million of money transferred from the industrious working or middle classes to rich speculators is enormous, and is thus a real cause of depression of trade. I think I am therefore quite justified in maintaining, that, although it is certain that the aggregate wealth of the country has been steadily increasing all these years, still that wealth has been becoming more unequally distributed, and that inequality is the direct cause of a large proportion of depression of trade.

Depression of Trade in America.

Now I did not mention it at first, but I may mention now, that the reason is very clear why the depression which affected us should affect all other great commercial countries of Europe and America. It is because all the causes which I enumerated as producing depression of trade as regards our foreign commerce would affect all those other countries just as well—that is, they have produced a real impoverishment of the peoples who were customers both of ourselves and other manufacturing countries. Therefore the causes acted in the same way on France, Germany, and America as they did with us, to the extent that their manufactures went abroad to other countries.

But there have been some special causes affecting America which account for the remarkable fact that, notwithstanding the advantages they possess in their enormous territory, and the great energy and enterprise of the Americans, they have still suffered from this depression perhaps as much as we have done. The reason is to be found in the fact that with them this last evil of speculation is greater and far more gigantic than even with us. Everybody has heard of the "corners" in America, by which a lot of speculators get hold of the whole trade of the country in a certain article, creating a monopoly which they manipulate for their own purposes. This has been applied to almost every industry. But the most destructive cause of depression in America is the successive railway manias which they have had. The first was from 1867 to 1875. There was a continuous railway mania during those years,—a mania for making railways in America. In that period 40,000 miles of new line were made, and in the one year 1872 no less than 7,000 miles of new railway were made. That coincides with the culminating point of our prosperity, and a large part of the iron for these lines was sent from England. The greater part of these railways was made merely for speculative purposes, and was very largely unproductive. The shareholders were often ruined, and consequently the exact effect was produced in America that was produced in our country by the limited liability mania. This railway mania, after a lull, broke out again in America a few years ago, in 1880, and in 1882 no less than 11,500 miles of new railway were made. It has been estimated by one of the most able statisticians in America, that this increase of the railway system went on four times as fast as the increase of the produce to be carried on the railways. That clearly shows that most of these railways have been failures—so much money thrown away, and those who lost it must have been impoverished. Here then you have a very widespread and enormous cause of impoverishment, and therefore of depression of trade in America. In fact, we hardly need to go further.

Then, again, as to millionaires in America, I do not know that they are greater in number, but they exceed us in the gigantic sums they possess. While our millionaires seldom have more than two or three millions, the American millionaires often possess ten and twenty millions. And of course the result is still more clear. All this money must have been obtained out of the purses of the community, and to that extent the labourers who produced it are so much worse off than if the money had gone into their own pockets instead of into the pockets of the millionaires.

There is yet another source of poverty in America which we have not to so great an extent in this country, and that is the "rings" that sometimes get possession of municipalities in the States. We have all heard of that wonderful "ring" in New York which got possession of the municipality, and plundered the whole community. They kept it up for years by wholesale bribery. That is a thing we do not hear much of in this country, but we may be sure that what was done so boldly in New York was imitated in other towns, and the result may perhaps be seen in the municipal debt piled up in America far beyond what it is in this country. The municipal debts of this country are held to be a great and growing evil, and help to occasion depression of trade. But in America it is worse. An estimate was given in an American paper some time ago; it may not be correct, but it gives perhaps a fair approximation. It compared American with English municipal debts. It compared the fourteen chief cities in America with fourteen large English towns, leaving out London, and it was found that the average taxation per head in America was fourteen dollars, whereas in England it was only seven dollars; and that while the municipal debt in America was forty-one dollars per head, in England it was only twenty dollars. In addition to that, it was stated that the area over which this municipal indebtedness extended was greater in America than in England; that small towns in America—the very smallest towns in the country—are often burdened with debt, to a much greater amount in proportion than the large towns. It has often puzzled people why America should have suffered from this depression, but I think the few facts I have here given afford a sufficient clue to it.

Depopulation of the Rural Districts.

I now come to what I consider to be by far the most important part of our subject, because it is that with which we are in the closest relation, and which is, I believe, the most direct cause of widespread poverty—rural depopulation. This rural depopulation has been going on for probably a very long time, but it was not seriously noticed till ten or twenty years ago. Before that date many of the counties seemed to be stationary in population, but in 1861 it was noticed that a few counties had not increased, but rather diminished, during the preceding ten years, in 1871 seven or eight had decreased in population, and in 1881 fifteen counties had decreased. But besides this decrease in certain counties, the census returns give very accurate and detailed information as to where this depopulation occurs, and to some extent how it occurs.

The whole of England is divided into registration districts and registration sub-districts. These registration sub-districts are about two thousand in number, and consist of an aggregation of parishes, roughly speaking not very unequal in size, and probably not very unequal in population. In towns they are, of course, much smaller in area. The increase or decrease of each of these registration sub-districts is given in the census, and I took the trouble to go through the tables and take out all the cases of decrease, and I found that there has been a decrease over a very large number of these sub-districts. The general result is, that over about half the area of England and Wales there was actually less population in 1881 than in 1871. But you must remember that the population of the country has been going on steadily increasing all that time. In the ten years the population of the whole country has increased fifteen per cent., and that is exclusive of those who have emigrated, so that the actual rate of increase of the population is somewhat more than that. Then, again, it is perfectly well known that the rate of increase—what we may call the natural increase

—of dwellers in the country is somewhat higher than that of dwellers in the towns; the birth-rate is higher, and the death-rate lower.[1] Therefore it is a very low estimate to consider that what may be called the normal increase of people dwelling in the country is seventeen per cent. Therefore the area that is actually decreasing will not represent the whole of the area from which people have migrated into the towns; they have also migrated from all those areas in which the population has not increased so much as it would normally have increased. That is, if in any area there is less than seventeen per cent. of increase of population since 1871, it is perfectly certain some of the people must have gone out of that area; and if we add to those which have actually decreased the areas in which the population must have migrated in order to make the increase so little as it is, then we shall find that in only about one-fourth of the whole country has the population increased to its normal amount. This increasing area consists almost wholly of the great towns and the residential districts around them, while all the rest of the country has been becoming more or less depopulated. The amount of the decrease of rural population is a distinct question. I find that the actual depopulation that is the diminution of inhabitants for the ten years in these decreasing sub-districts, amounts to three hundred and eight thousand. Then I take the amount the population of these areas ought to have increased in ten years at seventeen per cent., and that added to the actual decrease gives an effective diminution of nearly a million from this decreasing area. Then adding to this the emigration from the area of small increase, I find that in the ten years the people who have migrated out of the country districts into the town districts, with their natural increase in the same period, amounts to about one million and a quarter.

[1] See Dr Stark, in Tenth Report on Births and Deaths in Scotland, quoted by Darwin in his *Descent of Man*, p. 138.

The Effects of the Depopulation.

Now let us consider what are the results of this migration from the country into the towns. The greater part of those people who have migrated are not necessarily agricultural labourers. About one-third are agricultural labourers, and the remainder are what you may call villagers—people who carried on trades and occupations of various kinds in villages and small towns. The causes that led to the labourers migrating affected them also, and they migrated to a still larger extent, and the result is to be seen in a most striking fact which has been brought forward among others to prove the prosperity of the country, and that is the enormous increase in the import of certain articles of food. Most of you know—at all events it is a well known fact—that country labourers and many other rural inhabitants are fond, when they have the chance, of keeping pigs and poultry, growing potatoes and other vegetables. Now it is a most singular thing that if we compare the years 1870 and 1883 there is an enormous difference in the imports of these articles of food. It is so great that it seems almost impossible; but the figures are taken from official papers. In 1870 we imported less than a million—860,000—cwts. of bacon and pork, whereas in 1883 it had risen to 5,000,000 cwts. Of potatoes there were imported 127,000 cwts. in 1870, and 4,000,000 cwts. in 1883; of eggs in 1870, 430,000,000, and in 1883, 800,000,000. Now 1870 was in the midst of our period of prosperity; we were supposed to be all well off; wages were high, and men were all in full work. But 1883 is in our period of depression and distress, and it is actually maintained by Mr. Giffen and other statisticians who put forward these figures to show the prosperity of the country, that we consume enormously more when our trade is depressed than we did during the period when it was most prosperous! It appears to me, on the contrary, that these facts are due to a decreased production of food, caused in part by the continuous emigration of people out of the country into the towns ; and that means a diminished production of wealth for the country, and a great increase of pauperism and misery in the towns where these people go.

Evidence of the Increase of Destitution.

It is very difficult to get direct evidence of this, but there is one piece of indirect evidence—though it may be almost called direct—which I adduced some years ago, but can never find answered or explained in any way consistent with that increase of prosperity of the masses which is so persistently alleged. In the reports of the Registrar-General for London—and he takes in an enormous area called Greater London—he gives the deaths in workhouses and hospitals each year. In order to arrive pretty fairly at what may be called the destitute who die in these institutions, I have taken the deaths in the workhouses and one-half of the deaths in the hospitals. In 1872 they amounted to 8,674, or 12·2 per cent. of the total deaths; in 1881 to 13·132, or 16·2 per cent. of the deaths. Now I want to know, if the masses of the people of London and its suburbs were better off, or even as well off, in 1881 as in 1872,[1] why did 30 per cent. more of them die in destitution? If we take the proportion of deaths to those living, we find this increase of 4,458 deaths of the destitute in these ten years means the addition of 107,000 to the destitute poor of London! Now all this, which shows a real and dreadful increase of poverty, necessarily means depression of trade. If there are 100,000 more destitute persons in London now than there were ten years ago, there are so many less customers for the staple products of the country. Then, again, if we turn to another country—the sister country Ireland— we find that still more remarkable and still more distressing events have occurred. There the population has decreased half a million since 1870, and during the same period the emigrants have amounted to 883,000, so that

[1] The year 1872 is taken because 1871 was the year of the great epidemic of smallpox, when the number who died in workhouses and hospitals was abnormally large.

though the population has gone on slightly increasing, the increase has been far more than counterbalanced by the enormous number of emigrants; and you must remember that the emigrants are mostly men in the prime of life. Those who are left behind are the women and children and the old and the weak. We cannot wonder, therefore, at the increase of poverty and pauperism in Ireland. That increase is measured very well by the cost of poor relief. In 1870 the relief cost £814,000; in 1880 it cost £1,263,000—an increase of 50 per cent. on the cost of the poor, with a decreasing population! There, again, is a most tremendous cause of the depression of trade. You have got a much smaller population in Ireland, and a population very much poorer than it was, and that necessarily results in a depression of trade, because we supply Ireland with most of the manufactures she consumes.

Causes of Rural Depopulation.

It is, however, not sufficient to know the facts of this rural depopulation, but we must say a few words on its causes. These causes have been pretty clearly made out by little bits of evidence that have been found here and there in the reports issued by the last Agricultural Commission. We find it clearly stated by these official reporters that a considerable body of the farmers of England have been ruined by excessive rents. For many years past they have been paying rent out of capital, hoping for better times. Notwithstanding bad harvests and bad seasons, they have kept struggling on as long as they could by means of partial remissions from their landlords, but a large number have been utterly broken down, and have been obliged to give up their farms. The farms have not found fresh tenants, because the landlords will not let them, except on exorbitant terms and with the usual onerous conditions, and consequently a large number of landlords all over the country have been turning their lands from arable into pasture. The reason they do this is, that they can then obtain a return with a minimum of outlay and risk. When they have turned arable land into pasture, the annual produce is not above one-tenth of the value that it was before, but it is obtained with considerably less than one-tenth of the outlay. The consequence is that it means profit to the landlord; but it also means ruin to the country.[1] It is one of the causes, perhaps the chief cause, of the great exodus of population that I have been pointing out to you. It is estimated that for every hundred acres of land thus converted from arable into pasture two labourers must be discharged; and as at least a million acres of land have been so converted between 1873 to 1884, that means that 20,000 labourers and their families were discharged for this one cause alone. Along with them, of course, went numbers of tradesmen who depended on them for their support; and mechanics and others who were employed by the farmers and in the villages have also left, partly for the same reason, and partly because it has become more and more the custom for large farmers to get all their work done and machinery repaired in manufacturing centres rather than in the villages by the local workmen.

Now the amount of food lost to the country by this change from arable to pasture is enormous. I have taken the estimates made by two or three of the most authoritative writers. They give the average produce of arable land at £10 5s. per acre, and they also give the average produce of pasture land at £1 9s. per acre; consequently there was a loss of £8 16s. on every acre converted. That means nearly £9,000,000 of loss to the country by this 1,000,000 acres that we know from official returns have been changed from arable to pasture, and the change is believed to be going on to this day far more rapidly than ever.

But there is another cause of rural depopulation. Just now the landlords are trying to persuade the country that they are very glad to let poor men have land, but hitherto it is notorious that they have always refused to let them

[1] It is stated by Hume in his *History of England*, "that in the year 1634 Sir Anthony Roper was fined £4,000 for depopulation, or turning arable land into pasture land, under the provisions of a law enacted in the reign of Henry VII." Cannot this most just law, which has probably never been repealed, be put into operation now?

have it on any reasonable terms. This is very well known to be the rule, and to have been a chief cause of this terrible exodus of labourers from the country to the towns. In addition to this they will give no security to the farmers for their improvements. They treat the farmers in every respect exactly as they treat the labourers. If they do offer the labourers land—as they are doing now that there is a deal of excitement on the subject—they never give it except on what are prohibitory terms—that is, as yearly tenants, and without any security whatever for their labour and improvements.

Now the report of the Agricultural Commission, to which I have already referred, contains some remarkable evidence as to the results obtained in those few cases where landlords really do their duty, and treat the land as a trust rather than as property only. There are two or three landlords in the country who have done so, and in every case where such landlords' estates are referred to in these reports, it is invariably stated that there is no depression in agriculture, that the farmers are well off, the labourers are well off, and all are contented. That is remarkably the case in parts of Cheshire and Suffolk on Lord Tollemache's estates. Lord Tollemache is almost the only landlord in the country who not only gives his farmers voluntarily perfect security of tenure, but also gives every labourer as much land as he can cultivate, at a moderate rent, and on an equally secure tenure; and, what is more remarkable, he encourages outsiders of decent character—anybody, in fact, who likes—to come and settle on his estate. He offers land to build a house, and a few acres in addition on which to keep a cow, at a low rent. The result is that on his estate everybody is well off; the farmers are contented, the labourers are contented and prosperous. The farmers say they have the best of labourers to work for them, utterly disproving the common assertion that if you let a labourer have land he will not work for the farmer. At the same time the labourers and the farmers find customers in those persons who have come to live on the land, and small communities

are thus formed which are to some extent self-sufficing.
When we get a community of that kind, consisting of
various classes, all living together, but scattered about on
the land, they all tend to support each other. Each one
finds employment or assistance from the other. There is
a market at hand, and we do not see that absurd system
of sending all the butter and poultry to a place a hundred
miles away, while a person who lives a mile from the
farmer is obliged to get his poultry and butter from the
town. That is what they call economy of production, but
it is certainly waste in distribution.

Results of Peasant Cultivation.

The amount of loss involved by this driving the
labourers from the country to the towns is also brought
out very strongly by the evidence of a Tory landlord, who
has repeated it several times, and I will take it therefore as
correct. In Buckinghamshire Lord Carrington has land
which he lets out in lots to labourers. He has about
eight hundred of these allotments already in the hands of
labourers and others, and he has stated publicly that of these
allotments the average produce is £33 an acre *more than
the produce of the same land in farms.* Therefore, as far as
these allotments are concerned, there is a positive gain to
the country on every acre of land to the extent of £33 a
year. Some years ago, in 1868, when produce was not
nearly so valuable as it is now, there was a Government
Commission on the employment of women and children
in agriculture, and it obtained evidence that the average
produce of such allotments all over the country was £14
an acre more than that of farms. Then, again, there is
a curious piece of evidence recently given by an English
clergyman (Rev. C. W. Stubbs), also living in Buckinghamshire, who has a large amount of glebe lands,
which he lets out to labourers in acre or half-acre
allotments, and it is a noticeable fact that, the land of the
district being pretty good wheat-land, the labourers all
grow wheat upon their allotments. They have been
doing so for nine years, and Mr Stubbs has kept an

accurate account of the produce they get, and although it is constantly asserted that it is impossible to grow wheat on a small scale, yet these allotments produce £4 10s. more an acre than all the surrounding farms of Buckinghamshire. And what is more, he finds that the labourers' produce per acre is higher than that of the best scientific farmers in England; so that actually the poor labourer, working by himself on his own plot of land, can produce for us more wheat per acre than the most scientific farmer with all his skill.[1] Take these estimates together—£33 per acre, £14 per acre, and £4 10s. per acre, and that gives an average of net gain to the country of £17 for every acre of land cultivated by poor men in small quantities compared with the same land cultivated by farmers in large quantities. Now just think what a gain that would be to the country if the people, instead of being driven from the rural districts for want of land, had been encouraged to remain and cultivate the land for themselves. I have calculated the average gain at £17 an acre. But if, to avoid any exaggeration, we lower this, and say only £10 per acre, and if we suppose that out of the fifty millions of acres of cultivatable land—a considerable part of which is now going out of cultivation—only twenty millions of acres were cultivated by poor men in this minute and careful manner, and that they obtained £10 per acre of increased produce, that would give us £200,000,000 a year of extra wealth produced by poor men, and almost every penny of that £200,000,000 would be spent on the manufactures of the country.

Now that, in my opinion, indicates the method by which we may finally get rid of this terrible depression of trade, which is still increasing and is likely to increase, because we have been hitherto falsely guided by the political economists and by the great manufacturers, the speculators, financiers, and others. We have always been led to believe that our one line of business was manufacturing, that we were to be the manufacturers of the

---

[1] See *The Land and the Labourers*, by Rev. C. W. Stubbs, 1884. Swan Sonnenschein and Co.

world; and while we have been going on in this line, utterly neglecting agriculture and the land, forbidding people to use it, and driving them into the towns so as to increase the numbers of the unemployed and thus keep down wages, other nations have not been standing still, and are now competing with us in all the chief markets of the world.

There is a great deal of talk about finding fresh markets, but these would be open to all the competing countries, and would not supply our increasing population with fresh outlets for work; and therefore I maintain that the only real and substantial mode of getting rid of the depression of trade, is to utilise thoroughly that enormous store of wealth which exists in our neglected fields and our miserably cultivated soil.

Summary of the Argument.

I will now briefly summarise the points here brought forward. First of all, the enormous foreign loans led to an abnormal and unnatural increase of our trade, and then to a depression which was exaggerated and increased by the impoverishment of the people who had to pay the interest on these loans, and it must be remembered that they had to pay for millions which they never received, that never came into their country but were absorbed by the financiers in the cities—they had to pay and are still paying all this with interest upon it. Then we have the enormous increase of speculation in our cities, favoured by every act of the legislature and by every custom of the country, and as the result we have the concentration of wealth into fewer and fewer hands, and consequently a proportionate diminution of wealth that ought to be distributed among the people. We have also the dreadful increase of war expenditure; and lastly, the evils directly produced by the system of landlordism in this country—a system which gives a comparatively small body of men power to determine whether the land shall be used or abused, well cultivated or producing less than half of what it ought to produce and will produce—a system which drives the people away from the country into the towns, and turns into paupers men who would, if they were permitted freely to use the land on fair terms, produce an enormous increase of food, the prime necessity of a nation's existence, and by their prosperity cause such a demand for our manufactures as we have never known in this country before. And this evil is caused, and this good prevented, by the direct or indirect action of landlords under our vicious land system.

I maintain, therefore, that these are the real fundamental causes of the depression of trade, because every one of them, as I have shown, tends directly to the impoverishment of the great masses of the people, who are our best customers. Every one of them can be shown either to have begun about the period when the depression showed itself, or to have become greatly intensified about that period, and therefore as a whole they have worked together to produce this enormous and long-continued and increasing depression of trade.

Remedies for the Depression of Trade.

The remedies, of course, are some of them difficult, some of them comparatively easy. If you see and understand what I have endeavoured to make you see, that anything like a system of foreign loans bolstered up by the Government of this country is radically bad and immoral, then you ought to urge upon your representatives that in no way whatever should the Government lend its power or its influence to compel the oppressed populations to pay these loans or the interest upon them.

Another step will be to stop all aggressive wars on any pretence whatever. I consider in the present state of the world that there is only one class of wars that are justifiable or will be justifiable for us, and that will be a war to help a weak nation when oppressed by a stronger power. It is a singular thing that this is the only kind of war likely to do us good even in our trade, for it would protect for us our customers as well as bind them to us by the bonds of gratitude; but it is the kind of war that we never in any circumstances have undertaken.

Then, again, if we see clearly and distinctly that whatever facilitates the growth of abnormal wealth in the few is bad for the rest of the community, we certainly should favour all those steps which would render it more difficult to accumulate such wealth. It would take too much time now to go into all the measures which I think would be advisable for that purpose. One thing, however, would be certainly advantageous, though I am afraid it will never be done, and that would be to repeal the Limited Liability Act. I believe this Limited Liability Act has been a greater curse to the country than any Act of Parliament ever passed, because it as much as says, with the authority and voice of the Government to the people—You may enjoy the benefit and all the advantages of commercial prosperity by simply subscribing your money towards these companies. How are the people at large to know which are good and which are bad? The mere fact that such an Act was passed was an invitation to the people of the country. They accepted the invitation, and for each one who has benefited by doing so a score have suffered.

The last thing, and perhaps the most important of all, is to abolish the monopoly of land in this country. I believe no half measures will do any good here. The only thing will be to declare by law that the whole of the land shall revert to the State for the benefit of the people, but that no individual so far as is possible shall suffer any loss during his lifetime or during the lifetime of any of those who have reasonable expectations from him. If that were done no landowner would have a right to complain. He would receive an income probably as great as he has now for the rest of his lifetime and for the lives of all his children, while the nation would have the use of the land, and by allowing every man to have as much as he could cultivate, at fair rents and with complete security, would lead to the production of an amount of wealth probably two or three times greater than is now derived from it. This increased wealth would be earned by men who are now poor or pauperised; and as it would almost all be spent in home manufactures, it would in the most direct and speedy manner restore the prosperity of the country and abolish the Depression of Trade.

CHAPTER XIII

A REPRESENTATIVE HOUSE OF LORDS

A FEW years back, Mr. Labouchere introduced a Bill into the House of Commons declaring that, after January 1, 1895, the House of Lords shall cease to exist. But it is hardly possible that such a Bill can become law, either in this Parliament or in any of its successors for the next half century, since it would require that the Peers should commit political suicide, and this they would hardly do unless an almost unanimous public opinion compelled such a course, and they considered it more dignified than submitting to actual expulsion. There is, also, as Mr. Labouchere himself acknowledges, a preliminary difficulty, in a very wide-spread impression, even among Liberals, that a second chamber is necessary, combined with an extreme diversity of opinion as to how the second chamber should be constituted. It is evident, therefore, that the abolition of the House of Lords would by no means solve the problem, but would only lead to interminable discussions on the more difficult part of the question—what kind of chamber to substitute for it. The stoppage of all useful reforms by any attempt to remodel our constitution in such a revolutionary spirit would be exceedingly unpopular; and would probably involve a longer struggle and more expenditure of parliamentary energy than the effort we recently made to give Ireland permission to manage her own affairs. It may, therefore, be worth while to consider whether there is not a method by which a House of Lords may be retained in such a form as to render it a truly representative Upper Chamber, thus making it acceptable to most Liberals, and even to many Radicals; while, by preserving its ancient name and prestige, and by giving it both greater dignity and a more important part in legislation than it now possesses, the proposed reform might be upheld as truly conservative, and receive the support of the majority of the Conservative party.

It is clear that any such fundamental reform of the British Constitution as is now advocated by advanced Liberals should proceed on the lines of evolution rather than on those of revolution. Instead of abolishing the House of Lords we must modify, reform, and elevate it; and we must do this in such a manner as, on the one hand, to bring it into general and permanent harmony with the House of Commons; while, on the other hand, it is rendered so select, so dignified, so representative of all that is best in the British Peerage, past, present, and to come, that a seat in the Upper Chamber will become a more coveted honour than the insignia of the Garter, a higher dignity than a ducal coronet. It is, I think, essential to the successful carrying out of any such great reform that it should be initiated in the House of Lords itself, and simply accepted or rejected by the House of Commons. The discussion of its principles and methods should take place in the country at large, rather than in Parliament. The peers must be well informed as to the character and amount of change that will satisfy the people and bring about that substantial harmony between the two branches of the Legislature that is essential to good government; and it is with the hope of contributing towards the peaceful settlement of this great question that I now propose to set forth what appear to me to be the main principles on which such an important reform should be founded.

The two great anomalies of the present House of Lords are, first, its hereditary character; and, secondly, the presence in it of the bishops of the Church of England, who thus have a voice, and often a very important influence, in making or rejecting laws which affect the whole population. Both hereditary and ecclesiastical legislators are now felt to be wholly out of place in the parliament of a people which claims to possess both political and religious freedom. They have, during the last half century, been tolerated rather from the difficulty of getting rid of them, than from any belief in the value of their services; and it has long been seen, by all but the most bigoted Conservatives, that something must soon be done to bring the Upper House into harmony with modern ideals. In these concluding years of the nineteenth century our hereditary House of Lords is an anachronism. It may be said that our hereditary Sovereign is also an anachronism; but there is this great difference—that the peers systematically use their power to prevent or delay popular legislation, which the Sovereign, at the present day, never attempts to do.

It is clear, then, that any real and effective reform of the House of Lords must, in the first place, abolish the hereditary right to legislate, and must also exclude the bishops, as such, from any share in law-making. This, of course, does not affect the hereditary succession to the peerage, which may continue at all events for the present; but it would be most advisable to discontinue the creation of new hereditary peerages. Instead of these, life-peers should be created, but always as a mode of indicating distinguished merit, whether exhibited by services to the country at large, by philanthropic labours, or by exceptional achievements in the fields of science, art, or literature. The object of creating these life-peers should be, to raise the character and dignity of the peerage, and thus to afford material for the selection of a new House of Lords, which should be worthy of its historic fame and be in every way fitted to take a leading part in legislating for a free and civilised people.

Although all Liberals, and many Conservatives, will agree that the mere fact of succession to a peerage does not afford any sufficient guarantee of the possession of those qualities which should characterise the legislator, yet most of them will admit that the peerage as a whole does afford some good material from which to choose legislators, and this material may be indefinitely increased, both in quantity and quality, by the creation of life-peerages as above suggested. A peer is, at all events, an English gentleman. Many peers belong to families whose names are household words in our history, and these may well be supposed to have the real interests of their country at heart, and to be influenced more or less profoundly by the good old aristocratic maxim, *Noblesse oblige*. Most of them have had the best education our universities can afford, and have added thereto that wider education derived from foreign travel and from association with men of eminence at home and abroad. They have the means and the leisure to make themselves personally acquainted with the results of various forms of government, and especially with those of our Colonies, where free institutions are working out the solution of many political and social problems; and if the duty of legislation was conferred upon them, not as an accident of birth but by the free choice of their countrymen, and as an indication of popular confidence in their integrity and their special acquirements, they would probably devote themselves with ardour to the work. We know already that they do not lack either intellectual power or the special faculties of statesmen, and we could ill spare men of such attainments as the Marquises of Ripon and Salisbury, the late Duke of Argyll or the Earl of Rosebery, from the great council of the nation. Let us, then, briefly consider on what principles the new House of Lords should be constituted, what should be the qualification of its members, and how they should be chosen.

Constitution of the Representative Upper Chamber.

The first point to be considered is—what should be the constitution of the new House of Lords. And here it seems to me to be important that this House should be distinguished from the House of Commons, not only by the preliminary qualification of its members as peers, but by representing local areas considered as separate units, and therefore without regard to the population of the areas : just as the Senate of the United States represents the component States of the Union as units, each State returning two senators, irrespective of population. In our counties or shires we possess a series of such areas which in many respects correspond to these component States. Each of them has a very ancient individuality; many of them were British, Celtic, or Saxon kingdoms; and most of them preserve to this day distinctive peculiarities of speech or of customs. And the feeling of county unity or clanship survives, as seen in the friendly rivalry of county cricket and football clubs and of the volunteer forces; while birth or residence in the same county often constitutes a bond of sympathy between strangers who meet abroad. And this individuality of our counties is likely to be increased rather than diminished by the further extension of local self-government, offering fields for social experiment and for healthy rivalry in all matters involving the interests and well-being of their populations. It must always be remembered that our counties are not modern arbitrary divisions, but extremely ancient territories, often differing greatly in physical features, and, to a corresponding degree, in the character, interests, and occupations of their inhabitants. There is, therefore, ample reason for treating them as equal units, and giving to each an equality in choosing members of the Upper House.

The counties of the United Kingdom, reckoning the three ridings of Yorkshire as separate counties, are almost exactly a hundred in number; and, giving to each two representative peers, we should have a house of about two hundred members—amply sufficient for all purposes of legislation, but almost too large to be chosen from the limited number of existing peers, who are a little over six hundred. This difficulty, however, might be easily obviated by making all knights and baronets of the United Kingdom eligible for election to the House of Lords, those elected to be thereupon created life-barons, thus preserving the titular character of the House, while offering a more ample field for the selection of men of real eminence.

Provision might also be made for the admission of two representatives of each of our self-governing Colonies, those chosen also receiving titular honours. The presence of such Colonial lords would be of immense advantage, both as initiating a legislative union of the Empire, and in bringing to bear Colonial experience on our home legislation. There would probably be less objection to Colonial representation in the Lords than in the Commons; and when the time comes (if it ever comes) for a complete federation of the British Empire, the process would be greatly facilitated by this preliminary step towards a closer union. This, however, is not essential to the constitution of a new Upper House, although it presents advantages which should ensure for the proposal a full and careful consideration.

The next point to be considered is the preliminary qualification for membership, and, on this point, I hold very strong opinions. It has always seemed to me that the adoption of the minimum legal age which qualifies a person to hold property and to occupy the simplest public offices, as being sufficient also to qualify for choosing the national representatives or for being chosen as a legislator, is a very great political blunder. With us, most men of twenty-one have only just finished, and many have not yet finished, their education, whether intellectual or industrial; while few persons at that age have given any serious thought to politics, have made any study of the duties and rights of citizens, or have had any real experience to guide them in forming an independent judgment on the various political and social questions of the day. In this respect, most savage and barbarous nations set us a good example: with them, it is the elders who rule; and the very name of chief is often synonymous with "old man." The most suitable age to be fixed as that of political maturity should certainly not be below thirty, while I myself consider forty to be preferable.

But in the case of members of the Upper House, who are to represent the mature wisdom and experience of the nation, there can, I think, be no doubt that forty should be the minimum qualifying age. Some such limitation is especially necessary in order that the conduct, the character, and the attainments of the candidates may have become known to the electors, and this can hardly be the case at a much lower age than forty. By that time it will be seen whether a man has made any effort to qualify himself for so high a position, either by historical or legal study, or by having devoted himself to a practical inquiry into the results of the various political, economic, or social systems of other civilized communities. No one would wish to have such a House of Lords as is here suggested degraded by the presence of men who make use of the great opportunities they have inherited for mere selfish purposes, and whose highest pleasures are luxury or sport; or of such as are imbued with the prejudices and vices, rather than with the virtues and true nobility, of their ancestors. Instead of these we should seek for men who are able to show a good record of knowledge acquired or work done, and whose ability and character are known to be above the average.

Mode of Election.

Taking, then, the actual peerage, together with all knights and baronets of the United Kingdom who shall have attained the age of forty, as constituting the body from which the new House of Lords is to be chosen, the next point to be considered is the mode of selection. We may first set aside the method of election by their fellow peers (as in the case of the present representative peers of Scotland and Ireland) as being quite inadmissible, since it would perpetuate many of the evils to obviate which reform or abolition is demanded. The election must be a popular one, and the most obvious suggestion is that the existing constituencies of each county should choose its representative peers. There are, however, many objections to this. It would, in the first place, involve much of the expense and excitement of another general election; and, secondly, it is doubtful whether the average elector would be in a position to judge of the qualities and comparative merits of the several candidates. It would, therefore, be advisable to limit the voters to a body better able to make a wise and deliberate choice, and such a body will be found in the members of the several Town and County Councils, together with those of the District and Parish Councils now in existence. The members of these four classes of councils will constitute in each county an electoral body which will be truly representative of the people, since it will have been chosen on the widest and most liberal franchise to which we have yet attained. It will be sufficiently numerous and independent to avoid all suspicion of cliquism or wire-pulling, and it will be sufficiently intelligent and sufficiently interested in public affairs to make a sound and wise choice of members to sit in the Upper Chamber of the Legislature. Of course, as a rule, the representative peers for each county would be chosen from among candidates owning property in the county and residing in it, since these would be best known to the voters. But as, in some few cases, none of the residents qualified to be candidates might come up to the required standard of eminence and ability, it would probably be advisable to leave the choice of the electors entirely free.

A great advantage of such a mode of election as is here proposed would be, that it could be carried out with the minimum of trouble and expense, and without any of the publicity and excitement of an ordinary election. The clerks to the several councils would send the names and addresses of the councillors to a central office; each of them would receive by post a voting paper, which they would return in the same manner. No canvassing would be permitted, since the acts and general conduct, rather than the verbal promises, of the candidates would decide the elector's choice. Such elections would offer an excellent opportunity for a trial of the method of proportional representation advocated by John Stuart Mill, and in a modified form by Mr. Courtney and Lord Avebury. This system would ensure that, where the two political parties are not very unequally divided, the minority would obtain a representative. Each party in the county would, therefore, feel itself to be fairly treated, and the House of Lords would thereby acquire an amount of stability which would invest it with that character of a regulating power which an Upper House ought to possess.

Some persons may object to each county, however small, electing the same number of representative peers, and may urge that proportionate population should be the basis of representation as in the case of the House of Commons. But this is rather to mistake the purport of the mode of election here suggested—which is, not that the elected peers should be held to represent the counties in their local interests, but as a means of selecting the best possible Upper House, by the vote of an intelligent and popularly chosen electorate spread over the whole country, and likely to be personally acquainted with the merits or defects of those local residents who are qualified to be chosen as representative peers. For this purpose, the councillors in a small or thinly populated county would be at no disadvantage; on the contrary, it is quite possible that they might make a wiser choice than those which are most densely populated.

Advantages of the Plan.

The scheme now very briefly set forth claims to be, not only a good scheme in itself, but probably the best compromise which, under existing conditions, is possible. The more thoughtful and more influential among the peerage must see that the people of the United Kingdom will not much longer submit to a body of hereditary law-givers which not only has the power to defeat legislation earnestly desired by the majority of the voters, but which often exercises that power. They must also feel that the position of the present House of Lords is not a dignified one, while its record of service to the country will make but a sorry show in the pages of history. Almost every great reform which has been effected in this century, whether in ameliorating the severity of the criminal laws, in removing disabilities attendant on religious belief, in opening the universities to the people, or in the abolition of protective duties on the necessaries of life, has been at first strenuously opposed by the Lords, and ultimately adopted under pressure of public opinion, or for the purpose of forestalling political opponents. In very few cases, on the other hand, has the Upper Chamber initiated beneficent legislation or far-sighted policy, which has been ultimately approved by the people and accepted by the House of Commons. Yet in an Upper House which really deserved the name we should naturally look for guidance in the matter of those more important reforms which are essential to real progress, especially such as would tend to bring about a more equable distribution of the constantly increasing wealth of the nation among the masses of the people, thus diminishing and ultimately destroying that seething mass of misery and starvation which still persists among us, and which is the condemnation of our boasted civilization. An assembly which truly answered to its title of "noble" should be above the personal interests and petty prejudices which influence those who, in various ways, are engaged in the struggle for wealth or for mere existence.

The House of Lords, as it now exists at the end of the nineteenth century, is not only an anomaly but an utterly indefensible anomaly, and one wholly opposed to the spirit of the age. In the proposal now submitted to public consideration, a means is indicated of bringing it into harmony with modern ideas while preserving its historical continuity and constituting it so that it may be an aid, instead of a clog, to the wheels of progress. Will the Lords recognize the critical nature of their position, accept reform as inevitable and as the only alternative to destruction, and themselves initiate that reform? If they do so, in no hesitating or niggardly spirit, but fully recognizing that a body claiming power to legislate for Englishmen *must* be representative, and *must* be elected either directly or indirectly by the people, then it is probable that even the ever-growing Radical party would willingly accept such a reform. They would be wise to do so; because they would thus obtain a legislative chamber probably as good as any that could be obtained after a lengthy and profitless struggle; and, further, because a chamber such as is here suggested is of a nature to admit of continual improvement, and would necessarily develop as the nation developed, always keeping, as it should do, in the van of advancing civilization. When titles are given only for life, and are bestowed exclusively as recognitions of merit or of exceptional ability and integrity, there will grow up among us a true aristocracy characterized by the highest intellectual and moral qualities, while the old aristocracy of birth will be less and less esteemed, except in so far as it possesses similar characteristics. Educated public opinion will, from time to time, indicate the men who should be made eligible for election to the Upper Chamber, and no Ministry will then dare to advise the Sovereign to bestow this honour on the unworthy, or as a reward for mere political support, thus lowering the standard of those who are eligible for election by the people's local representatives. If, further, it was the rule that each of the great political parties should give titular honours to not more than a fixed number in each year, the balance would be kept even, and at successive elections each party would have an equal range of choice.

There are many matters of detail which it is not necessary to discuss at present. Among these are—the term for which the Lords should be elected; whether they should change with a change of Ministry, or by a portion retiring at intervals; whether the judges, when they leave the bench, should be *ex-officio* members or should be eligible for election. These are matters of minor importance, which will be easily settled when the main principles of the scheme of reform are decided on. These more fundamental points may be summarized as follows: (1) The limitation of the number of members in the new House of Lords to about two hundred; (2) The extension of the range of choice to knights and baronets; (3) All titular honours in future to be granted for life and only in recognition of distinguished merit; (4) An age-qualification of about forty years; (5) Representation of counties as units, by two members of each; (6) The constituency to consist of all the members of the County, District, Town, and Parish Councils in each county.

I now submit this scheme of reform, first, to the leaders among the existing peers, to whom it offers an honourable mode of escape from a difficult position, not by an ignominious surrender to popular demands after a long struggle which they know must terminate in defeat, but by voluntarily recognizing their anomalous character as hereditary legislators in an otherwise representative government, and by themselves initiating the reconstruction of the Constitution which at no distant date is inevitable. By so doing they may preserve the continuity of the aristocratic Upper Chamber, add greatly to its dignity and power, and give to the world the too rare example of a privileged class voluntarily resigning such of its privileges as are inconsistent with modern civilization. This I conceive to be true conservatism.

To the Liberal and Radical parties—and I am myself an extreme Radical—I submit my scheme as one that will remove all the evils and anomalies of the present House of Lords, transforming it into a representative chamber of the very highest character, which must always be in harmony with advanced public opinion as expressed by the whole body of its freely elected local representatives. Such a House of Lords would be really capable of fulfilling what is supposed to be the special function of an Upper Chamber—that of calm and judicial consideration of such measures as were the result of sudden waves of prejudice or passion, either in the population at large or in the House of Commons. It would thus appeal from Philip drunk to Philip sober, and would give expression to deliberate and permanent, rather than to hasty and temporary, public opinion. To secure a body capable of doing this—and I cannot see how it could be more effectually done than by some such method as I have here suggested—would be, in my opinion, a measure of radical, yet safe and judicious, reform.

CHAPTER XIV

DISESTABLISHMENT AND DISENDOWMENT : WITH A PROPOSAL FOR A REALLY NATIONAL CHURCH OF ENGLAND

THE wide-spread agitation for the disestablishment and disendowment of the English Church calls for more notice than it has hitherto received from those who, while agreeing with the necessity for some such movement and the abstract justice of its main object, do not look upon the existing Established Church merely as a powerful sect whose prestige and influence are to be diminished as soon as possible and at almost any sacrifice.

At the various meetings in favour of disestablishment (some twenty years back), little or nothing was said as to the details of the proposed or desired legislation; no scheme was formulated as to a practicable and beneficial mode of applying the national property now held by the Church, or of preserving and utilizing for national objects the parish churches, cathedrals, and other ecclesiastical buildings spread so thickly over our land, and which constitute a picturesque and impressive record of much of our social and religious history for nearly a thousand years. The only thing we have to guide us as to the aims and objects of these agitators is a constant reference to recent legislation in the case of the Irish Church, and we are therefore left to infer that some very similar mode of dealing with the English Church, its property and its buildings, is what these gentlemen have in view. But if this be so, it is surely the duty of all who have the social and moral advancement of their country at heart, and are uninfluenced by sectarian rivalry, to protest against any such scheme as in the highest degree disastrous. It may be thought by many that this agitation cannot possibly succeed in gaining its object for a very long time, and that it is useless to discuss now what shall be done at some indefinite and distant future. But this may be altogether a mistake; gross abuses do not now live long, and when an agitation is started as powerfully and influentially as this one, supported as it will undoubtedly be by the great mass of the operative class, and made a party cry at future elections, the end may not be very far off. We may then find it too late to introduce new ideas, or to persuade the Nonconformist leaders of the movement to give up their special programme, however injurious some portions of that programme may be to the best interests of the country.

My object in this chapter is, therefore, to urge upon all independent liberal thinkers to lose no time in taking part in this movement, laying down at once certain principles to be adopted as an essential condition of securing their support; and I propose, further, to show a practicable mode of carrying out these principles so as to produce results in the highest degree beneficial to the whole community.

The main principle that should guide our action in this matter, I conceive to be, that existing Church Property of every kind is National Property, and that no portion of it must, under any circumstances, be alienated, either for the compensation of supposed or real vested interests, or to the uses of any sectarian body; and further, that the parish churches and other ecclesiastical buildings must on no account be given up, but be permanently retained, with the Church property, for purposes analogous to those for which they were primarily established—the moral and social advancement of the whole community.

That the property now held by the Established Church is national property, is generally admitted; and also that the Church, as represented by a body holding particular religious opinions, can have no permanent vested interest in that property, although the individuals of which it is composed may have life-interests; and the case of the Irish Church should be a warning to us to look far enough ahead, and prepare for the inevitable change so much in advance of any immediate political necessity for it that we may allow all individual vested interests to expire naturally, and so have no need to make special compensation for them. In Ireland every kind of vested interest was brought forward, and it was even claimed that, as every clergyman had a chance of obtaining a better living, or of becoming a bishop, he should be compensated accordingly ; and that every member of the Church had an actual vested interest in its maintenance during his life. It was because all legislation had been put off till it could no longer be delayed, that these interests had to be considered, and the result was, that a sectarian Church was permanently endowed with a large amount of national property. But any such necessity of compensation for vested interests of individuals may be obviated by a little foresight, and by legislating sufficiently early to allow everyone to retain his rights and privileges in the Church *during his lifetime*. All individual vested rights would thus be satisfied, and it is probable that they would not interfere with the complete establishment of a new system at a comparatively early period, because a transition state is always an unsatisfactory and an unpleasant one, and long before half the individual lives had expired, and perhaps in the course of a very few years, the change might be voluntarily effected.

While legislation was proceeding in the case of the Irish Church, it was made sufficiently clear that it is almost impossible suddenly to abolish any such great national institution, and to find any suitable mode of applying the surplus property without grievous waste, or so as to be really beneficial to the community; and it was almost felt to be a means of getting out of a difficulty that every shadow of a vested interest should be fully compensated, and the inconveniently large amount to be disposed of reduced to manageable proportions. I believe, however, that in the case of England no such difficulty exists, and that the whole of the Church revenues may be applied in such a manner as,—Firstly, to retain all that is most useful in the organization of the existing Church of England; secondly, to extend its sphere of usefulness almost indefinitely; thirdly, to remove all cause for the ill-feeling with which it is viewed by Nonconformists, and by the members of other religious bodies; and lastly, to create, without violent change, a great national institution, which shall always be up to the highest intellectual level of the age, and be a means by which the moral and social advancement of the whole nation shall be permanently helped forward. In order to show how these desirable results may be obtained, it is necessary first to say a few words as to the status of our existing clergy, and the importance of the functions they fulfil.

The Church of England, as a religious body, owes much of its power and influence in society to its venerable antiquity; to its intimate association with our great Universities; to its establishment by law and its position in the Legislature; and to its possession of the cathedrals and parish churches, which from time immemorial have been the visible embodiments of the religion of the country. The clergy of the Church of England owe their chief influence for good in their respective parishes to their connection with these permanent and often venerable buildings; to their being the official representatives of a law-established religion, to their being the recognized heads, either officially or by courtesy, of many local organizations for charitable purposes, for education, or for self-government; and, though last not least, to their social position, their intellectual culture, refined manners, and moral character. It must, I think, be admitted that an institution which provides for the residence in every parish of the kingdom of a permanent representative of the best morality and culture of the age—a man whose first duty it is to be the friend of all who are in trouble, who lives an unselfish life, devoting himself to the moral and physical improvement of the community, who is a welcome visitor to every house, who keeps free from all party strife and personal competition, and who, by his education and training, can efficiently promote all sanitary measures and healthful amusements, and show by his example the beauty of a true and virtuous life—that an institution which should really do this, would constitute an educational machinery whose influence on the true advancement of society can hardly be exaggerated.

But in order that such an organization should produce all the good of which it is capable, it is above all things essential that it should keep itself free from sectarian teaching, and from everything calculated to excite religious prejudices. So long as there is but one religious creed in a country, or if the dissentients form a small and uninfluential minority, the ordinary clergy may possibly effect much of the good here indicated; but with us this has become impossible, owing to the adoption of a fixed creed by the Established Church, and to the multitude of opposing sects, equal in political influence, and perhaps superior in the number and enthusiasm of their adherents. The earnest Nonconformist cannot look with satisfaction on a man who is unjustly paid by the nation to teach doctrines which he firmly believes to be erroneous; while the conscientious and well-informed sceptic can hardly respect one who is not only often inferior to himself in mental capacity as well as in acquired knowledge, but who professes to believe and continues to teach as fact much that modern science has shown to be untrue. The clergyman, on the other hand, too often considers that every dissenting chapel in his parish is an evil, and looks upon every Nonconformist minister as an opponent.

The time seems now to have come when we shall have to get rid of the anomaly and the injustice of devoting an elaborate organization and vast revenues to sectarian religious teaching, while we loudly proclaim the principle of religious freedom in all our legislation. In order to get rid of an Established Church which is behind the age, there are men who would not hesitate to break up the whole institution, destroy or sell the churches, and devote the revenues to support free schools or hospitals.

Such a step would, I believe, be an irreparable loss to the nation; and I propose now to consider what means can be adopted to preserve this great organized establishment, which has grown, with the nation's growth, and has from time immemorial formed an essential part of the body politic, and to separate from it everything that can impair its efficiency or check its healthy development. I claim for every Englishman a share in this great property, devoted by our ancestors to the relief of distress, the protection and advancement of the people, the example of morality and virtue, the teaching of the highest knowledge of the age, and the inculcation of doctrines which were once universally accepted as absolute truths of the first importance for the welfare of mankind. I claim that it shall be preserved to our successors for analogous purposes, and that it shall be freed from association with all sectarian teaching, and from everything that can impair its value. Let it be reformed, not destroyed.

The Proposed National Church.

I will now proceed to show how it can be so reformed, and how it may be made a means of national advancement more efficient than all ordinary educational machinery, because its sphere of action will be wider, and because it will carry on a higher education than that imparted by schools, not for a few years only, but throughout the entire life of all who choose to profit by it. I will first sketch out what I consider should be the status and duties of the man who will take the place of the existing clergyman as the head and representative in every parish or district of the National Church.

First, as to his designation; he might be termed the Rector, a name to which we are already accustomed, and which does not necessarily imply a religious teacher. He should be chosen, primarily, for moral, intellectual, and social qualities of a much higher character than are now expected. Temper and disposition would be carefully considered, as his usefulness would be greatly impaired if he were not able to gain the confidence, sympathy, and friendship of his parishioners. His moral character should be unexceptionable. He should be specially trained in the laws of health and their practical application, and in the principles of the most advanced political and social economy. His religion should be quite free from sectarian prejudices, and his private opinions on religious matters would be no subject for inquiry. He should, however, be of a religious frame of mind, so as to be able to work sympathetically with the clergy of the various religious bodies in his district, and excite in them neither distrust nor antagonism. He must have a fair knowledge of physiology, and of simple medicine and surgery, of the rudiments of law and legal procedure, of the principles of scientific agriculture, and of the natural-history sciences, as well as of whatever is considered essential to the education of a cultivated man.

He should not be allowed to undertake the care of a parish till thirty years of age, and only after having assisted some rector in parish duties for at least five years.

The duties of the parish rector would comprise, among others, all those of the existing clergyman, *but he would never conduct religious services of any kind*. The parish church, with its appurtenances, would however be under his entire authority, in trust for the whole body of parishioners, to be used for religious services by all or any duly organized religious bodies, under such arrangements as he might find to be most convenient for all. Any religious body should be able to claim the use of the church as a right (subject to the equal rights of other such bodies), the only condition being that it should possess a permanent organization, and that its ministers should be an educated class of men, coming up to a certain standard of intellectual culture and moral character. The State might properly refuse the use of the churches to those sects whose ministers are not specially trained or well-educated men, on the ground that the public teaching of religion among a civilized people is degraded by being placed in the hands of the illiterate, and that such teachers are likely to promote superstition and increase fanaticism.

The rector might himself lecture in the church on moral, social, sanitary, historical, philosophical, or any other topics which he judged most suitable to the circumstances of his parishioners. He would also allow the church to be used during the week for any purpose not inconsistent with the main objects of his position, but always having regard to religious prejudices so long as they existed, his first duty being to promote harmony and good-will, and to gain any object he might think beneficial by persuasion rather than by an abrupt exercise of authority. His knowledge of law, and his position as *ex-officio* magistrate, would enable him to settle almost all the petty disputes among his parishioners, and so greatly diminish law-suits. He would be an *ex-officio* member of the School Board, and of the governing body of any other public educational institution in his district. It would be his duty to see that new legislative enactments were brought to the notice of the persons they chiefly affected, so that no one could offend through ignorance. He might, if he pleased, visit the sick, if his services were asked for, but this would be altogether voluntary. It would be an essential part of his duty to be on good terms with the ministers of all religious sects in his district, to bring them into friendly relations with each other, and to induce them to work harmoniously together for moral and educational objects.

With a sphere of action such as is here sketched out, the rector of a parish would have far more influence for good than the existing clergyman can possibly have. The position would be one of weight and dignity, and would be, I believe, in a high degree attractive to some of the best men in the country. The choice of men to fill it would be indefinitely wider than it is now, since no special religious beliefs would be insisted on. The educational qualifications being at once broad and high, and the appointment offering a wide field for useful labour, a sphere would be opened for a class of able men who, while they are imbued with the purest spirit of philanthropy, are too conscientious to teach religious doctrines they cannot themselves accept.

Some years ago, a proposal for a nationalization of the Church of England was made by Lord Amberley, in two very striking articles in the *Fortnightly Review*. These attracted much attention at the time, but do not seem to have produced any permanent impression. That proposal contemplated, if I remember rightly, perfect freedom of doctrine in the Church of England, and some power of modifying the formularies, while retaining the duty of conducting religious service and of preaching as at present.

It was probably felt that the difficulties of carrying out any such scheme were insuperable, and the advantages doubtful, since it involved some form of election or veto by the majority of the parishioners, or some mode of getting rid of a clergyman whose doctrines were greatly disliked. The Church would thus remain as sectarian as ever, but it would be a varying instead of a uniform sectarianism; and the necessary uncertainty of tenure would at once diminish the clergyman's influence for good, and render it more difficult to induce the best men to undertake the duties.

It seems to me to be an important and valuable feature of my plan, that it renders the rector's tenure of office for life almost certain, since the only causes (other than voluntary retirement) for his displacement would be immorality, or the fact of his making himself generally disliked by his parishioners. But the careful education and selection of the candidates, and the perfect freedom in the choice of the profession, would render either of these events of very rare occurrence.

No man, who held any special doctrinal tenets so strongly as to make him intolerant of others, would choose a profession in which he would be compelled to recognize and work harmoniously with the clergy of all denominations; nor would one who felt himself by nature unfitted to associate familiarly with all classes, and make himself their friend and counsellor, undertake an office in which it would be his chief duty to do this. We may fairly anticipate, then, that our rectors of the future would be of as high a character as our judges are now, and that there would be as little necessity for the retirement of the one from his honourable duties as there is for that of the other. This would induce better men to seek the office, and would render them far more capable of effecting beneficial results than if they were mere temporary occupants, liable to be ejected by the votes of a majority of parochial schismatics.

How the New System may be Introduced.

If no hasty and irretrievable step is taken, there seems no reason why the change from the existing state of things to something like that here sketched out, might not be gradually effected without any interference with vested interests. The new rectors would take their places wherever vacancies occurred, after the expiration of the time allowed for the disestablished Church to reorganize itself; and there need be no interference with the right of presentation to livings, the desirability of which as positions of social importance would be increased by the new arrangements. Some official recognition of the appointment would be required, and the stringency of the qualifications, both as to education and character, would render any abuse of this kind of patronage impossible. It seems highly probable that many clergymen who feel their present position more or less irksome, owing to their being obliged to read and teach much that they cannot accept as truth would gladly resign their positions as ministers of a disestablished Church in exchange for that of rector in the National Church. Such men would be quite at home in their new position, for the wider duties of which many of them would be admirably qualified. Of course there would have to be some high officers fulfilling the duties of bishops, or inspectors over the rectors; and over the whole a Supreme Board, or a Minister of Public Instruction; but these are matters which would offer no difficulty in an institution of which the main features are so well marked out.

It has now, I trust, been shown that it would be possible to remodel the framework and machinery of the Church of England as by law established, so that it should become, in connection with the various voluntary religious bodies—which, while retaining their perfect freedom of action would be to some extent associated with it—a real and highly efficient National Church; and further, that this could be done without infringing any existing right, while it would, on the other hand, confer on every section of the community the right, from which they have long been debarred, of an equal share in the use of national buildings and in all the benefits that may be derived from a proper application of the national property. It now remains to answer, in anticipation, a few of the more obvious objections that may be made to this proposal; to discuss briefly a few important details; and to point out some of the advantages that would almost certainly result from its adoption.

Objections Answered.

The first objection that will probably occur is a financial one. It will be asked how the existing endowments of the Church can be increased so as to make the position of Rector worth the acceptance of men of the required high standard of ability? The answer to this is to be found in the fact of the excessive inequality, both as regards area and population, of our parishes. In the north of England they are said to average six or seven times the size of those in the south, and we shall find that more than half of the parishes in England and Wales are far too small to require the exclusive services of a rector. A judicious system of union of small parishes, and approximate equalization of endowments, will entirely overcome the financial difficulty. A few facts and figures will make this plain. Some thousand of parishes have an area of from 5,000 to 12,000 acres, and even the largest of these are not too extensive for the supervision of an active and energetic man, while those of 4,000 or 5,000 acres and an average rural population would be comparatively easy work. But an examination of about 200 parishes, taken alphabetically in two series, shows that there are, as nearly as possible, one half of our parishes which do not exceed 2,000 acres and have less than 1,000 population, the average population of these being less than 400 by the last census. Of the thirteen thousand parishes or places in England and Wales which form distinct ecclesiastical benefices, no less than 62 or 63 per cent. have under a thousand inhabitants. The average value of all the benefices is about £307 a year, but this value is by no means in proportion to area or population, for the average of those parishes whose population is under 1,000 is still about £275 a year.

A careful examination of the circumstances of these parishes, as regards area, means of communication, and increasing or decreasing population, would enable us to combine them, so that the number of rectors required would be little more than one-half that of the existing incumbents. About one-fourth of the parishes whose population is less than a thousand could most likely be attached to others with a population somewhat exceeding that number, while the remainder might be formed into groups of two, three, or four parishes. This would result in a total reduction of about 45 per cent. A further reduction might be made in towns, where three or four parish churches might almost always be placed under the control of one rector, because, although the population might be large, many of the duties he would have to fulfil in rural districts would be performed by existing establishments, such as corporations, mechanics' and other institutions, and ministers of religion; and his chief duties would be to protect and preserve the churches for the use of the various religious bodies, and to promote harmonious action among them. The average endowment might thus be nearly doubled, and in addition there would be the vacant parsonages and glebes, the rents of which might form part of the income of the rector of two or more combined parishes. We thus arrive at a nominal average endowment of about £600 a year, while the actual inequalities are enormous; and we have to deal with a large number of advowsons which are private property fully recognized by the law. But this need not interfere with an approximate equalization of livings. Just as in other cases of far less momentous reforms, land or house property has to be given up for public uses, the owners receiving just compensation, so must the owners of advowsons be dealt with.

In cases of the union of parishes, the several patrons

might either exercise their right of nomination jointly or alternately, or one might pay a sum to the other for exclusive possession. If they failed to agree to either of these alternatives, the joint advowson must be sold by public auction and the proceeds equitably divided between them. Equalisations of endowments might be treated on a similar principle. In every case they might be effected by taking a definite sum, say £100 per annum, from one living and adding the same amount to another. The owner of the advowson which is increased in value might either pay a sum to be determined by arbitrators, to the owner of that which is diminished, or the advowson which is increased must be sold, and the proceeds divided equitably as before. It would be advisable to leave some inequalities in the value of rectories, and while none should be under £300 or £400 a year, a few might remain as high as £1,000, in important districts, to which men of special abilities would alone be appointed. The revenues now devoted to episcopal and cathedral establishments have not been reckoned as sources of increased rectorial incomes, although, whatever system of supervision might be adopted, it is probable that a considerable surplus from these revenues would remain. It may, perhaps, be further objected, that the country could not supply six or seven thousand men of the requisite ability and character, in addition to the clergy of the disestablished Church, who would continue in existence as an independent body. But we must consider that the new men would be only required in gradual succession as livings became vacant; and, as it is almost certain that no voluntary establishment would be able to appoint resident clergy in the thousands of small parishes with a very scanty population, the total number of men required for the service of the national and the disestablished Churches would not perhaps be very much greater than at present.

Although the power of nominating rectors now possessed by private persons is not proposed to be interfered with, candidates would have to pass a much more rigid examination, and to furnish much better evidence of temper and moral character, than is now required; and they would have to submit to the probation of five years' service under a rector, which would sufficiently test their capacity and suitability for the office. All livings now in the gift of Government or of public bodies should be thrown open to public competition by annual examinations, the details of which need not now be considered.

It will doubtless be further objected, that the scheme now advocated is Utopian, and aims at an ideal perfection which could not be realized even were public opinion ripe for any such revolution; and also, that it will be repulsive to the feelings of a large number of persons by placing religion and religious teachers in a subordinate position. To this I would reply, that a few years ago, before the Irish Church had been disestablished, and when household suffrage and the ballot were still ideal propositions which our Parliament would hardly seriously discuss, any such proposal as the present one would have been thoroughly Utopian: but I cannot admit that it is so now. The body which has set up the cry for disestablishment and disendowment of the Church of England is a more powerful and a more united one than that which inaugurated any of the other great reforms; and the probabilities seem to me to be great that they will attain their object in less than half a century. If so, it is not Utopian to discuss the subject in all its bearings; and although my scheme may aim at an ideal perfection which it is not in existing human nature perfectly to attain, the question to be considered is whether this ideal is a just, a true, and a noble one; if it is so, we shall assuredly do well to keep it in view and so legislate as not to prevent our successors from ever attaining it. Neither do I believe that such a scheme can be in any way degrading to religion; it will, on the contrary, keep up a connection between religious teaching and the State, and by dealing out equal justice to all creeds, will go far to do away with that sectarian animosity which more than anything else really degrades religion. As knowledge and true civilization spread more widely, it is to be expected that religion will become more and more a personal matter, without necessarily losing any of its influence on the human mind;

and an organization which provides for the diffusion of those moral and social teachings which are the highest products of the age, must necessarily aid in the development of that religion which is the truest reflex of man's higher nature.

Advantages of the Scheme.

It now remains only to point out a few of the advantages which would result from the adoption of the scheme here advocated.

It will be generally admitted that, were the English Church to be disestablished and disendowed, the Church buildings to be devoted to sectarian or secular uses, and the Church property applied in almost any way that can be suggested (other than that here proposed), a void would be left in the social organization of the country that could not be easily filled up. The clergy of rival sects, all equal and equally without authority in the eye of the law, could not possibly fulfil the various social and moral functions even of the present established Church, still less could they ever attain the standard of usefulness which could be easily reached by men in the position I have indicated in the Church of the future. What that standard might soon become it is not only difficult to exaggerate, but difficult even adequately to realize, because no institution equally well adapted to produce great results has ever before existed. If we were to say that its beneficial influence upon society would be equal to that produced by the whole of our best literature, many would at first think it an exaggerated estimate. But a little consideration would, I think, convince them that it is on the contrary far too low. For literature only reaches certain defined and very limited classes, consisting largely of men who least require the lesson it conveys, while the great mass of the population know no literature, or only that of the cheap newspaper; and the teachings of modern science and philosophy, as well as the instruction to be derived from history and biography, would be to many of them as startling as the revelation of an unknown world. Most of these would be reached by the

National Rectors, whose duty and pleasure it would be to convey to the minds of their parishioners, in interesting and instructive series of lectures, some idea of the beauties of literature, of the marvels of science, and of the instruction to be derived from the example of great and good men. Is it possible to foresee the ultimate effects of such teaching, as a supplement to our new system of National Education, carried out systematically, not in our great towns only, but in every country parish, not by the occasional visits of itinerant superficial lecturers, but continued week by week, year by year, and from one generation to another, by a body of the best educated, the most earnest, and the most practical teachers the country can produce.

Men of this stamp would be able to influence all classes for good; they would aid in introducing the best methods of agriculture and of household economy; they would be the men to see that sanitary inspectors and School Boards did their duty; they would take care that in their district no common lands were wrongfully enclosed, no public paths stopped up, and generally no injustice done to those who did not know, or could not enforce, their legal rights. Not coming into competition with any class of men, and not exciting any sectarian or religious animosity, the National Rectors might be in our age all that the monks and abbots were in the best monastic days—and much more—respected by the rich, loved by the poor, feared by the evil-doer, centres of culture and of morality throughout the land; by their example their teaching and their assistance helping on the higher civilization, and thus fulfilling the noblest function that can fall to the lot of any body of men.

But besides these direct benefits to society, which such an institution would be naturally expected to produce, there are others of hardly less value which would incidentally flow from it, and a few of these I should wish to touch upon. One of the results of the extreme competitive activity of modern life, and of the somewhat commercial character of our institutions, is, that there are exceedingly few positions open to men of high intellectual culture and scientific or literary tastes, such as will leave them sufficient leisure to devote themselves to original research in their favourite pursuits. But the position of a National Parish Rector would supply this want in the most complete manner. From their liberal education and special training, and the high intellectual standard required for the appointment, a large proportion of them would be men of exceptionally active and powerful minds. They would have a good elementary knowledge of modern science and philosophy. Their duties, though numerous, and in the highest degree important, would not, as a rule, be laborious, and would leave them a considerable amount of leisure—and leisure with such men necessarily implies occupation. Some would devote themselves to science, some to experimental agriculture or horticulture, some to history, philosophy, or other branches of literature ; and we may fairly conclude, that from the body of six or seven thousand National Church Rectors, we should have a very large accession to our original thinkers and general workers—a class of men who not only reflect glory on their country, but more than any others help on the work of human progress.

It has been already suggested that the rectors would be able to see that Sanitary Inspectors and School Boards did their duty ; but I think we may go further, and say, that over a large portion of the rural districts no sanitary or educational legislation will be efficiently carried out till some such body of men is called into existence. Their value, too, can hardly be exaggerated as a means of obtaining trustworthy information on the working of any new law affecting our social relations, and especially those connected with pauperism. The narrow education, imperfect training, and sectarian prejudices of so many of the clergy of the Established Church, prevent their opinions having much weight, either with the public at large or with the Government. But the National Rectors would be in a very different position. Their education and special training would render them well fitted to consider such questions in all their bearings, and their perfect independence would give weight to their opinions ;

while their means of obtaining accurate information would be much greater than that of any visiting inspector, who can seldom detect abuses which can be temporarily concealed, or which only occasionally become prominent.

These are some of the incidental advantages (and many others might be adduced) that would follow the establishment throughout the country of such a body of men as has been indicated; but I lay no stress upon these as arguments for the proposed change, compared with the direct and unparalleled advantage of establishing a truly National Church, in which every Englishman, whatever be his religious opinions, shall have an equal share; and of abolishing for ever, so far as it is possible to do so, all causes of religious animosity. I would also claim a favourable consideration for this proposal, because it is a settlement of the question that would adapt itself to any possible future change in the religious beliefs of the community, and would therefore be permanent. Whether sects increased or diminished in number, and whether religion or secularism should ultimately prevail, an institution that should provide for the teaching of the best morality of the age to those most in need of such teaching, and that should aid in producing harmony and good-will among all classes of society, would never become obsolete.

In conclusion, I would most earnestly press upon all unprejudiced thinkers to consider the essential conditions of this great problem, not my imperfect exposition of it. Let them reflect that they are actually in possession of an elaborate organization, and an ample property, handed down to us by our forefathers, with whom it did at one time fulfil many of the high functions which I wish to restore to it. We have suffered it to remain in the hands of a narrow religious corporation, which in no sufficient degree represents either the most cultivated intelligence or the highest morality of our age, and which, by its dogmatic theology and resistance to progress, has become out of harmony both with the best and the least educated portion of the community. The question that now presses upon us is, shall we suffer this grand institution and these noble revenues to be irrevocably destroyed, or shall we bring them back to the fundamental purposes they were originally intended to fulfil, and which the conditions of modern society—its terrible contrasts of profuse wealth and grinding poverty, of the noblest intellectual achievements with the most degrading ignorance, of the most pure and elevated morality with the lowest depths of vice —render perhaps of more vital importance to our national well-being than at any previous epoch of our history?

Shall we preserve and re-create, in accordance with the principle of religious liberty, or shall we utterly abolish our great historic National Church?

CHAPTER XV

INTEREST-BEARING FUNDS INJURIOUS AND UNJUST

> "The millionaire is, and as long as he is allowed to exist, always will be, a useful member of society; because he produces more wealth in comparison to the amount that he exhausts than any other member of society. . . . The richer a man is, the greater is the proportion of his savings to his income. Take a man with a fortune of £20,000,000, and an income of £1,000,000, of which he could not very well spend more than £100,000 a year. Now if this fortune was owned by 10,000 persons instead of one, they would have £100 a year each, of which they would probably spend £90. Therefore, their savings would amount in the aggregate to £100,000, while the multi-millionaire's savings from the same capital would be £900,000. Therefore the community which had the multi-millionaire would grow richer at the rate of £800,000 a year, at compound interest over the community that had divided *his property* up."—(Bradley Martin, jun., in *Nineteenth Century*, "Is the Lavish Expenditure of Wealth Justifiable?" p. 1029, Dec., 1898.)

THE passage quoted at the head of this chapter shows such an extraordinary misconception as to the real nature of wealth that I propose to devote a few pages to a discussion that will, I think, show the fact to be the very reverse of that maintained by Mr. Bradley Martin, jun. He confounds money with wealth; the increase of money, or its equivalent—claims on the earnings of other people —as increase of wealth. I maintain on the contrary that the result of numbers of rich men saving a large portion of their incomes every year, instead of making a community or a nation richer makes it poorer—makes it a smaller producer of real wealth—causes the bulk of its people to work harder and fare worse. Those who will read the following pages carefully, will, I think, admit that my conclusion is correct, and that it is by moving in the very opposite direction, so as to bring about the diminution rather than the increase of great individual money-wealth that the real wealth and well-being of the whole community is to be attained.

Interest-bearing Funds a Danger to the Community.

The evil effects of wealth-accumulation, as now understood, are by no means limited to those cases where it is accumulated in abnormally large amounts by individuals, but are not less real or less important when more widely distributed and devoted, as it so often is, to supporting a considerable section of the population in idleness and luxury during their whole lives. For this class of persons are not only, so far as they are idle, an incubus on the community, but, so far as they are luxurious, they become a far greater source of evil in withdrawing large bodies of labourers from the production of useful wealth and keeping them employed in useless or even injurious work. The thousands and tens of thousands whose lives are spent in the manufacture and sale of luxuries, ornaments, nick-knacks, and worthless toys, as well as the majority of those permanently occupied as domestic servants to the wealthy, or in connexion with horse-racing, yachting, and other amusements of the upper classes, represent not only so much lost productive power but are themselves a heavy burthen, since they are all supported by the productive labour of others.

In an ideal social state no one would live idly and luxuriously on wealth produced by another. No one would be able to obtain surplus wealth, and no children would be brought up to a life of idleness, but would be taught that they are debtors to society for all that they receive and must therefore perform their fair share of useful work.

In order to make true social progress we must always keep some such ideal in view; and this brings me to what I consider to be at the very root of the question—the evil of all institutions which permit or favour the paying of (nominally) perpetual interest, income, or profits, on any invested capital.

Real and Fictitious Wealth.

Nothing is more certain than that wealth—real wealth—is continually used up and destroyed, and as continually reproduced by fresh labour. All articles of food, of clothing, of furniture, and even tools, machines, dwellings, books, works of art, and ornaments, are either wholly or partly used up day by day or year by year, and as continually reproduced; and it is those which are most continually consumed and reproduced which are, pre-eminently, beneficial wealth. If, then, a man acquires a large surplus of this real wealth beyond what he can consume himself, it must be either profitably consumed by others or be wasted by natural decay. In either case it soon ceases to exist. But our fiscal and legislative arrangements enable a man to change this *perishable* wealth into securities which bring him a *permanent* income —an income supposed to be perpetual, but at all events lasting long after the wealth of which it is the symbol, and which is supposed to produce it, has totally disappeared. Incomes thus derived constitute a tax or tribute on the community, for which it receives nothing in return, and may truly be termed fictitious wealth.

To show that this is so, and at the same time to exhibit clearly not only the evil but the inherent absurdity of these arrangements, let us consider a few indisputable facts. Every year much surplus wealth is accumulated by individuals and is invested as reproductive capital. This invested capital goes on producing an income which, it is supposed, is and ought to be permanent; and as more and more surplus wealth is continually produced and invested, it is evident that the permanent incomes thus derived will continually increase, and thus in each successive generation a larger and larger number of persons will be enabled to live in idleness on incomes derived from this invested capital. But this is a state of things that evidently carries with it its own destruction, since a time must come when the number of idle persons living on " independent incomes," and the aggregate of those incomes, will be so great, that the working portion of the population will be ground down to penury in the attempt to pay them, and this will more quickly come about from the tendency of a large idle and luxurious class to withdraw labour from beneficial work to the production of useless luxuries. The end will inevitably be the worst kind of revolution, brought about by the determination of the labouring poor no longer to support the burthen of an ever-increasing class of idle rich.

How to Abolish Fictitious Wealth.

The only cure for this state of things (to which we are steadily drifting) is for our Legislature to acknowledge the principle and act on it, that all capital expenditure must be repaid (if at all) by means of terminable rentals or annuities (including interest and repayment of capital), in a limited term, such term never much to exceed the average duration of one generation—say 30 to 40 years; while all permanent works of general utility, such as railways, harbours, docks, canals, gas and water-works &c., shall, when the capital is thus repaid, revert to the State, to the Municipalities, or to other freely elected local authorities, and be thenceforth administered for the public benefit. By this measure alone a considerable portion of the permanent investments, which now enable wealth-producers to provide for the support of the next generation in idleness, would be abolished; but there would remain the largest and the least defensible of all—the national debts of civilized nations. It is now generally admitted to be wrong for one generation, or even for one government, to borrow money for its own purposes (generally for the most wasteful and injurious of all purposes—war) and leave the debt as a burthen on its successors.

How to Extinguish National and other Debt.

Confining ourselves to our own National Debt, I maintain that it is an unmixed evil, as well as a cruel injustice to the present generation of workers; and I further maintain, that the least injurious mode of abolishing it would be to declare, that after a fixed date the Government will not allow transfers of Stock (except in cases of inheritance) but will pay the dividends to the holders at that date for their lives and for the lives of any direct heirs living at the time they make their will or die, after which all payments will cease, and the community will at length be released from the oppressive and unjust burthen of taxation it has so long borne.

Of course this proposal will be met by the cry of confiscation, repudiation, and other ugly terms; but let us look at it a little closer before we decide as to its justice and inadmissibility. It will hardly be denied that the Legislature may, on grounds of public policy, and after ample notice to allow of any desired sales and transfers, limit the payment of the interest or principal of its debt to the then representatives of its original creditors. Having gone so far we have to inquire further what injury will be done to the holders of stock, if this income is limited to their lives and that of their living heirs. It is admitted that the unborn can have no claims to a special provision, while the fundholder can have no affection for, or interest in hypothetical beings who may never exist. It is evident, therefore, that the supposed injury is purely imaginary, and could not be estimated at the value of the smallest coin of the realm; whereas, the payment of the debt in full, supposed to be the only honest course, would appreciably injure the fundholders themselves, as well as the whole nation.

For, let us suppose it to be determined that the National Debt shall be paid off within, say, thirty years, and that increased taxation to the necessary amount is thereupon raised annually. The immediate effect would be to raise the price of consols, which would soon rise very much above par owing to their affording the only safe and easily realizable temporary investments for great capitalists, financiers, &c., so that all persons whose stock was paid off would have to buy in again at a loss in order to secure a safe income, which would be then permanently diminished. If we add to this loss the heavy burthen of taxation, of which they would have to bear their share, the fundholders (who we suppose had rejected the scheme of terminable annuities), would find out their mistake, and begin to ask themselves why they should suffer in order that certain unborn individuals might live in idleness at the expense of the nation. The supposed honest plan of going on paying interest for ever, is really dishonest, because it perpetuates a heavy burthen on the whole community, not for any real benefit to any existing portion of the community, but for the injurious and immoral purpose of providing that even in unborn generations certain selected individuals shall be able to live idle lives at the expense of their fellows. The alleged confiscation, on the other hand, is really the honest course, because, while providing that no living individual shall suffer either in purse or vicariously by the suffering of those dear to them, it provides for the abolition of an unjust burthen which we have inherited from evil times, but which will become still more unjust and harder to bear the longer it exists.

How to Deal with the Land.

Having thus, we will suppose, got rid of all such permanent means of investment as railway-stocks, gas shares and the public funds, there would remain only the land of the country. But this, as I show in the four succeeding chapters of this volume, is the most injurious to the community of all forms of permanent investment of capital, since it secures to the capitalist not only the general power due to wealth, but enables him to absorb in the form of increased rents a large portion of the nation's surplus wealth-production, and gives him also direct power over the health, the happiness, and the lives of his fellow citizens, who must live upon the land on his terms so long as he is allowed to possess it and deal with it as he pleases. For these reasons alone, all possession of land except in limited quantities for personal occupation, should be prevented, and then the last remaining means by which a permanent and ever increasing income can be obtained from accumulated wealth which is, in its very nature, transitory and perishable, would have been taken away, and the result would, I maintain, be wholly beneficial to the community.

Some Objections Answered.

Among the objections that will be made to this proposed reform it will no doubt be said that on the system of short terminable payments to cover interest and repayment of principle, capitalists would not lend money and consequently no good national or local work would be effected. But the fact is that they could be much better effected without borrowing money, whenever the proposed works were such as would be remunerative, while all other necessary works could be done in the same way, but of course would have to be paid for by some kind of public or private contributions. The following experiment shows how easily this may be done.

In the island of Guernsey some years ago a market-place was much wanted, and the Government of the island having determined to build it, issued notes, inscribed "Guernsey Market Notes," for £1 each, and numbered from one to four thousand, £4,000 being the estimated cost of the market. With these notes the Government paid the contractor, the contractor paid his men, and the men bought all the necessaries they required, as the notes were a legal tender in the island. They were used to pay rent, to pay taxes, and for all other purposes. When the market was finished, it immediately produced a revenue, and this revenue was applied to redeem the notes; and in ten years all were redeemed, and henceforth to the present time the market returns a considerable revenue to the Government of the island, which goes to reduce taxation; and all this was done without borrowing any money or paying any interest.

Now here is a principle, applied on a small scale by a small self-governing community, which is capable of a very extensive application. All remunerative public works could be executed by some such method; while if it is urged that some works, like sanitary improvements, are not directly remunerative, it may be replied that this is usually because the benefit of such works is allowed to be absorbed by individuals instead of accruing to the community. This is because individuals possess the land in our towns and cities, and every sanitary improvement effected at the public expense increases the value of this land. In fact, no public improvement of any kind can be made in a city without increasing the value of the land, so that there is a double motive in urging on costly, and, perhaps, unnecessary improvements—jobs are effected by financiers and contractors, while the owners of land know that, however much the ratepayers may suffer, *they* are sure to be benefited. Here is surely another indication that the land of every municipality, or other local community, which grows in value owing to the increase and the expenditure of the whole population, should belong to the community and not to private individuals.

The subject, however, which we were more particularly considering was the doing away with those funds and investments by which money is made to produce a perpetual income. Now, when, as in Guernsey, there was no permanent debt created and no interest paid, there was no "stock" to speculate in and no income derivable from it. Here, then, we have a double advantage over the usual mode of creating interest-bearing debts, which indicates that we have discovered an important principle which is applicable to almost every case of public improvement. Let us take the case of railways, for example. These are usually constructed under legislative acts, empowering a company to take the necessary land to build the line and to work it for the profit of the shareholders. This plan has led to the greatest possible amount of mischief. Lines have been made where not wanted; speculation to an enormous extent has been encouraged; huge monopolies have been created; shareholders by thousands have been ruined; while the thing least considered has been the general interest. During the last great American railway mania it has been estimated by Mr. Atkinson that railway-construction went on four times as fast as the increase of produce to be carried by the railways, thousands of miles of railways being made long before they would be wanted, involving loss in a great variety of ways, and being, in fact, one of the causes of recurring depression of trade.

If, on the other hand, no such power had been given to companies, but, when public opinion in any State or country demanded a particular line of railway it had been constructed by means of *Railway Bonds* created for the purpose, bearing no interest and serving as legal currency within the State till they were all redeemed and paid off out of the profits of the line, then no speculation would have been possible. It would have been no one's interest to build unnecessary and unprofitable lines, because so soon as this was done the bonds of the particular line would have little chance of being redeemed; and as they would be a legal tender, they would soon be all paid in as taxes, and the Government—that is, all the taxpayers—would have to bear the loss. This would check further railway-making for a time, and thus prevent useless expenditure in the interest of speculators and contractors.

On the other hand, every railway that returned any profits at all would steadily redeem its bonds, and then the whole of the future profits would go to reduce taxation or to make railway travelling free. It would thus be the interest of every one that no railways should be made that were likely to be worked at a loss, because that would lead to a depreciation of the bonds, and thus be a loss to the whole community. But it would be equally every one's interest that all really useful and necessary lines should be made, because, besides the direct benefit, the bonds would be quickly redeemed and the profits of the line would enable the general taxation to be reduced. Water-works, gas-works, public parks, new streets, and all similar improvements could be executed on a similar principle, the only safeguard required being that no large improvement should be undertaken in any town or district till the preceding one had been completed and had begun to redeem its bonds out of its genuine profits or proceeds.

It has now, I think, been made clear how all public works and public improvements may be effected by public *credit*, properly so called, instead of by public *debt*, involving far less risk of loss, no permanent charge on the community, but leading, on the contrary, to a continuous reduction of taxation, and cutting away the very foundations of the system by which the financier and speculator are now enabled to plunder the working people.

Concluding Observations.

Returning to our main subject, some may think that by thus checking great accumulations of capital by individuals the country would be impoverished; but the fact would be exactly the reverse, since the accumulation of real capital would be greatly facilitated. For that large class which now makes its wealth by financial operations and speculations—a mere form of gambling—would find its occupation gone, and would be forced to turn its attention to genuine industrial pursuits. Instead of money being used, as it so largely is now, as a mere instrument to make more money by pure speculation, it would have to be invested in true reproductive wealth—machines, tools, buildings, roads, bridges, ships, &c., and thus the whole country would be enriched and benefited.

When the changes here indicated had been effected, capital could be profitably invested only in some form of agriculture, manufacture, or commerce; and, while the wealth of the community would thus be indefinitely increased, the accumulation of excessive wealth by individuals would become almost impossible, since, there is a limit to the number of industrial concerns that can be profitably managed by one man. As the race of hereditary idlers would no longer exist, the total production of wealth would be much increased from this cause, while the free use of land in small quantities by a large proportion of the population would render labourers less dependent on capitalists than now, and thus lead to a more equable distribution of wealth than now prevails.

My object in this brief chapter has been to call

attention to a principle of great importance which appears to have been overlooked by political economists and ignored by legislators, namely,—that the system which enables a permanent and sometimes an increasing revenue to be derived from nominal wealth long after the real wealth it is supposed to represent has ceased to exist, is wrong in itself; and that, like all wrongs, it inevitably leads to suffering. One of the evil results of this system is that it affords the main, if not the only, support to millionnaires and to hereditary plutocrats.

I have further endeavoured to show that we may remove all the sources whence vast revenues can be derived by individuals without any productive exertion on their part, not only without injury but with the most beneficial results to the community. Lastly, I believe, that it is in some such view as to the economic error and moral wrong of deriving permanent incomes from perishable wealth, that we shall find the true solution of the problem of the antagonism between capitalists and labourers now everywhere agitating the civilized world.

CHAPTER XVI

HOW TO NATIONALIZE THE LAND: A RADICAL SOLUTION OF THE IRISH LAND PROBLEM [1]

> "Land is not and cannot be property in the sense that movable things are property. Every human being born into this planet must live upon the land if he lives at all. The land in any country is really the property of the nation that occupies it; and the tenure of it by individuals is ordered differently in different places, according to the habits of the people and the general convenience.
>
> "To treat land, with the present privileges attached to the possession of it, as an article of sale, to be passed from hand to hand in the market like other commodities, is an arrangement not likely to be permanent either in Ireland or elsewhere."—J. A. FROUDE, in the *Nineteenth Century*, September 1880, pp. 362, 369.

THE Irish Land League proposed that the Government should buy out the Irish landlords (at an estimated cost of two hundred and seventy millions), and convert the tenants into a peasant proprietary who were to redeem their holdings by payments extending over thirty-five years. That a scheme so impracticable as this—and even if practicable so unsound and worthless—should be put forth by a body of educated men, who had, presumably,

[1] This article appeared in the *Contemporary Review*, November 1880, and it is reprinted here because the principle of separating the inherent value of the land from the improvements, as a means of obviating the need for any "management" by the State or Municipality, was I believe first enunciated in it, and led in the following year to the formation of the Land Nationalisation Society, which, with its offshoot, the Land Restoration League, have done much to spread correct views as to the fundamental importance of its proposed solution of the Land question. The article has therefore, in some degree, an historical value.

studied the subject, is a noteworthy fact, and one which shows the importance of a thorough and fearless discussion of all questions relating to the tenure of the land, in order that we may arrive at some fundamental principles on which to base our practical legislation.

The total neglect of the study of this most important subject is further illustrated by the way in which the daily press widely promulgated, either without criticism or with expressed approval, an objection to the Land League's proposal which is more absurd than that proposal itself, inasmuch as it involves and rests upon an oversight so gross as almost to constitute a true "Irish bull." Mr. W. J. O'Neill Daunt, an old colleague of O'Connell, was the author of this remarkable piece of criticism, the most important part of which, and that which has been quoted as so especially crushing, is as follows:

"There are, roughly speaking, about half a million of tenants in Ireland. But there are about five and a half millions of people in the country. Suppose the half million of tenants are established as peasant proprietors, what is to be done with the claims of the remaining five millions? Have they not a right to say to the peasant landocracy, 'You are only one-eleventh of the nation. Why should one-eleventh grasp all the land? Our right to the land is as good as yours. We will not permit your monopoly. We insist on getting our share of your estates.'"

But neither Mr. O'Neill Daunt himself, nor the writers who approvingly characterized his letter as "remarkable," and his criticism as "pertinent," can have given five minutes' real thought to the matter, or they must have seen the absurdity of their remarks. For surely the half million of tenants have wives and families, and reckoning the children at three and a half per family (which is rather higher than the average for the whole country), we arrive at a tenant population of two and three-quarter millions, or about half the total inhabitants of the island. And what will the other half consist of? There are the landlords, the clergy, and other professional men, the army and navy, the members of the court and officials, the manufacturers, the merchants, and all the mechanics and shopkeepers of the towns. What then becomes of the "five millions" who would cry out against the "half million" monopolizing the land? Would the wives and the children of the new peasant proprietors cry out against their husbands and fathers? Would the manufacturers of Belfast or the shopkeepers of Dublin suddenly want to turn farmers, merely because the same people who now cultivate the land as tenants then cultivate it as owners, or prospective owners, having paid its full value? The whole objection thus vanishes, as a mere "Irish bull," which the English press adopted and circulated as if it had been sound logic and good political argument!

Some other objections stated by Mr. Daunt are, however, more valid. The whole rental of the land during the thirty-five years would necessarily go to the London Treasury, and as it would be the repayment of a loan, distress and eviction must follow non-payment of rent, just as it does now. More important, however, is the consideration that so soon as the new proprietors had acquired the fee simple of the land (or even before), the buying of land by the more wealthy, and the selling of it by the poorer, will, inevitably, begin again. The land will be mortgaged by the poor or improvident, and the wealthy will again accumulate large estates. Then, absentee landlords and discontented tenants, rack-rents, agents, middlemen, evictions and agrarian outrages will all arise as before, till some future Government will again be asked to advance money to buy out the new landlords, and transfer the land to those who will at that time be the tenants. It is evident then that no such proposal as that of the Land League would be more than a temporary palliative applied at an enormous cost, and that we must seek in a different direction if we would effect a radical cure. That direction, is, I believe, indicated by the remarks of Mr. Froude placed at the head of this article, and which fairly represent the views of many advanced thinkers. Hitherto, no practical mode of carrying such ideas into effect has been hit upon, and they have accordingly been relegated to the limbo of "unpractical politics." But this defect is not inherent in the views themselves; and I now propose to show in some detail how all the difficulties in their application may be overcome, and the land of Ireland, or that of any other country may be gradually, but surely and permanently, restored to the great mass of the people who desire to cultivate it, without injustice to any of the present landowners, although the operation will be effected entirely without cost. This is undoubtedly a bold statement, but, before rejecting it as absurd or impracticable, I beg for the reader's careful and unprejudiced consideration of the propositions I shall endeavour to establish, and the definite scheme that will be set forth.

A General Principle of Legislation.

My proposal is mainly founded upon a very simple proposition, which I think will be admitted, and which, if not capable of logical demonstration can yet hardly be disproved. This proposition is, that whatever acts may be done by an individual without injustice or without infringing any rights which others possess or are entitled to claim in law or equity, then acts of a similar nature may be done by the State, also without injustice. In judging of the validity of this proposition, we must remember, that an individual may be actuated by purely personal motives; may be influenced by passion, by pride, or even by revenge, and yet may not go beyond what always has been admitted to be his right; while the State will, presumably, be guided in its action by a desire for the public welfare, and cannot possibly, in the particular cases here contemplated, be influenced by those lower motives which often affect the individual, and yet have never been held to impair either his legal or his moral rights.

The proposition here generally stated appears to me to be so nearly in the nature of a political axiom as to require no attempt at a formal demonstration. It will be time enough to defend it when good, or at least plausible, reasons have been given why it should not be accepted. I will now proceed to its application in the present inquiry.

The right to transfer land (or other property) by will, to any successor not insane or criminal, has been allowed by most civilized nations to some extent, and by ourselves with hardly any limitations. A British landowner may leave his property to be divided among his family, or to any single member of his family. If he has no family he may leave it to any relation or to any friend; and he is not said to be unjust if he passes over many of his relatives and bequeaths his land either to a personal friend, or to some man of eminence, or to benefit some public institution or charity, or for any analogous purpose. Even his own immediate family—his sons and daughters, his parents, or his brothers—have no legal claim on his land, if he chooses to leave it to a more distant relation, or to a friend, or to a charity; but public opinion does, in such a case, condemn his action as more or less unjust. But whenever the choice is between remote relations and some public purpose or even personal friendship, public opinion rather applauds his freedom of choice, and it is never allowed that the more or less distant relatives who may be passed over have any right to complain of injury or robbery because the land was not left to them, even if they were the actual heirs-at-law and would have received it had the owner died intestate.

Now comes the first application of my above-stated proposition or axiom. If the personal owner of land does not rob or injure a distant relative (even if he be the heir-at-law) by making a will and otherwise disposing of his land, neither can the State be justly said to rob or injure any one if, for public purposes, it alters the law of inheritance so as to prevent the transfer of the land of intestates to any persons who are not near blood relations of the deceased. The exact degree of relationship that may be fixed upon is not of importance to the principle, except that it must not be so narrowly limited as to interfere with what Bentham termed "just expectation." A son or a brother certainly has such just expectations, while the expectations of a third cousin or a great-grand-nephew can hardly be so termed. For the sake of illustrating the principle, let us suppose that the limit of inheritance to the land of an intestate is fixed at what may be termed the second degree, that is, that it shall not pass to any more remote relative than an uncle, first-cousin, or grandchild, but when none of these exist shall devolve to the State for public purposes. No one can deny that the State could justly make such a law, when laws which disinherit acknowledged children because they are illegitimate, as well as all a man's legitimate daughters and other female relatives, have been long upheld as both just and expedient!

Before going further I may as well state, that for the purpose of the argument in this paper I assume that settlements by which land can be tied up and life interests created for several generations, as well as the law of succession to the eldest son or nearest male heir, do not exist, as it seems pretty certain that they will be abolished long before any such radical reform as that proposed in this chapter will come on for discussion in the Legislature.

It may, however, be objected, that if the law of inheritance were altered as above suggested it would produce little effect, because it would afford an additional incentive to the owners of land to dispose of it by will. But it is a fact that much stronger incentives—such as the fear of leaving daughters destitute—has not prevented men from dying intestate; and it may be argued on the other side, that in those cases in which a landowner had no near relatives, and all power of entailing an estate had ceased, the inducement to make a will at the earliest possible period would be very weak indeed, and thus a certain number of estates would continually lapse to the Government.

Application of the Principle.

It must be admitted, however, that the quantity of land thus annually acquired by the State would be inconsiderable, and would not be sufficient to produce any important amelioration of the condition of the country. We must, therefore, proceed to the second and far more important application of our general principle, to which what has hitherto been proposed is merely the introduction.

The interest of a landowner in his property is of two kinds, commercial and sentimental, and these together constitute its value to him. He claims, and possesses, the right to deal with it as he pleases during his life, and to bequeath it to any successor at his death. For the State to interfere with either of these rights would be an injury for which he should be compensated. The British landowner has, however, been allowed to extend his sentimental interests to an indefinite extent, by leaving his property *in trust* for certain purposes, which trust the law has enforced for generations, or even for centuries, after his decease. It is now very generally admitted that this is impolitic and unjust in the case of any property, and especially so as regards land. It is felt that each generation should have absolute possession of the land and goods that have descended to it, and should not be hampered in the use of them by the dictates of the dead who cannot possibly be able to judge what is best for a new and, in many respects, differently circumstanced population. This question is far too extensive to be discussed here, and I refer my readers to Sir Arthur Hobhouse's volume, *The Dead Hand*, in which they will find abundance of facts and arguments demonstrating the absurdity and the evil consequences of allowing the dead still to hold property; while the enormous mischief produced by entails and other life-interests in landed estates, has been fully exposed in Mr. J. Boyd Kinnear's useful work, *The Principles of Property in Land*. I accept it, then, as an established principle, that the present owner of land should be allowed to bequeath it to any successor he may choose, but that he should have no power to restrict that successor in his use of it. He may *recommend*, or make known his wishes as to the use of it; but the State is unjust to the living if it allows the dead to *command*, and then enforces their commands on posterity.[1]

Having thus established the only right of transmission from one generation to another which ought to be recognized by the State, we see that the "expectation" or

---

[1] The subject of "Trusts" has been discussed in chapter X. of he present volume.

"sentiment" of a landowner, as to the continued possession of his estate by his descendants, is so liable to be traversed by his successors that he can hardly be said to have any right or property in it beyond the first or second generation. It is true that in many cases the estates have passed from father to son for centuries; but this is a rare exception, and has probably only been secured by the law of primogeniture and the power of entail. When these are abolished, and no man can influence the succession of his land beyond one generation, it is clear that the value of this "sentimental possession" in land will rapidly become a vanishing quantity as we pass beyond the first few generations. For each successive possessor will be free to sell or bequeath it as he pleases, and this freedom will certainly lead to the breaking up of estates, and render all calculations or expectations as to their possible owners three or four generations hence altogether futile.

It may be admitted, however, that the desire to transmit property to the second or third generation, or to the families of any living person in whom the owner may be interested, is a legitimate sentiment which, though not proper to be forcibly carried into effect by the Government, should yet not be checked, or its realization be rendered altogether impossible, by any act of the Legislature, except to obtain important benefits for the whole people. The more limited desire or sentiment, that a personally occupied estate, such as an ancestral house, farm, or grounds, should long continue in the family, is decidedly one to be encouraged and aided in its realization, as keeping up the love of home and country, and having a generally good moral and social tendency; and, as will be seen further on, this is fully recognized in the scheme we are here developing.

But the power of unlimited transmission of land, in a fixed line, not as a home to be occupied and personally enjoyed, but solely as a source of wealth and social influence, has been shown to be contrary to public policy; and here, therefore, our main principle will come into operation—namely, that whatever may be done legally and equitably by individuals, may also be done by the

State. Now, any individual owner has the power of diverting the transmission of land into another direction than that desired by the previous owner. He may do this in accordance with his personal wishes, his necessities, or even as impelled by his vices, and no person has a right to claim compensation for any supposed injury or injustice in consequence of his not inheriting it. The State, then, may properly claim and exercise a like power for important public purposes; but in order that "just expectations" may not be interfered with, nothing should be done to prevent an estate from descending in due course, at least as far as the grand-children of any existing owner; and, if we go one step further and say that the law shall not be altered so as to affect even his great-grand-children, we certainly extend the principle as far as any one can reasonably claim on the ground that his "sentimental interests" ought to be respected. It is, therefore, proposed that a law shall be enacted by which all landed property in Ireland *shall legally descend for three generations beyond the existing owner and then pass to the State.* It has been already shown that this will not infringe any individual right or privilege that ought to be permitted to landowners, or even any sentimental interest that they really possess; neither, as will be shown further on, will it, in all probability, appreciably diminish the market value of their property during the lifetime of any existing owner or heir-at-law.

In all those cases in which land does not pass from father to son or daughter, but collaterally to brothers, uncles, cousins, or other persons, as well as in all cases in which it is sold or given away, each separate transfer is to be counted as equivalent to one succession in the direct line of descent—the general statement of the new law being, that land will be allowed to pass to three successive owners other than the actual owner at the time of the passing of the Act, and will, on the decease of the last owner, become the property of the State.[1]

[1] This was a first tentative application of a great principle which I have considerably extended since, especially in Chapter XXVIII. of this volume.

Before considering how the land so acquired should be dealt with in order to realize the greatest good to the community, and avoid all the evils that result directly and indirectly from absolute individual ownership, I would call attention to the advantage of the very gradual acquisition of the land which the mode here advocated would ensure, so that the necessary machinery for dealing with it might be adequately tested, and much valuable experience gained, before the bulk of the land became national property. By means of the law of intestacy, as already explained, a few estates would at once drop in; while from the law which limited the future transfers of land to three in number, other estates would lapse in the course of a very few years, and afterwards in gradually increasing numbers, just as the more perfect State and local organization and modified habits of the people became better adapted to utilize the changed conditions of tenure.

Proposed Land-tenure.

I will now proceed to explain in detail the exact manner in which the land so acquired should be held by the people, in accordance with the general principles already laid down; and in doing so, I shall endeavour to show that it is possible to give full satisfaction to every just sentiment of ownership of the land, to every desire for family permanence, to every home feeling and local attachment, which it should be a primary object of Government to maintain and restore. The encouragement and extension of such sentiments and influences is of the highest importance to the real well-being of the community, and it is one of the greatest objections to the present system of land-tenure that, by leading to vast accumulations of land in the hands of comparatively few individuals, it has more and more destroyed these beneficial influences, by condemning the bulk of the population to the mere temporary occupation of house and land, and has thus made us what an earnest and talented writer has well termed " A Dishomed Nation."[1]

[1] See Rev. F. Braham Zincke, in *Contemporary Review*, August 1880.

HOW TO NATIONALIZE THE LAND

My proposal will best be understood, and its numerous advantages explained, by taking an illustrative case, and showing exactly how it would work. Let us suppose, then, that owing to a rapid succession of deaths a gentleman has come in unexpectedly as the third successor to an estate, and therefore having only a life interest in the land. The estate consists, perhaps, of a house and extensive pleasure grounds, of a home farm, and of, say, a dozen surrounding farms. This gentleman has a family of sons and daughters, and he wishes his eldest son to continue to live on the estate, which, we will suppose, has been long connected with his family. At the death of this last freeholder the whole *land* of the estate becomes public property; but anything *on* the land or which has been added to its value by the preceding three owners, remains the property of the heirs, and every future holder of the land will have an indefeasible *tenant-right* to everything they may acquire, besides the land itself, and also to every addition or improvement of whatever kind they themselves make to it.

Soon after the passing of the law we have here advocated, a general valuation of all the land of Ireland will have been made, every separate field, plot, or holding being estimated according to its *inherent comparative value* as dependent on soil, subsoil, aspect, climate, elevation above the sea, vicinity to towns or markets, means of communication, and all other facts and conditions, *not given to it by any preceding owner*, but dependent either on natural qualities and surroundings or on the general development of the country. The annual value thus estimated will be the State "ground-rent" or "quit-rent;" and, as may be decided on from time to time, either the whole or some fixed portion of this ground-rent will be payable by every holder of land which has ceased to be private property. This "ground rent" will, of course, be very much lower than the lowest rent ever paid by a tenant to a landlord on the old system; but even this will probably never have to be paid in full, except in the earlier stage of the transition from public to private ownership; and whatever proportion of it is decided on by the Government to be payable will be uniform over the whole country, and will only be raised or lowered for State purposes, or as a substitute for oppressive or injudicious taxation, so that it will be impossible that any favouritism should be shown to particular individuals or particular localities.

So much being premised, we will return to our illustrative case of the estate whose last private owner has just died. In due course the heirs will come into possession of so much of the land as the last owner personally occupied, at the "ground rent" determined by the general valuation, which will be open to inspection in every parish, and whose amount will thus have been long known to the heir. If he decide to continue to reside in the house and occupy the home farm he may do so, with the same certainty and security as if he were still the freeholder and the "ground rent" were merely an enlarged land-tax; and he will also be able to transfer the occupation to his son or successor, or to sell his "tenant-right" to any one so as to obtain the market value of any improvements he may make in the estate. He may, if he likes, pull down houses or fences, cut down trees, plant or remodel in any way he pleases; for in doing this he is only improving or injuring his own saleable or transferable property. One thing, however, he must not do, and that is to sublet or mortgage the land or tenant-right, it being a principle of State policy (carried into effect by the Act already referred to) that no one must hold land except from the Government direct, and must not, except under certain defined conditions, subject it to any claims which would destroy or interfere with the security for the ground-rent payable to the Government. This is the very essence of the proposed system of land-tenure; since, if it were not adopted, the same accumulation of land in the possession of individuals that now prevails might again occur; tenants would again be subject to prohibitory stipulations; and that perfect freedom and unfettered ownership essential to the full development, both of the capacities of the soil and of those good moral and social effects which such ownership is calculated to produce, would be again destroyed.

Supposing, however, that the heir or heirs did not wish to occupy the estate, but wanted to realize and divide their property, they could freely sell the tenant-right, including everything that was upon the land, either by private contract or public auction, and the purchaser would at once become the holder of the land under the State.

As regards the other farms, which had been rented out by the last owner, each tenant in actual occupation at the time of his death would have the right to continue undisturbed in his holding, thenceforth paying the fixed ground-rent to Government, and purchasing the tenant-right from the heirs of the last owner of the land. If a private agreement could not be arranged between the parties, owing to exhorbitant demands by the owners of the tenant-right, the tenant should be empowered to claim that the amount payable should be determined by an official valuer, who should take as a basis of his valuation the difference between the "ground-rent" and the average net rent actually paid for the preceding five years, calculated at a moderate number of years' purchase, dependent on the state of repair of the premises and general condition of the farm.

Should the tenant not be able to pay this amount, authorized public associations of the nature of our building societies or, failing these, the Government or the municipality might advance a certain proportion of it on security of the tenant-right, repayment to be made by equal installments for a limited period. This, of course, refers only to these cases in which the tenant does not already possess the tenant-right. But when all the buildings and improvements on the farm have been made by the tenant, or have become his by purchase or in any other legal or equitable way, then he will have nothing to pay to the last owner of the land, but will at once become a holder under the State, at a very greatly reduced rent, with absolute certainty of tenure for himself and his heirs, and with perfect security as to the possession of whatever further improvements he may make upon the land.

It may here be objected that, as in the scheme of the Land League, the country would be impoverished by the whole rental of the land being paid to the English Treasury, and thus leaving the country. But this need not be so, because there is a radical difference between the two cases. In the Land League scheme the tenants would be paying interest and repaying part of the principal of a loan, and the money so paid would not, of course, be again available for any local purpose; but in the case we are now considering, the ground-rent paid by the tenant would be so much clear gain to the State, and should therefore be applied to the remission of local and general taxation in equitable proportions. But, for some considerable time after the scheme came into operation, large funds would be required, to be employed, by way of loan or otherwise, to enable the poorer class of tenants to build themslves decent houses, to make roads and fences, to stock the farms, and generally to bring the holdings into a reasonably good state of cultivation and improvement, which has been altogether impossible under the system of absentee landlords, middlemen, and exorbitant rents. It must be claimed as a special merit of this scheme of land reform, that it would provide ample funds for such a truly national purpose as the raising of a whole people from a chronic state of pauperism, only relieved by emigration or by the depopulation caused by famine and disease; and we may be sure that whenever the Legislature becomes sufficiently liberal and far-seeing to enact such a law as is here advocated, it will be generous enough to empower the National Land Commission (or whatever body may be created to carry the law into effect) to apply the funds at their command in any way that may best further the great object of raising the peasantry of Ireland into a condition of independence and well-being.

Aid of this kind would of course be strictly limited to repairing the obvious physical evils which the old system had brought about. When once the lowest class of tenants were placed in such a condition as to enable them to cultivate and improve their holdings with a fair prospect of success, the rest must be left to the influence of a sense of secure ownership, and the possession of a tenant-right under far more favourable conditions than was ever asked for or thought possible, even in Ireland. Of course there will always be a few men so utterly thriftless, idle, or incompetent, as, under the most favourable conditions, to come to ruin. For such there is no help, and they must be left to sink to the condition of unskilled day-labourers.[1] But there is no reason to think that men of this stamp will be much more numerous in Ireland than elsewhere; and we may fairly expect that under such extremely favourable conditions of tenure as this scheme would give them, the Irish agriculturist, on whatever scale, would work with the same devotion and energy as in any other country where there is complete security that the result of every hour's additional labour will be to increase the permanent value of his own property, and thus add to the well-being of himself and his family.

It has often been urged that no system of State-ownership of the land ought to be adopted, even if practicable, because it would be impossible to avoid jobbery and favouritism by the officials who would have the power of letting the Government lands; and the objection has been thought to be very serious, even by those who see all the evils inherent in unrestricted personal property in land. But it will be evident that no such objection applies to the plan here advocated, because no State official, or Government officer whatever, would have anything to do with letting the land, and could not possibly favour one person more than another even if he were disposed to do so. This arises from the fact, that in all enclosed and cultivated lands, the "tenant-right," or that portion of the land's value which has been given to it by preceding holders, will have a personal owner By virtue of this ownership of the "tenant-right," he has an indefeasible title to hold the land, subject only to the payment of the National "ground-rent;" and as this "tenant-right" will be a marketable commodity, and one without the posses-

[1] In Chapter XXV. it is shown that even this class might be rendered industrious and self-supporting.

sion of which the land itself cannot be held, it follows that no enclosed land will ever be given up till the actual holder finds a purchaser for his "tenant-right," when that purchaser at once becomes the new holder, and as such becomes liable for the ground-rent, just as the new tenant of a house becomes liable for the "house-tax." So far, then, as regards the transfer of land from holder to holder, Government or Government officials would have no more to do with it than they have in the transfer of land or houses now, though the new owner or tenant now becomes responsible for the land-tax or the house-tax to the Government.

Even in cases of intestacy, with no relatives within the degree required by the law, it would only be the land itself that it is proposed should pass to the State, the houses or other property upon it, and generally the "tenant-right" of it, being treated as personal property, which would follow the other property of the deceased. In most such cases the value of this tenant-right would have to be realized for division among the heirs. It would, therefore, be put up to auction and sold to the highest bidder, and the purchaser of it would thenceforth be liable for the State ground-rent. Under no circumstances, then, would Government have anything to do with letting the land, except in the case of default of payment of rent. Should this remain unpaid for a certain fixed period, the tenant-right would have to be sold to defray it. The purchaser would become the new holder, and the balance of the purchase money, after paying the arrears of rent, would be handed over to the ejected tenant.

The only cases in which Government would have the unfettered disposal of the whole of the land would be in the case of commons, moors, and unenclosed tracts generally as well as land which had been kept uncultivated and used for sporting only. Along with other landed property, this would of course fall in to the State in due course, and would have to be dealt with in a variety of ways, depending upon special local conditions. Some might, and probably would, be kept as common land in perpetuity, for the use of the surrounding occupiers and the enjoyment of the public generally. Where extensive tracts of moor, bog, and mountain prevail, as in many parts of Ireland, the reclamation of some of this might be encouraged by granting definite portions rent-free for a certain term of years, and at a low ground-rent afterwards, on condition of enclosure and cultivation; but in all these cases the letting should be public—by auction or tender, and such as to allow of no chance for jobbery or favouritism.

The question of private dwelling-houses in towns remains for consideration, and would have to be decided in accordance with the same general principles as govern the occupation of land generally—namely, that the occupier or holder of any land from the State, should reside on or near it, and be the real owner of the fixed property upon it. Everything would therefore be done, as town and village lands fell in, to facilitate the acquisition of houses by all classes of the community. Ground-rents would be fixed at a low rate, proportioned somewhat to the character and density of the population; while the first acquisition of the houses would be rendered easy to the purchaser of the tenant-right by fixing the official valuation (to come in action on the failure of private agreement with the heir of the last owner), at a small number of years' purchase of the average rental or rateable value. Legalized companies might also be allowed to advance money for such purchases; but in all such cases a sufficient margin would have to be left to cover the possibility of loss if the tenant were ejected and the house sold for payment of ground-rent, which would always be a first charge on the property.

There would, however, remain a considerable number of persons who require temporary abodes, and these might be accommodated in two ways. There would, first, be large buildings let out in lodgings either in flats or otherwise; while in localities where numerous small houses already existed, persons specially licensed might be allowed to hold the land on which a number of these stood, on condition that they personally superintended and managed them, were responsible for their repair and sanitary condition, and made the letting and supervision of house property their personal business. If no house owner of this kind was allowed to employ agents (except temporarily) in the entire management of such property (just as the holder of a farm would not be allowed to live at a distance, and manage it entirely by deputy), the wants of the public would be adequately supplied, while the evils now arising from the occupation of temporary houses, the operations of speculative builders, and the system of building leases, would be reduced to a minimum.

Probable Results of Land Nationalization.

Having thus sketched the main features of the system of the Nationalization of the Land here advocated, let us endeavour to trace out some of its probable effects, both while the operation was in progress, as well as after its completion; and in doing so we shall be able to consider some of the objections that will inevitably be brought against it.

And first, as to the effect of such a scheme on the value of land. It will no doubt be alleged that the passing of the Act here proposed would immediately lower the value of all landed property, and thus do a direct injury to existing landowners. As, however, anticipations of the effect of certain changes of legislation on the value of land have almost always been falsified by the result, we may well refuse to put much faith in similar prophecies now. If the change here advocated should come into effect, any purchaser of land before the Act passes will be sure of absolute possession for three generations; while, if he purchases after the Act has passed, his prospective possession will extend to only two generations, the purchase itself forming one transfer. Now, as such a measure will only be passed, after long discussion and agitation and repeated failures, till at last it is seen to be inevitable, it is probable that there will, towards the last, be a kind of rush to get land before the law is changed; and this might enhance its value considerably, and compensate for any slight subsequent fall. Then, after the Act has passed, estates will very soon begin to drop in, and these will be altogether withdrawn from the land market so far as investment is concerned. This will diminish the supply of saleable land, and will thus tend to keep up the high prices previously attained. Again, we must remember that to the majority of purchasers of land absolute possession for two generations (or two transfers) after themselves would be practically the same as a theoretical perpetuity of ownership; for the present perpetuity of ownership of freehold land is, in most cases, imaginary, as no man can possibly tell what will become of it in the third generation after his decease; or, at all events, he will not be able to do so when entails are abolished, and this abolition of entails is always taken for granted as having occurred before the present scheme comes into operation.

Another important consideration is, that all the land of the country will be equally affected; and it is a great question whether any such change of the law *could* lower in value *all* the land of the country, while its population continued to increase. If some districts were excepted and retained their land as freehold, while others came under the operation of the new law, no doubt there would be some difference of value produced, though even then it would not be much; but as all land would be at first on an equality in this respect, and the alteration of tenure would be so remote that its effect would be more sentimental than real, it is a question whether the continually diminishing supply of land would not for a considerable time keep up its full market value. When we pass on to the second or third generation after the new law had come into operation, the question becomes still more complicated, and it is not easy to say whether there would be even then any important fall in value. For by that time so much of the land of the country would have gone entirely out of the market as a possible investment (being held by personal occupiers under the State), that all other classes of securities, such as railway debentures, tramroad and telegraph shares, colonial and municipal bonds, and Government stocks, would be in great demand, and, therefore, increase in value. This would certainly react upon land; and as this could still be purchased for one life, with the option of continuation at the very moderate State ground-rent, it is possible that the demand for the poorer classes of land for occupation and improvement, and for more favourable sites as residences, might still keep it up to nearly the full value it had when freehold. Even if there were a considerable depreciation, this would be to a large extent, compensated by the diminution of local and general taxation that would by this time have been effected by means of the ground rents which had already fallen in to the Government; and it is not at all improbable that, with a nominally lower value of their land, the landowners who remained in Ireland might, owing to the peace and general prosperity of the country, and the diminished taxation, be really better off than they are at the present day.

Evils of Free Trade in Land.

Let us now pass on to another question. It is a favourite dogma of some reformers that all the evils of the present system would be got rid of by what they term " free-trade in land." They seem to think that, if all obstacles to the sale and purchase of land were abolished, if entails of all kinds were forbidden, and the conveyance of land made as cheap and expeditious as it might easily be, the chief obstacle that now exists to the growth of a body of peasant proprietors would be got rid of. This notion appears to me to be one of the greatest of all delusions. The real obstacle to peasant proprietorship or small yeoman farmers in this country is the land-hunger of the rich, who are constantly seeking to extend their possessions, partly because land is considered the securest of all investments, and which, though paying a small average interest, affords many chances of great profits, but mainly on account of the political power, the exercise of authority, and widespread social influence it carries with it. The number of individuals of great wealth in this country is enormous, and, owing to the diminution of the more reckless forms of extravagance, many of them live far below their incomes and employ the surplus in extending their estates.

The probabilities are that men of this stamp are increasing, and will increase, and the system of free-trade in land would serve chiefly to afford them the means of an unlimited gratification of their great passion. With such men for competitors in the market, who will ever be able to buy land for personal occupation and cultivation as a business? Such a course will become more and more impossible; and nothing seems more likely to check and render difficult the growth of a peasant proprietary than free-trade in land, with the unlimited power of accumulation by wealthy individuals which such free-trade will render still easier than before. This increased accumulation will inevitably exaggerate the numerous evils of absentee proprietorship, such as management by agents, restriction of agricultural processes, discouragement of improvements, the preservation of game, the system of short building leases,[1] and a pauperized class of agricultural labourers; and thus, although the abolition of restrictions on the transfer of land is a valuable reform, and receives my hearty support, it is yet utterly powerless to ameliorate the evils inherent in the unlimited possession of the soil of the country by individual owners, either as a money investment or as a source of political and social power.

The advocates of the views here opposed seem to have overlooked two fundamental facts—that the land of a country is the great essential of human existence; and, that, being fixed in quantity and incapable of increase, absolute freedom to buy and sell it must result in a monopoly, and in giving absolute power to the rich who possess it over the poor who do not—a power which, in civilized countries, is checked by public opinion and by special legislation, but is nevertheless always incompatible with the well-being of a free people.[2]

[1] I am informed that some landowners will now only let their land on building leases for eighty or even sixty instead of the usual ninety-nine years, and when they have the monopoly of fine sites land is actually largely taken on these onerous terms.

[2] In Mr. Froude's remarkable paper on Ireland, in the *Nineteenth Century* for September last, he gives the following case of (probably ignorant) abuse of power, apparently from personal knowledge. He says:—"Not a mile from the place where I am now writing, an estate on the coast of Devonshire came into the hands of an English

The scheme I have here developed destroys the monopoly of the land—the very life-blood of the nation—by any class, while it allows for the freest interchange and the most unrestricted use of the land, and for the most

duke. There was a primitive village upon it occupied by sailors, pilots, and fishermen, which is described in *Domesday Book*, and was inhabited at the Conquest by the actual forefathers of the late tenants, whose names may be read there. The houses were out of repair. The duke's predecessors had laid out nothing upon them for a century, and had been contented with exacting the rents. When the present owner entered into possession, it was represented to him that if the village was to continue it must be rebuilt, but that to rebuild it would be a needless expense; for the people, living as they did on their wages as fishermen and seamen, would not cultivate his land and were useless to him. The houses were therefore simply torn down, and nearly half the population was driven out into the world to find new homes. A few more such instances of tyranny might provoke a dangerous crisis."

This is a sufficiently striking case of the evils of landlordism which gives a rich man the power to tear the poor man away from his ancestral home. Can we really boast of our freedom when even centuries of occupation give these poor seamen no right to live on their native soil? But even this, bad as it is, is as nothing compared with the wholesale misery we have caused by forcing our land-system upon a large portion of India. This is what a Bengal civilian (quoted in the *Statesman* for September 1879, p. 329) states to be the present condition of the unhappy peasants of Bengal: "The zemindar and ryot are as monarch and subject. What the zemindar asks the ryot will give; what the zemindar orders, the ryot will obey. The landlord will tax his tenant for every extravagance that avarice, ambition, pride, vanity, or other intemperance may suggest. He will tax him for the salary of his ameen, for the payment of his income tax, for the purchase of an elephant for his own use, for the cost of the stationery of his establishment, for the payment of his expenses to fight the neighbouring indigo-planter, for the payment of his fine when he has been convicted of an offence by the magistrate. The milkman gives his milk, the oilman his oil, the weaver his cloths, the confectioner his sweetmeats, the fishermen his fish. The zemindar fines his ryots for a festival, for a birth, for a funeral, for a marriage. He levies black mail on them when an affray is committed. He establishes his private pound, and realizes five annas for every head of cattle that is caught trespassing on the ryot's crops. . . . . These cesses pervade the whole zemindari system. In every zemindari there is a naib (deputy), under the naib there are gumashtas (agents), under the gumashta there are piyadas (bailiffs). The naib exacts a perquisite for adjusting accounts annually. The naib and gumashtas take their share in the regular cesses; they have other cesses of their own. The piyadas when they are sent to summon defaulting ryots, exact from them four or five annas a day. It is in evidence before the Indigo Commission that in one year a zemindari naib, in the district of Nuddea, extorted ten thousand rupees from his master's ryots. . . . .

perfect free-trade in land compatible with the liberty, the progress, and the free development of the whole community. There can be no conceivable *use* of the land for which it would not be available under the new régime. The absolute freedom of sale of tenant-right to intending occupiers of land would provide for experimental cultivation in any direction. The capitalist who wished to devote himself to farming on a large scale would first purchase the tenant-right of some large farm, and gradually add to it the surrounding farms as they came into the market, or as he could persuade their owners to sell them by liberal offers. Farms would often be broken up, and the tenant-right to single fields or small plots sold separately whenever there was a demand for such lots, and thus the industrious labourer or the retired tradesman would be able to obtain portions suited to their respective wants. Spade husbandry on small peasant properties, and huge machine-cultivated farms like those of Western America, would have an equal chance of trial; each district would gradually merge into that style of husbandry which suited it best, and in no case would there be any hampering restrictions to check its progress. This would be real free-trade in land as opposed to its present monopoly by the rich, and would lead to the freest and most perfect development of the agricultural resources of the country.

This system of cesses has eaten, like an incurable disease, into the social organization of the country. An energetic Government might have grappled with the question, and succeeded in abolishing a system which, though forbidden by law, yet flourishes in undisturbed luxuriance; yet no one raises a hand on behalf of the ryots, no one speaks a word in their interest. . . . . It seems almost as though they were doomed never to be emancipated from their present degrading life."

The result is that the ryots exist always on the verge of starvation. They were once, it must be remembered, direct holders of their land under the Government. But Lord Cornwallis and the then Home Government of India handed them over to a body of tax collectors (the zemindars) as tenants, thinking that our English landlord system, perfect in the eyes of a landlord Government, must be best for all the world. The result has been, that, "under British rule, the soil of India has either passed or is fast passing into the power of land speculators and money-lenders, while the ancient landowners have been converted into half-starved, poverty-stricken serfs on the fields which were once their own."

Effect of Free Access to Land on the Well-being of the People.

It is very difficult to foresee, and perhaps impossible to exaggerate, the influence of such a state of things on the real well-being of the community. Judging from what is known to be the effect of extended land-ownership in other countries, in stimulating industry, in diminishing crime, and in abolishing pauperism, and knowing the love of country people for their home and its associations, we may surely anticipate that the land would soon exhibit the effects of such favourable conditions of life in well-cultivated fields and gardens, comfortable houses, and a well-clothed, well-fed, and contented population.[1]

But, it will be said, what is now proposed is a revolution, and a revolution more portentous than any the world has yet seen, since it would inevitably lead to the complete extinction of the territorial aristocracy, a class which has hitherto formed an important—perhaps the most important—part of every community raised above the savage or nomad condition.

This is very true. The change proposed *is* indeed a great and a fundamental one. But the question before us is, not its greatness or its radical character, but simply whether it would be beneficial to the community as a whole, and, if beneficial, whether it could be effected with-

---

[1] Mr J. Boyd Kinnear in the work already referred to says:—"Who does not see how much happier England will be when, instead of one great mansion surrounded by miles beyond miles of one huge property, farmed by the tenants at will of one landlord, tilled by the mere labourers, whose youth and manhood know no relaxation from rough mechanical toil, whose old age sees no home but the chance of charity or the certainty of the workhouse, there shall be a thousand estates of varying size, where each owner shall work for himself and his children, where the sense of independence shall lighten the burden of daily toil, where education shall give resources, and the labour of youth shall suffice for the support of age. Changes like these cannot, indeed, be created; they must grow. But our business ought at least to be to permit their growth." Mr. Kinnear thinks that free trade in land will permit their growth. I have already shown how extremely improbable this is, while it might even exaggerate many of the existing evils: whereas the plan here proposed *necessarily* brings about the state of things which is allowed to be so highly beneficial.

out injustice and without danger. I think I may claim to have shown that the last question may be answered in the affirmative. On the plan here sketched out, the change might be effected, either without injury to any individual other than a possible, but by no means certain, depreciation of his property—and if such a depreciation did occur, and could be valued, there would be ample available funds to award compensation.[1] Moreover, there is no finality advocated for the actual proposals here made, but only for the general principle. If it should be estimated that the termination of absolute property in land after three generations would be really injurious to existing landowners to any appreciable extent, then a higher number might be fixed upon. But, on the other, hand a smaller number would give the existing generation a greater interest in passing such a law.

To many it will no doubt be almost impossible to realize a state of society in which there were no great landowners, no country gentlemen living wholly or mainly on the rent of land. They will picture to themselves the country relapsing into a state of semi-barbarism, the parks turned

[1] In taking such extreme precautions against any interference even with the "sentimental interests" of existing landowners, I have been actuated by a desire to deal fairly with every class, and to give the least possible offence to a powerful vested interest. That I have gone even further in this direction than is required by simple justice, in a case where the sentimental interests of individuals conflict with those of the community, is shown by the following remarkable statement on this very point, by the late Nassau W. Senior, a writer who will certainly not be accused of being an extreme Radical. In his Essays on *Ireland* (vol. i. p. 3) he says:—"Nor can any interest, however lawful, be considered property as against the public, unless it be capable of valuation. And for this reason : if incapable of valuation it must be incapable of compensation, and therefore, if inviolable, would be an insurmountable barrier to any improvement inconsistent with its existence. If a house is to be pulled down, and its site employed for public purposes, the owner receives full compensation for any advantage connected with it which can be estimated. But he obtains no *pretium affectionis*. He is not paid a larger indemnity because it was the seat of his ancestors, or endeared to him by any peculiar associations. His claim on any such grounds for compensation is rejected, because, as the subject-matter is incapable of valuation, to allow it would open a door to an indefinite amount of fraud and extortion ; nor is he allowed to refuse the bargain offered to him by the public, because such a refusal would be inconsistent with the general interest of the community."

into sheep farms, the mansions into farmhouses, and the noble pleasure-grounds into market-gardens. But they entirely overlook the fact that the real wealth of the country will certainly be greater than ever, and that every mansion now existing, and many additional ones, will still be occupied, possibly with less of display and magnificence, but often with more of taste and high cultivation. The mere fact that thousands of educated men who now live comparatively idle live on the rents of their ancestral estates will be converted into workers of one kind or another, must surely be a source of additional wealth and power to the nation. During the transition state (all checks in the way of entail having been abolished) the surplus revenues from the land, or the proceeds of its sale, will be gradually invested in other ways. Some of it will go into genuine industrial enterprises of various kinds; while it is not improbable that the personal management and improvement of a great agricultural estate (of which a park and mansion may still form the central part) will come to be considered as the proper and most honourable occupation for the descendants of the landed aristocracy. The great landowner and country gentleman of that day, with an estate of many thousand acres, employing hundreds of labourers and supporting thousands of cattle, using the best machinery and manures, developing in every possible way the productive capacity of the soil, and as proud of the health, comfort, and well-being of his men as of the breed and condition of his horses and his oxen, would certainly not be an unworthy successor of the great landowner of to-day, who receives his rents from half-a-dozen counties, and possesses mansions which he never inhabits and estates which he never visits but for purposes of sport.

The impossibility of having any land except for personal occupation would render it necessary that agriculture should be studied as a part of every gentleman's education, in order that whatever land he had around his country-house, whether park or home farm, might be not only a source of pleasure but of profit; and this wide extension of agricultural knowledge would certainly become a source of wealth to the country which it is impossible to estimate.

Although sporting will necessarily be a far less important feature in country life than it is now, there is no reason to think it would altogether cease. The wealthy could devote as much land as they pleased to the preservation of game for their own or their friends' amusement, or sporting might take other forms more suited to the altered state of the rural population, in which both wealth and intelligence would be more widely distributed than at present. Even if the present land-system were to continue unaltered, we could hardly anticipate that the growth in population and changed habits and ideas of two centuries hence would leave the customs of the country, as regards field sports, what they are now; and it is therefore quite unnecessary to conjecture further what might happen to them under such extremely changed conditions as are here anticipated.

Although the legislature and the press may alike ignore it, there is undoubtedly growing up among the more intelligent of the working classes, as well as among a large body of independent thinkers, a profound dissatisfaction with the actual state of things as regards private property in land. They see that its possession or enjoyment by any but the wealthy is yearly becoming more and more difficult, and that its accumulation in the hands of a few owners is opposed in many ways to the public welfare. They see wide areas of common lands enclosed, to the pauperization of the needy labourer, and the further enrichment of the wealthy landowner; while a system of obsolete laws framed by, or in the interests of, the so-called "lords of the soil," are now being everywhere strained against any free and adequate enjoyment of their native land by the great mass of the people. They find themselves often shut out from the downs and moors and picturesque mountains, almost all of which are said to have private owners who may, and often do, enclose them; while the very rivers and streams, which ought to be as free for the enjoyment of all as the winds of heaven or the light of the sun, are everywhere being monopolized for the exclusive pleasures of the rich. Even the beautiful country lanes, with their wide margins of grass, and banks often shaded with trees and adorned with wild flowers —lanes which afford the purest delight to the constantly increasing population of our towns and villages, and which are often the only examples of picturesque Nature within their reach, are now constantly being stolen from them by the owners of the adjacent fields who (regardless of decisions in our highest courts) fence in the narrow roadway in order to add a few perches to their land; while everywhere we find what were once pleasant footpaths either stopped altogether or shut in by obstructive fences. This land-monopoly of the rich pursues the mass of the people even to their homes, since they are obliged to live in crowded, badly built, and often unhealthy houses, because so many landowners will only grant land on building-leases and at high ground-rents, in order to enrich their unborn, and perhaps unworthy successors, at the expense of the health, the comfort, and the freedom of the present generation. And, lastly, they see that these great landowners are, as a class, the opponents of all progress, the upholders of cruel and obsolete game laws, and that they possess legal powers and privileges virtually giving them a command over others which in a free country no class of citizens ought to possess. This widespread feeling of discontent manifests itself in Ireland in land-leagues and tenant-right associations, and in other more destructive forms; while in England there is a very general but as yet undefined belief that the true remedy for the evil is to be found in the Nationalization of the Land.[1] The danger is, that all reform should be opposed

[1] Proposals for the nationalization of the land have received comparatively little attention by modern writers, because it has been *assumed* that *present* owners must be paid its full value, and this was clearly impossible. Thus in the recently published work of Mr. J. Boyd Kinnear, in a chapter on this subject, we find such statements as the following:—"Under the broader form of the proposal the first step is that the State shall purchase the land from its *present owners*, either compulsorily or by agreement, but in either case paying *its full value*" (p. 106). And again:—"There is, of course, no objection in principle to the State taking possession of all the land in the realm, on the understanding, *which is always included in the proposal*, that it shall

too long, and the people, in whose hands political power now rests, should at last insist upon some hasty and ill-considered remedy which, while bringing ruin on many, should only afford a temporary and imperfect cure of the disease.

The present writer had his attention forcibly drawn to this great question about forty years ago, by the perusal of Herbert Spencer's demonstration (in his *Social Statics*) of the immorality and impolicy of private property in land, and since that time he has endeavoured to make himself acquainted with what has been written on the subject, and by means of constant thought and discussion to arrive at the true solution of the problem. This he believes he has at length done. The difficulties that surrounded the subject were many and great. It was necessary, firstly, to find a means of transferring the ownership of the land from individuals to the State without taking anything away from existing owners, or infringing any right, real or sentimental, which they actually possess; secondly, to devise a new tenure of the land which should combine all the incalculable advantages of safe possession and transmissible ownership, together with the full benefit of every improvement and increase of its value, while guarding against the recurrence of unlimited landed estates, absentee landlords, life-interests, subletting, building leases, restriction on improvements, and all the other evils which accompany our present system; thirdly, to avoid the dangers which have been hitherto believed to be inherent in State-Landlordism— jobbery, favouritism, waste, and the creation of a vast addition to State patronage; and, lastly, to render the land

compensate the present owners." These quotations from one of the latest and best informed writers on the land question sufficiently prove that previous writers have not seen how the land may became State property without paying for it and yet without injury to any one; and this is the very essence of the question which determines its practicability. My plan not only does this, but it also, as already shown, completely removes the difficulty of State-Landlordism by retaining the tenant-right as saleable and heritable property—a system which has been in actual operation on Lord Portsmouth's estates in Ireland for more than half a century, and with the most beneficial results.

a productive and practically inexhaustible source of national income, and to bring all these changes about in a gradual and almost imperceptible manner, by the action of a few simple principles embodied in law, so that society may have ample time to adapt itself to the new conditions; while, during the process of adaptation, successive generations may grow up to whom they will bear the aspect of being as natural, as orderly, and as beneficial, as private ownership does to most of the present generation.[1]

All these essential conditions of a true system of land-reform are embodied in the scheme now briefly explained. Although not really injurious to existing landowners, it is not expected that it will meet with any support from them, since it has not been framed in their exclusive interest, but with a view to the well-being of the entire population. It will, no doubt, be said that the title of this chapter is misleading, since the arguments I have used are equally applicable to England as to Ireland. This is

[1] Although any change of the nature here proposed will no doubt be fiercely opposed by most landowners, and will perhaps not be admitted to discussion in Parliament for many years, yet changes more directly affecting vested interest in land have been made in the present century. Mr. Nassau Senior tells us in the work already referred to (*Ireland*, p. 8), that—"Until January, 1834, no person could inherit the freehold property of his lineal descendants. On the death of a person possessed of such property, intestate and without issue, leaving a father or mother, or more remote lineal ancestor, it went over to his collateral relatives. In 1833 this law was totally altered. In such cases the property now goes to the father or mother, or remoter lineal ancestor, in preference to the collaterals. The brothers, uncles, nephews, and cousins, of lunatics, or of minors in such state of health as to be very unlikely to reach the age at which they could make a will had, until the 3rd and 4th Will. IV. cap. 106, was passed, prospects of succession so definite, that in many cases they would have sold for considerable prices. All these interests, though lawful and capable of valuation, have been swept away without compensation. And it was necessary that this should be done; the old law was obviously inconvenient, and to have attempted to compensate all those who, if the principle of compensation had been admitted, must have been entitled to it, would have involved such an expense as to have rendered the alteration of the law impracticable." This precedent is very valuable, because no such calculable vested interests occur in the present case, while the political and social importance of the change, and its beneficial effects on the bulk of the community, are vastly greater. The discussion, therefore, becomes limited to the question whether the proposed change would be a beneficial one.

quite true. The principles laid down are of universal application, but the time and the mode of applying such principles are matters of expediency. The land-question in Ireland is a burning one. It is a source of chronic discontent and disaffection, and is likely to be so in spite of all the patchwork remedies that may be applied to it. More than anything else it maintains the antagonism of the Irish representatives in the British Parliament, an antagonism which has unhappily too much justification, and which, so long as it exists, will be a drag on the wheels of the legislative machine, and thus be directly injurious to every British subject. The solution of the Irish land-question is, therefore, urgent. It is of importance to every one that it should be settled on a sound and permanent basis, and this can never be the case unless the true principles of land-tenure are discovered and acted upon. The present scheme is, therefore, proposed to be applied, in the first instance, to Ireland alone. It is claimed that its very gradual operation—which to some will appear an objection—renders it far safer and more likely to be effective than more heroic measures, while its discussion need not interfere with any remedial legislation which successive Parliaments are willing to enact. It is further claimed that it is founded on principles of abstract justice, and that, while respecting all existing rights and possessions, it will ultimately abolish that system of unlimited property in land which was founded originally by conquest, oppression, or rapine, and which, although perhaps useful in a transition stage of civilization, is incompatible with our national well-being, or with the general happiness and advancement of the community.

To the independent Liberals of Great Britain and to the long-suffering Irish nation I now submit my proposals, asking only for a careful perusal, an unprejudiced consideration, and a searching criticism.

CHAPTER XVII

THE "WHY" AND "HOW" OF LAND NATIONALIZATION [1]

IN *Macmillan's Magazine* (for July, 1883) an article appeared on "State Socialism and the Nationalization of the Land," from the pen of the late Professor Fawcett, in which he referred to two books as having more especially drawn attention to this question—one of these being my own volume on *Land Nationalization*, the other, Mr. Henry George's well-known *Progress and Poverty*. In consequence of the wide circulation of the latter work, Professor Fawcett thinks it important to examine carefully the proposals there advocated, and he proceeds to do so, though, as it seems to me, far from "carefully," since he starts many difficulties which would never arise under Mr. George's proposals, and entirely ignores the vast mass of fact, argument, and illustration, by means of which the radical injustice of private property in land, and its enormous and widespread evil results, are set forth and demonstrated. With the treatment of Mr. George, however, I do not here propose further to meddle; but as Professor Fawcett has quoted the title of my book as one of those which have drawn attention to the subject, while he deliberately ignores every fact, argument, and proposal contained in it; and as the press has very widely noticed

[1] This article first appeared in *Macmillan's Magazine* (August and September, 1883). It is now reprinted, because it discusses and replies to many of the popular objections against Land Nationalization which are still used, and explains details which are not touched upon in the preceding chapter. A few verbal alterations have been made in order to make it more intelligible at the present day.

and praised this article as demonstrating the futility and impracticability of land nationalization, I gladly seize the opportunity afforded me of stating the other side of the question. This is the more necessary because the readers of Professor Fawcett's article will certainly carry away the impression that my proposals are substantially the same as those of Mr. George, and that a criticism of the one will apply equally to the other; whereas not only are they absolutely distinct and unlike, but those first advanced by myself have commended themselves to a considerable number of advanced thinkers who previously held nationalization to be impracticable, and have led to the formation of a Land Nationalization Society, which has now been twenty years[1] in existence, and is gradually but surely aiding in the formation of a distinctively English school of land reformers. These facts, to which Professor Fawcett's attention has been specially directed, surely required that some notice, however brief, should be given to them in an article written expressly to instruct the public on this great question.

In order to place this problem fairly before my readers within the limits here assigned to me, it will be necessary to omit the consideration of some of its aspects altogether, and to treat others very briefly. The fundamental question undoubtedly is, the right or the wrong, the justice or the injustice, of private property in land. And then follows the question of results; right and justice lead to good results—to happiness and general well-being; wrong and injustice as surely lead to bad results, and their fruits are moral evil and physical suffering. We have to inquire, then, what are the actual results of modern landlordism? and thus confirm or modify the conclusions we have reached from general principles. Finally, we have to consider how we can best carry into effect right and just principles so as most certainly to reap the reward of moral and physical well-being. This really exhausts the subject. The historical inquiry—how private property in land arose, what changes it has under-

[1] Founded in March, 1881.

gone, the results of legislation by landlords, and usurpation by kings, the story of royal grants, confiscations, and inclosures, are all exceedingly interesting, and will be found to support and strengthen at every point the argument from principle and from results; but it is not essential to a comprehension of the main question, and we shall therefore omit it here, referring our readers to such works as Mr. Joseph Fisher's *History of Landholding in England;* Thorold Rogers' *Six Centuries of Work and Wages;* *Our Old Nobility* (originally published in the *Echo* newspaper); and Dr. W. A. Hunter's lecture on "The Land Question" (*Mark Lane Express,* January 8th, 1883), for condensed information on this branch of the subject.

First, then, we have to inquire whether private property in land is just and right; and here we find ourselves at once in conflict with the great body of Liberals and land-law reformers, who advocate, as their sole panacea, free trade in land. For the foundation of their doctrine is, that land should be treated as merchandise; that it is *right* for individuals to own it absolutely and in any quantity; that it is *good* for great capitalists to add farm to farm, and to build up great estates; that land should be bought and sold as easily as iron or railway shares. We nationalizers, on the other hand, say that all this is fundamentally wrong. We maintain that land should not be treated as merchandise, for the following reasons :—

1. Because it is absolutely essential for all productive industry, while it is the first necessity of human existence; therefore those who own it will, as a whole, possess absolute power over the happiness, the freedom of action, and the very lives of the rest of the community.

2. Because it is limited in quantity, and tends therefore to become the monopoly of the rich—a monopoly which will surely be intensified by free trade, which will render it easier than now to accumulate large estates, and will thus make the landless people still more surely than they are now the virtual slaves of the landlords.

As Professor Fawcett and Mr. Shaw-Lefevre have both shown, the land hunger of the rich is insatiable; and, as is well put by the *Edinburgh Review*—" It stands to reason that if the sale and purchase of land were perfectly easy and free, those persons would buy most land and give the best price for it who had most money to buy it with."

The Right or Wrong of Landlordism.

To determine whether private property in land is right and just, and compatible with the well-being of the whole community, it will be well to glance briefly at the true foundations of property, and the admitted rights of a free man. Property is, primarily, that which is obtained or produced by the exertion of labour or the exercise of skill. In this a man has a right of property, to use, to give away, or to exchange. This is a universally admitted right which forms part of the very foundations of society, and many eminent writers maintain that it is the only way in which private property can justly arise. Property is, however, usually admitted in any natural product *found* by an individual and obtained without labour; but this kind of property has never the absolute character of the former kind, since if the thing found is not abundant and is essential to life or well-being, the individual right to its exclusive possession is not admitted. The single good spring of water on an island, a single group of fruit or other useful trees, a single pond or stream containing abundance of fish, are not allowed to be appropriated by the first discoverer to the exclusion of his fellow men.

Property in the results of a man's labour has no such limitations; it is usually hurtful to no one, and with free access to natural agencies and products, and freedom of exchange between man and man, is beneficial to all. There is no other natural and universal source of private property but this—that every man has a right to the produce of his own labour; and hence, as land is not produced by man and is essential to man's life and happiness, it cannot equitably become private property.

Let us next look at the question from the point of view of the rights of individuals, as members of a society which upholds freedom as a fundamental principle of its existence. In such a society it will surely be admitted that every man has an equal right to live. Not, be it observed, a right to be kept alive by others; not a right to claim any part of the produce of others' labour, but, simply, freedom to support himself by labour, freedom from all obstructions by his fellow-men of his own freedom to labour. Not to have *this* freedom of action is to be a slave: and to this extent at least it will be admitted that all men are, or should be, equal.

But man cannot live without access to the natural products which are essential to life—to air, to water, to food, to clothing, to fire. If the means of getting these are monopolized by some, then the rest are denied their most elementary right—the right to support themselves by their own labour. But neither pure air, nor water, neither food, clothing, nor fire, can be obtained without land. A free use of land is, therefore, the absolute first condition of freedom to live; and it follows that the monopoly of land by some must be wrong, because it necessarily implies the right of some to prevent others from obtaining the necessaries of life.

Another consideration which shows the private ownership of land to be unjust is the fact (admitted by all economists), that the whole commercial value of land is the creation of society, increasing just as population and civilization increase. If one man had a grant of an uninhabited island or country, the size of Britain, the value per acre of all the land which he did not use himself would be *nil*. Rather than live alone he would give land to any one who would settle near him. And when others came he would sell them land, as it is sold in all new countries, for a mere trifle, while he could never enforce his rights over those who took possession of remote parts of his territory. But, just as the population increased the land would rise in value; till, when towns and cities had sprung up, and all the arts of civilized life were practised, and communications were established with every part of the world, a single acre might sell for £1,000 or £10,000. Who created this value? Not the original settler, but society. And this shows the absurdity of comparing, as some do, the occasional increase in the value of other property with that of land. In the case of everything which is the product of human labour, the tendency is for it to become cheaper as population increases and civilization advances. When the reverse occurs it is usually owing to exceptional conditions, or to the influence of some kind of monopoly, and it never applies to the necessaries of life. But with land the increase of value is universally coincident with and due to the growth of society, and the only fluctuations in this constant rise are owing either to monopoly and speculation forcing the price at a certain epoch above its natural value, or to restrictions on its free use by the people. Here again we bring out a broad distinction between the products of a man's labour which are and should be private property, and land, the gift of nature to man and the first condition of his existence, which should ever remain the possession of society at large, and be held in trust for the equal benefit of all.

One other consideration remains, and perhaps the most important of all as affording a demonstration of the necessarily evil results of unrestricted private property in land. If a portion of the community is allowed to appropriate the whole of the land for its private use and benefit, this appropriation necessarily carries with it the right and the power to appropriate the bulk of the products of the labour of the rest of the community, while it keeps down wages to a minimum rate just sufficient to maintain physical existence. Carlyle recognized this truth when he speaks of the poor widow boiling nettles for her only food, and the perfumed lord at Paris extracting from her every third nettle as rent. The late Professor Cairnes in many of his writings dwelt upon this consequence of landlordism. In one of his essays in the *Fortnightly Review*, he says:—

"The soil is, over the greater portion of the inhabited globe, cultivated by very humble men, with very little disposable wealth, and whose career is practically marked out for them by irresistible circumstances, as tillers of the ground. In a contest between vast bodies of people so circumstanced, and the owners of the soil—between the purchasers without reserve, constantly increasing in numbers, of an indispensable commodity—the negotiation could have but one issue, that of transferring to the owners of the soil the whole produce, *minus* what was sufficient to maintain in the lowest state of existence the race of cultivators."

But this result has been most clearly and forcibly demonstrated by Mr. Henry George, who makes it the very key-note of his book, and demonstrates it by a wealth of illustration and a force of argument which must be carefully studied to be appreciated. I can here only find space for an abstract of one of his illustrations.

Imagine an island, with no external communications, and but moderately peopled, in which the land was equally divided among all the inhabitants; and let us suppose that there was free trade in land as in everything else, just as desired by our most advanced politicians. After fifty or a hundred years let us look again at this island, and we shall certainly find the land most unequally divided; some will be very rich and have large landed estates, many will be very poor and have sold or otherwise parted with all their land. We may suppose there to be no wars, a pure government, few taxes, no state church, no hereditary nobility; yet inequality in ownership of land will have caused pauperism and virtual slavery. For, all must live on the land, and from the products of the land; therefore those who do not own land can only have the use of it or obtain its products on the terms of those who do own it. They are really slaves; for, in order to live, they must accept the landlord's terms and do as he bids them.

If landlords are rather numerous it will not *seem* like slavery, because the *forms* of free contract will be observed. But there can really be no free contract, because the landowners can wait, the landless cannot. They must work on the landowner's terms or starve. And thus, just in proportion as population increases, and the competition for land and its products, especially for bare food, becomes keener, the landowners will obtain a larger and larger share of the products of the soil—in other words, rents will rise, and wages will fall or remain stationary. Now let us introduce a fresh element — labour-saving machinery. This will enable more wealth to be produced with the same labour or with less; but it will not decrease the dependence of labour upon land. All the increased production of wealth will go to the wealthy—the landlords and capitalists, the landless remaining as poor as before. As the climax of this argument, Mr. George supposes the case of labour-saving machinery to be brought to absolute perfection so that all wealth will be produced by various forms of automata, without human labour. Then all wealth will belong to the landowners, for even standing room for houses and machinery cannot be obtained except on their terms, and the landless multitude must necessarily starve in the midst of plenty, or live as servile dependents on the landlord's bounty. Thus, private property in land—even were all other social and political evils removed—necessarily makes the many poor that the few may be rich; for it prevents free access to those natural elements without which man cannot live, and thus directly causes *poverty* and *pauperism*, and the long train of miseries and crimes that spring therefrom.[1]

By this preliminary inquiry we have shown—

(1) That private property in land can never justly arise, because land is not a product of human labour.

(2) That the monopoly of the land by a class is inconsistent with the fundamental rights of individuals in a professedly free country.

(3) That the whole commercial value of land is the creation of society, not of landlords and tenants, and should therefore belong to the community.

(4) That private property in land necessarily leads to the poverty and subjection of the many for the benefit of the few.

I therefore claim to have completely answered the

---

[1] If we trace the social condition of the United States from the early part of the nineteenth century to the present time, we see that all these changes *have* taken place as a result of increasing land-monopoly.

fundamental question with which I started, and to have demonstrated that, as a matter of principle, our present land-system is absolutely wrong, cruelly and perniciously unjust. Before proceeding to consider how far this conclusion is supported by the facts and results of modern landlordism, I cannot but remark on the absolute silence of Professor Fawcett on the whole question of right or wrong. He knows that this aspect of the subject is treated with wonderful force and most convincing illustration in Mr. George's book; he knows that nearly a hundred thousand copies of that book have been circulated among English readers; and yet he confines himself exclusively to Mr. George's practical proposals which have really nothing to do with the main question. Are we to suppose that he upholds the convenient doctrine that whatever is is right; that ethics need have no place in political teaching; and that the happiness or misery of millions is as nothing compared with the maintenance of the usurped rights and privileges of British landlords? He can surely not imagine that such a mode of treating this great subject can have the slightest effect as an antidote to Mr. George's teachings.

The Effects of Landlordism.

We now come to the important practical question, What is the outcome of modern landlordism? We have seen that, from various points of view, it is wrong in principle; the works we have referred to on the history of the subject show that it had its origin in force, and has since been largely maintained by confiscation and by unjust legislation; but we are so practical a people, that, if it can be shown that its results are good we should care little about principles or about history, but would be quite content to maintain a system which works tolerably well. And people actually do say that it works well. Press and Parliament are never tired of exclaiming—" See how rich we are! What a trade we do with all the world! Our system, which produces such results, *must* be all right!" But along with our great riches we have a mass of terrible poverty, and it is the opinion of disinterested writers that we are the most pauperised country on the globe.[1] Our public men continually assure us that pauperism is diminishing; or that at the worst it is stationary, while our population is increasing rapidly, and that it is therefore proportionally diminishing; and they base their statements on the official statistics of pauperism. I shall show however, that these are not trustworthy guides, and that there is good reason to believe that, during the very periods in which our aggregate wealth has increased most rapidly, pauperism has increased also in positive amount, and perhaps even in greater proportion than the increase of population.

If we take the official statistics of pauperism in England and Wales for the last thirty years, we find great fluctuations, but nothing like a regular diminution. Between 1849 and 1880 the numbers were lowest in the years 1853 and 1876–78, while they were highest from 1862 to 1873. The only years in which the numbers rose above a million were 1863–64 and 1868–71, and this was the very period when our commerce was increasing so rapidly as to excite the enthusiasm of our legislators, and when our prosperity was supposed to be greatest. The extremely irregular fluctuations of official pauperism render it possible almost always to choose some year, twenty, thirty, or forty years back, when it was higher than now, and thus show an apparent decrease of numbers; but if we take the whole period from 1849 to 1880 as one during which our commerce and wealth increased enormously, and all the industrial arts and means of communication made the most rapid strides, and take the average pauperism of the first twelve and last twelve of these years we find them almost exactly the same, thus—1849—1860, average paupers, 863,338; 1869—1880, average paupers, 864,398. Between

---

[1] As this has been denied without proof of its inaccuracy it will be well to quote the words of Mr. Joseph Kay, Q.C., author of *Free Trade in Land*, who says: "The French, the Dutch, the Germans, and the Swiss look with wonder at the enormous fortunes and at the enormous mass of pauperism which accumulate in England side by side. They have little of either extreme." And again he speaks of the astonishment of foreigners at "the frightful amount of absolute pauperism amongst the lowest classes."

the middle points of these two periods (1854 to 1874) the population increased about 23 per cent., and thus the proportional pauperism *appears* to have decreased considerably, though not at all in proportion to the increase of our aggregate wealth, which was at least doubled during the same period.

Several causes have, however, been in operation during this period which have led to the numbers of officially recorded paupers forming a less and less adequate indication of the total mass of pauperism in the country, so that even the small comfort derived from its supposed decrease in proportion to population may be denied us. In the first place there can be no doubt that the extent and efficiency of private charity all over the country have been steadily increasing, and that by its generous aid, large numbers have been saved from becoming paupers. Not only have old charities been better administered, but many societies have been formed for the systematization of private charity; while all over the country the clergy, and an ever increasing army of lady visitors, have aided the poor with advice and timely relief. It is impossible to estimate the amount of these various agencies, but it seems not impossible that they may have relieved the ratepayers from an amount equal to that due to increase of population during the same period; and this is the more probable when we consider the enormous increase of the wealthy middle class, and the increasing feeling that the poor have some moral claim upon the rich, leading to more and more liberality in every case of undeserved misfortune.[1]

But yet more powerful agencies have been at work tending to decrease the numbers of official paupers without any corresponding decrease of poverty and pauperism. For many years there has been a growing disposition to diminish out-door relief, and apply more generally the "workhouse test" which is the fundamental principle of our poor law. It is well known that there is such a wide-spread dislike and dread of the workhouse among the more respectable poor that many will rather starve that enter it, and as a

---

[1] Some of the evidence on this point has been given in Chapter XII. page 213), but more fully in my *Wonderful Century*, Chapter XX.

matter of fact many *do* starve who might have been well fed within its walls. Now, in the *Daily News* of April 18th, 1883, there was a remarkable article giving an account of the results of this change of system in some London parishes. It states that, ten years earlier, a severe reduction of out-door relief was commenced in Whitechapel and other parishes, till at the time of writing there was no out-door relief given in that parish, nor in Stepney and St. George's in the East, while the same process was going on in Marylebone and other parts of London, and to a less extent all over the kingdom. The article in question states the remarkable fact that this great change had produced practically *no increase of indoor paupers*. In Stepney, for example, out-door relief had been reduced by 7,000 in the preceding ten years, and there was no increase whatever of indoor paupers, the reason being (as expressly stated) that *organized private charity had taken the place of out-door relief.* Now, if there has been a reduction of 7,000 official paupers in one London parish without any proof of a corresponding decrease of real want and destitution, how utterly unmeaning and even misleading becomes the quotation of these official statistics as showing any real decrease of our pauperism. If we take as our guide the fact, that, in one of the worst and most poverty-stricken districts of the metropolis, organized private charity has been able to take the place of the relieving officer when a *total* cessation of out-door relief has been effected, we may be sure that it has been found quite equal to the much easier task of relieving those thrown on its hands by a very partial application of the same methods of dealing with the poor in other parts of the country. We may, therefore, fairly assume that the diminution of out-door paupers over the whole country during the last thirty years has been largely due to the stricter application of the workhouse test, and that those thus refused relief by the guardians have been aided and kept alive by more extensive and better organised private charity. If this is the case, the only official test of pauperism as actually increasing or decreasing will be found in the records of indoor relief, and these show numbers steadily increasing at a much greater rate than population! Thus in the thirty years from 1852 to 1882 the number of indoor paupers in England and Wales has continuously increased, from 106,413 in the former year to 188,433 in the latter, an increase of 83 per cent., while in the same period population only increased 45 per cent. The plain inference is, that confirmed pauperism—that which includes all the most degraded and the most hopeless of our poor—had been steadily increasing at a greater rate than our population, during a period in which our aggregate wealth had been doubled, and our commerce, of which we are so proud, had increased three-fold.

Before quitting this subject, it is well to point out that the way in which the number of paupers is estimated is most misleading, and gives no adequate idea of the real numbers. The tables show only the numbers relieved on the 1st January in each year, but it is estimated that the actual number of persons receiving relief during the year is nearly two and a half times this number, or about an average of two millions [1] for England and Wales, or two and a half millions for the United Kingdom. If we add to this latter number those who receive relief in the casual wards (which are not included in the official tables), and the very large numbers who depend wholly or partially on private charity for support, we shall perhaps bring the figures up to three and a half millions. But beyond this number of actual paupers loom a vast host of the poor who ever live on the verge of pauperism, and from whom the ranks of the actual paupers are constantly recruited, including whole populations, like the cottiers of the west of Ireland, and the crofters of the Highlands and Islands of Scotland, living in such a condition of perennial want that it only requires that most certain of periodical events —a bad season—to produce actual famine. If we add only one million for all these, we bring up the number of actual or potential paupers in this civilized, Christian and pre-eminently wealthy country to about four millions and a half, or about one in seven of the whole population!

This dreadful failure to distribute among our workers

[1] For details of this estimate see the present writer's *Land Nationalization, its Necessity and its Aims*, p. 3.

with any approach to fairness the enormous wealth which they alone produce, is rendered more disgraceful when we take account of the vast extension of labour-saving machinery during the epoch we are considering.

It is calculated that we now possess steam-engines of about ten million horse power, equal to a hundred million men always working for us. Reckoning six million families in the United Kingdom, we may say that every family has the equivalent of sixteen hard-working slaves, who are never idle and can always do a full day's work. What ought to be the result of all this labour, in addition to the grinding toil of all our working men and women? Should we not expect abundance of food and clothing for all, and ample leisure for the cultivation of the mind, and the enjoyment of the beauties of nature and of art? Instead of this, we have wide-spread, ever-present pauperism; crowded cities reeking with squalor, filth, drunkenness, and vice; a depopulated country; and, as a direct consequence of these two factors—streams polluted with wasted fertilising matter, destroying at once natural beauty, valuable fish-food, and human life. Everywhere we find wealthy people enjoying all the luxuries and refinements of a high civilization; but amidst them we also find masses of human beings living more degraded lives than most savages, and working harder and more continuously than most slaves.

In our preliminary inquiry we have shown that some such result as this *must* arise from absolute private property in land. It is surely a remarkable coincidence (if it be only a coincidence) that these results should actually occur, in such extreme and painful development, in the country where land is concentrated in the fewest hands, where the legal rights of landlords are the most absolute, and where, owing to the enormous aggregation of wealth, the divorce between those who own and those who cultivate the soil is the most complete. Let us endeavour to throw further light upon this question by an examination of the effects of our system in the special cases of Ireland and Scotland, in which the facts are both undisputed and easily accessible,

Landlordism in Ireland.

In Ireland we have the spectacle of landlords doing what they like with their own for three centuries, backed up by a landlord parliament which made any laws they thought necessary; and the result has been a country in continual rebellion and a people ever on the verge of starvation.

This chronic starvation has been imputed to any and every cause but the real one—to over-population, to idleness, to potatoes; the real and all-sufficient cause being that the mass of the population are crowded on small and utterly insufficient holdings of the worst lands at extravagantly high rents, which means that everything they raise besides enough potatoes to support life goes as tribute to the landlords.

Under such conditions, no population however limited, no industry however great, no agriculture however perfect, no soil however fertile, could save a people from poverty and recurring famines.

As Mr. De Courcy Atkins well puts it:

"Less than 2,000 persons own two-thirds of the land in Ireland, and out of its five or six million inhabitants there is no man of those who have tilled it and given it all its present value who owns one sod of its soil. For the land owned by these two thousand persons, many of whom are absentees, five hundred thousand families are competing, as the sole stay between them and starvation." (*The Case of Ireland Stated*. 1880.)

The Devon Commission, published in 1847, declared authoritatively that in Ireland everything on the land which gives it value—houses, buildings, fences, gates, drains, &c., had been made by the tenants, and were undoubtedly their own property; yet from that day onward for many years our Parliament allowed and even encouraged the Irish landlords to rob the tenants of this property by forms of law, and thousands and tens of thousands of Irish tenants were robbed accordingly! Yet more. In the four years succeeding the great famine, there were over two hundred thousand evictions; whole town-lands were depopulated, and their human inhabitants driven off to make room for cattle and sheep—houses, schools, churches, everything being destroyed. The results of this are still to be seen over a large part of Ireland, where the traveller seems to be passing through a land bereft of human inhabitants, but marked by abundant ruins. The *Daily News* special commissioner, writing from Mayo, in October, 1880, said :

"Tradesmen, farmers, and all the less wealthy part of the community still speak sorely of the evictions of thirty or forty years ago, and point out the grave-yards which alone mark the sites of thickly populated hamlets abolished by the crowbar."

The lands thus cleared were let in blocks of several square miles each to English or Scotch farmers for grazing farms, in order, as he tells us, that landlords "might get their rents more easily and more securely," even though they were sometimes less than those paid by the former inhabitants. And what became of these inhabitants? Let the Devon Commission, appointed by Parliament, and consisting mostly of landlords, answer the question :— "It would be impossible for language to convey an idea of the state of distress to which the ejected tenantry have been reduced, and of the disease, misery, and even vice, which they propagated in the towns wherein they have settled; so that not only they who have been ejected have been rendered miserable, but they have carried with them and propagated that misery. They have increased the stock of labour, they have rendered the habitations of those who received them more crowded, they have given occasion to the dissemination of disease, they have been obliged to resort to theft and all manner of vice and iniquity to procure subsistence; but what is perhaps the most painful of all, a vast number of them have perished of want." [1] Now, consider these horrible results produced in four years to a million of people; consider further, that the same kind of eviction, with its consequent misery and vice, has been going on in Ireland in varying degrees down to our own times, and that all this untold wretchedness, this cruel, heart-rending wrong, this vice, and crime, and pauperism, this disease and death, have been caused —not in a great war between nations struggling for

[1] *Parl. Rep.* 1845, vol. xix., p. 19.

supremacy, not to maintain any great principle of religious or civil liberty, but, "in order that landlords may get their rents more securely and more easily!" And now, whenever the people of Ireland, crowded into towns and on the poorest lands of the west coast, are again starving, the only remedy our landlord-legislators can propose is to ship them off by thousands to other countries, and thus increase and intensify that widespread hatred of English rule which is the natural and just punishment we are receiving for persistent injustice. These deserted villages are not to be again repeopled; the cattle and sheep must be still allowed to displace Irishmen; the "easy collection" of the landlords' rents must on no account be endangered; let everything go on as before, and when our consciences or our fears are aroused by the cry of too many starving Irishmen, let subscriptions be got up, and let the English people be taxed to ship off a few thousands of surplus paupers to Canada or Australia, and all will be well!

Here we see pure landlordism having its own way, and working out its natural and inevitable results, in the extreme case of ownership of the soil of the country for the most part by absentees and by aliens in race and religion. About this there can be no dispute. And if the absurd and totally unfounded cry of over-population is always to be followed by more emigration, there can be no end to the process. For even were the population reduced to one million of Irish peasant cultivators, that million would continue in exactly the same condition of misery and destitution as the present population, if they were confined to limited areas and were subjected to the extortions of the agents of absentee or alien landlords. Even before the famine the exorbitant rents and high taxes were paid chiefly by means of exported *food*, showing that the land of Ireland was able to support many more than the eight millions which then inhabited it. Yet now, when the population has been reduced to less than five millions, the same cry of over-population is raised —as it was a century ago when there were only two millions; and whether there be two or five or eight millions in the country, there will certainly be starvation and local over-population if the people are forbidden the free use of their native land, are confined to the least productive districts and to insufficient holdings, and if all the surplus produce above a bare supply of potatoes in average years is exported to pay rent! That more starving Irishmen should be expatriated while millions of acres which they once tilled are given up to cattle and sheep, is the condemnation of landlord government. That the chronic famine which has prevailed in Ireland for a century should still devastate it, is the condemnation of landlordism itself.

Landlordism in Scotland.

Let us now turn to another country, where the landlord power has had complete sway for a century, unfettered by any of the difficulties which are often alleged as the reason for its terrible failure in Ireland. In the Highlands of Scotland there has been no religious difficulty, and there has been no antipathy of race; the people have not depended wholly upon potatoes, and the country has certainly never been over-populated. Neither has there been any rebellion against authority; but the universal testimony of all who know them best is, that in the whole British dominions there exists no more intelligent, religious, peaceable and industrious people than the Highland peasantry. Yet here too, under the most favourable conditions, we find perennial destitution and famine, and a series of Royal Commissions seeking out that which is plain as the sun at noonday—the causes of want and misery among the tenantry of an enormously wealthy, and in their own territories almost omnipotent, body of landlords. The causes are, simply, that the native inhabitants have been driven from the inland valleys to the sea-coast to make room for sheep and deer. The terrible history of the Highland clearances is too long to go into in this place. Suffice it to state that more than two millions of acres once inhabited by human beings are now devoted to deer only, from which the noble Highland chieftains or their successors get much sport or large rentals. Seventy men at this day own half Scotland, and if they choose to complete what they have begun, and turn more fields and meadows into hunting-grounds, our existing law permits them to do so; while no amendment of the law as yet proposed by Liberal politicians would place the slightest check upon this iniquitous power.

The reader who wishes to know how the brave Highlanders have been treated by those who owe everything to them, and who should have been their protectors —their hereditary chieftains—should read Mr. Alexander Mackenzie's interesting work *The History of the Highland Clearances*, or the outline of the main facts in my own work *Land Nationalization*. Let us now see what are the conditions under which the Highlanders live, and consider whether under such conditions anything but poverty, discontent, and famine is possible. We learn from the various reports that have appeared in the daily papers, confirming the testimony of all previous impartial writers, that the Highland crofters are confined to miserably small holdings—the largest croft in Skye, for example, being seven acres; that the land is poor and the rent very high; that the landlords have continually encroached on the commons and mountains the use of which for grazing is essential to the crofter's existence. These have been usually taken from them, without compensation, to make either large sheep-farms or deer-forests; and in many cases they suffer without redress from the incursions of the deer which eat their crops, while they are not allowed to keep dogs to mind their own sheep (when they have any) for fear of disturbing these sacred deer, whose well-being and due increase are carefully attended to, even though it entails starvation on men and women.

Then, again, these great estates, often as large as continental kingdoms or dukedoms, are managed by agents and factors who represent an unknown and unseen landlord, and who are really despotic rulers, carrying out their own decrees under the penalty of eviction—a penalty as severe as that imposed by the law of England on hardened criminals. It was stated in the *Daily News*, a paper which is celebrated for the careful accuracy of its information, that on one estate (a generation back) a whole body of crofters were removed because they had good land which the factor wanted: and this is the more credible because many other cases are recorded in which the factors take farms from which the former holders have been evicted. Under the rule of the factors the people may be oppressed and pauperized, even with the most benevolent of landlords. Take the case of the late Sir James Matheson, who, in the year 1844, bought the extensive island of Lewis (as large as an average English county) and who is universally admitted to have been personally most benevolent and liberal. Yet under his paternal government tenants were ejected at the will of the factor, and extensive tracts turned into sheep farms and deer forests, and such cruel injustice was perpetrated for years that the people at length rebelled, and then only did their landlord know they had anything to complain of. The *Quarterly Review* declared that Sir James Matheson made wonderful improvements in Lewis, " pouring out money like water," and spending over £100,000 there, besides giving largely in charity; and the result was that soon after his death there was famine in Lewis, and the representatives of this wealthy and benevolent landlord were obliged to beg for subscriptions in the city of London to save the people from starvation! Nothing is said about the sheep and deer of the island; no doubt they were fat and flourishing and gave handsome returns, whereas men and women were encumbrances and had to be kept alive by charity! What a cruel satire is this. An enormously rich country, which taxes its people heavily under the pretence that it dispenses justice and gives protection to all; which is highly civilised and highly religious; and which yet upholds a system under which large masses of its subjects have no right to live but by the permission of landlords and their irresponsible agents!

I cannot here go further into this distressing subject, but must again refer my readers to the easily accessible sources of information I have quoted. I will only give one passage from a writer of full knowledge and authority, Dr. D. G. F. Macdonald, to show that my conclusions are not the result of prejudice or imperfect knowledge:

"I know a glen, now inhabited by two shepherds and two game-keepers, which at one time sent out its thousand fighting men. And this is but one out of many that may be cited to show how the Highlands have been depopulated. Loyal, peaceable, high-spirited peasantry have been driven from their native land—as the Jews were expelled from Spain and the Huguenots from France—to make room for grouse, sheep, and deer. A portly volume would be needed to contain the records of oppression and cruelty perpetrated by many landlords, who are a scourge to the unfortunate tenants, blighting their lives, poisoning their happiness, and robbing them of their improvements, filling their wretched homes with sorrow, and breaking their hearts with the weight of despair." (*The Highland Crofters of Scotland.* 1878.)

Here, then, we have reviewed the results of our land system. Persistent pauperism in the midst of boundless wealth in England—largely due to the great farm system so dear to English landlords and agents, to the consequent driving of labourers to the towns to seek a subsistence, to the utter divorce of the labourer from any right in his native soil, and to land and building speculation, making it the interest of landlords and speculators that people should be driven to live crowded together in towns rather than be scattered naturally and beneficially over the country.

In Ireland we see agrarian war, chronic famine, and a degraded population, while by the cruel evictions and forced emigration, and the long-continued robbery of the Irish peasant's improvements, a deadly enemy to our country has been established in the United States.

In Scotland, a religious, patient, educated peasantry have been forcibly driven from their native soil, while those which remain are pauperised, discontented, and famine-stricken.

Beneficial Results of Small Holdings.

In the preceding pages I have given a brief outline of the effects of landlord rule in England, Ireland, and Scotland, and I cannot but express my amazement that a writer like Professor Fawcett, who must have been fully acquainted with the whole of the terrible facts I have here only been able to hint at—who nearly twenty years earlier wrote so strongly on the pitiable condition of the British labourer—a condition afterwards changed if anything for the worse, since he had become more than ever divorced from the soil and was tending to a nomadic life—and who has depicted so forcibly the land-hunger of the rich, should yet, in his latest teaching, have no other remedy to propose for the evils due to unrestricted private property in the first essential of human existence, the soil, on and by which alone men can live, than to make that property still more easily acquired by the wealthy and still more absolute, by means of complete free trade in land!

It has repeatedly been said, and will no doubt be said again, that many of the admitted evils here pointed out are not due to our system of landlordism, but to various other causes, such as improvidence, over-population, and idleness. To this I reply, that even where these causes do exist they are but secondary causes, and are themselves due to the landlord system. Improvidence is the inevitable result of insecurity for the produce of a man's labour; over-population is always local, and is produced by men being forcibly crowded together and not allowed freely to occupy and cultivate the soil; while idleness is the last accusation that should be made against any of the inhabitants of these islands, whose fault rather is a too great eagerness for work whenever they can work for fair wages or with full security that they will reap the produce of their labour. I reply, further, that the close correspondence between the theoretical and the actual results, between deduction and induction, must not be ignored. An irresistible logic assures us that the possession of the soil by a few must make those who have to live by cultivating the soil virtual slaves; that it must keep down wages and enable the landlords to absorb all the surplus wealth produced by the labourer beyond what is necessary for a bare subsistence; while it indisputably prevents all who are not landowners from any use or enjoyment of their native soil except at the landowner's pleasure—a fact everywhere visible in the unnatural mode of growth of our towns and villages, where, with ample land suitable for pleasant and healthy homes all round them, people are forced to live in close-packed houses with the view of nature's beauties shut out and much of the discomfort and insalubrity of large towns carried into the country. The best way, however, to prove that these and many other evils are directly due to landlordism is to show by actual examples how constantly they disappear whenever the land belongs to those who cultivate it. This is the case in many parts of Europe; and although climate, race, laws, and the character and habits of the people may widely differ, we always find an amount of contentment and well-being strikingly contrasted with what prevails under the system of landlordism.

As regards Switzerland, Sismondi, in his *Studies of Political Economy*, gives full information. He declares that here we see the beneficial results of agriculture practised by the very people who enjoy its fruits, in "the great comfort of a numerous population, a great independence of character arising from independence of position, and a great consumption of goods the result of the easy circumstances of all the inhabitants." Many other writers confirm the accuracy of these statements. The observant English traveller, Inglis, speaks of the wonderful industry of the Swiss, the loving care with which they tend their fields and fruit-trees, and the complete way in which all the peasants' wants are supplied from the land; while, as a rule, pauperism and even poverty is unknown in the rural districts where peasant or communal properties prevail. The common objection that small proprietors cannot use machinery or execute any improvements requiring co-operation, is answered by the examples of Norway and of Saxony. In the former country, Mr. Laing tell us how extensively irrigation is carried on by miles of wooden troughing along the mountain sides, executed in concert and kept up for the common benefit. The roads and bridges are also kept in excellent repair without tolls; and he considers this to be done because the people "feel as proprietors who receive the advantage of their exertions."

Mr. Kay, a most unimpeachable witness, tells us that there is no farming in all Europe comparable with that of the valleys of Saxon Switzerland. After giving a picture of the perfect condition of the crops, the excessive care of manure, and other details, he adds:

"The peasants endeavour to outstrip one another in the quality and quantity of the produce, in the preparation of the ground, and in the general cultivation of their respective portions. All the little proprietors are eager to find out how to farm so as to produce the greatest results; they diligently seek after improvements; they send their children to agricultural schools in order to fit them to assist their fathers; and each proprietor soon adopts a new improvement introduced by any of his neighbours."

The late William Howitt, writing on the rural and domestic life of Germany, says:

"The peasants are not, as with us, for the most part totally cut off from the soil they cultivate—they are themselves the proprietors. It is, perhaps, from this cause that they are probably the most industrious peasantry in the world. They labour early and late, because they feel that they are labouring for themselves. The German peasants work hard, but they have no actual want. . . . The English peasant is so cut off from the idea of property that he comes habitually to look upon it as a thing from which he is warned by the laws of the large proprietors, and becomes, in consequence, spiritless and purposeless. . . . The German *Bauer*, on the contrary, looks on the country as made for him and his fellow-men. No man can threaten him with ejection or the workhouse so long as he is active and economical. He walks, therefore, with a bold step; he looks you in the face with the air of a free man, but a respectful air."

And Mr. Baring Gould, although showing how poor the peasant of North Germany often is, owing to the miserable system of each farm being cut up into scores or hundreds of disconnected plots, and his cruel subjection to Jew money-lenders, nevertheless thus compares him with the journeyman mechanic:

"The artisan is restless and dissatisfied. He is mechanised. He finds no interest in his work, and his soul frets at the routine. He is miserable and he knows not why. But the man who toils on his own plot of ground is morally and physically healthy. He is a freeman; the sense he has of independence gives him his upright carriage, his fearless brow, and his joyous laugh."

In Belgium, the most highly cultivated part of the country is that which consists of peasant properties and even M'Culloch, the advocate of large farms, admits that,—

"in the minute attention to the qualities of the soil, in the management and application of manures of different kinds, in the judicious succession of crops, and especially in the economy of land, so that every part of it shall be in a constant state of production, we have still something to learn from the Flemings."

In France, though the farming may not be what we call good, the industry and economy of the peasant-proprietors is remarkable, while their well-being is sufficiently indicated by the wonderful amount of hoarded wealth always forthcoming when the Government requires a loan. The connection of peasant-cultivation with well-being is apparent throughout France. Sir Henry Bulwer remarks that by far the greatest number of the indigent is to be found in the northern departments, where land is less divided than elsewhere, and cultivated with larger capitals. Mr. Birkbeck, noticing that in one district the poor appeared less comfortable, found, on inquiry, that few of the peasants thereabouts were proprietors; while in Anjou and Touraine, Mr. Le Quesne noticed that the houses of the country people were remarkable for their neatness, indicative of the ease and comfort of their possessors, and on inquiry as to the cause, was told that the land was there divided into small properties. So when the celebrated agriculturist and traveller, Arthur Young, noticed exceptional improvement in irrigation and cultivation, he is so sure of the explanation of the fact that he remarks:

"It would be a disgrace to common sense to ask the cause; the enjoyment of property *must* have done it. Give a man secure possession of a bleak rock and he will turn it into a garden; give him a nine years' lease of a garden and he will convert it into a desert."[1]

It hardly needs to adduce more evidence to prove the intimate connection of the sense of secure possession with industry, well-being, and content. But we must briefly notice one more example at our very doors, and under our

[1] For these and many other examples see Thornton's *Plea for Peasant Proprietors*.

rule—the case of the Channel Islands. The testimony of all observers is unanimous as to the happy condition of these islands, and to its cause in the almost total absence of landlordism as it exists with us. The Hon. G. C. Brodrick, in his *English Land and English Landlords*, says:

"If we judge of success in cultivation by the produce, we find that a much larger quantity of human food is raised in Jersey than is raised on an equal area, by the same number of cultivators, in any part of the United Kingdom. Not only does it support its own crowded population in much greater comfort than that enjoyed by the mass of Englishmen, but it supplies the London market out of its surplus production with shiploads of vegetables, fruit, butter, and cattle for breeding. Even wheat, for the growth of which the climate is not very suitable, is so cultivated that it yields much heavier crops per acre than in England; and the number of live stock kept on a given area astonishes travellers accustomed only to English farming. Nor are these only the results of spade-husbandry, for machinery is largely employed by the yeomen and peasant proprietors of the Channel Islands, who have no difficulty in arranging among themselves to hire it by turns."

Mr. Brodrick, like every one else, attributes this wonderful success to the land system of the country.

Lest it be now said that there is something in the climate or soil of these various localities, or in the character or habits of the people to which these favourable results are to be imputed, I must refer to a few crucial examples in which every other cause but difference of land tenure is eliminated, and which therefore complete the demonstration to which this whole argument tends. The first of these is afforded by Italy, over large portions of which there are still, as in the time of the Romans, *latifundia* or large estates farmed by middlemen, and cultivated by labourers or tenants at will. In his volume on *Primitive Property* the great economist, M. de Laveleye, speaks of the

"naked and desolate fields, where the cultivator dies of famine in the fairest climate and most fertile soil; such is the result of the *latifundia*. Economists, who defend the system of huge properties, visit the interior of the Basilicata and Sicily if you want to see the degree of misery to which your huge properties reduce the earth and its habitants."

Yet in the same country and under the same laws, wherever fixity of tenure or peasant properties exist, the utmost prosperity prevails. Again, let us hear M. de Laveleye:

"I know of no more striking lesson in political economy than is taught at Capri. Whence come the perfection of cultivation and the comfort of the population? Certainly not from the fertility of the soil, which is an arid rock. . . . Before obtaining the crops, it was necessary, so to speak, to create the soil. It is the magic of ownership which has produced this prodigy."

Now let us come back to our own country, where we shall find that exactly similar results are produced by similar causes. On the Annandale Estate in Dumfriesshire, plots of from two to six acres were granted to labourers on a lease of twenty-one years. They built their own cottages, having timber and stone supplied by the landlord, and these little farms were all cultivated by the labourer's family, and in his own spare hours. Now note the result. Among these peasant-farmers pauperism soon ceased to exist, and it was especially noticed that habits of marketing, and the constant demands on thrift and forethought brought out new powers and virtues in the wives. In fact, the moral effects of the system in fostering industry, sobriety, and contentment, were described as no less satisfactory than its economical success.

Again, on Lord Tollemache's estate in Cheshire, plots of land from two-and-a-half to three-and-a-half acres are let with each cottage at an ordinary farm rent, and the results have been eminently beneficial. It is remarked here too that the habits of thrift and forethought encouraged by cow-keeping and dairying, on however small a scale, constitute a moral advantage of great importance.

One more example must be given to show that even in Ireland the laws of human nature do act the same as elsewhere. Mr. Jonathan Pim, in his *Condition and Prospects of Ireland*, gives an account of how the rugged, bleak, and sterile mountain of Forth, in Wexford, is sprinkled with little patches of land, many of them on the highest part of the mountain, reclaimed and enclosed at a vast expense of labour by the peasant-proprietors, who have been induced to overcome extraordinary difficulties in the hope of at length making a little spot of land their own.

"The surface was thickly covered with large masses of rock of various sizes, and intersected by the gullies formed by winter torrents. These rocks have been broken, buried, rolled away, or heaped into the form of fences. The land when thus cleared has been carefully enriched with soil, manured, and tilled. These little holdings vary from half an acre to ten or fifteen acres. The occupiers hold by the right of possession; they are generally poor; but they are *peaceable, well-conducted, independent, and industrious; and the district is absolutely free from agrarian outrage.*"

A volume might be filled with similar cases, but more are unnecessary here, for the evidence already adduced or referred to is absolutely conclusive. Wherever there are great estates let on an insecure tenure, we find in varying degrees the evils here pointed out. On the other hand, wherever we find men cultivating their own lands, or lands held on a permanent tenure at fixed rents, we find comparative comfort, no pauperism, and little crime. And as this is exactly what a consideration of the immutable laws of human nature and economic science has demonstrated must be the inevitable result, we have fact and reasoning, induction and deduction supporting each other.

The Remedy.

Having now cleared the ground by an inquiry into principles and a survey of the facts, we come to the practical question—Can any adequate remedy be found for these widespread and gigantic evils?

The common panacea of the Liberal party, "Free-trade in land," must surely now appear to my readers to be ridiculously inadequate. It was tried in Ireland by the Encumbered Estates Act, and it so aggravated the disease that a Liberal Government has now been forced to stop free-trade in land altogether in Ireland, and fix rents by act of parliament! It has always prevailed in America, yet many of the evils of land monopoly are very great even there. It exists in Italy, yet great estates prevail, and the tiller of the soil starves in the midst of abundance as with us. It will do nothing for the poor evicted crofters or the famine-stricken cottiers of Ireland. It will not cause the tracts now occupied by sheep and deer to be given up again for the use of men and women; but it will allow rich men, more easily than now, to make more deer-forests and sheep-farms if they choose. It will not help the labourer, the mechanic, or the shopkeeper, to a plot of land where he requires it. It will not give back the land to the use of the people who want it most, and who, as the universal experience of Europe shows, are always benefited by it both physically and morally. Let us then appeal to first principles and simply follow their teaching.

We have seen (1) that private property in *land* cannot justly arise at all; and (2) that its results, except where small portions are personally occupied and tilled, are always evil. Hence we conclude that the land of a country should be the property of the State and be free for the use and enjoyment of the inhabitants on equal terms. In order that every one may feel that sense of property in the land he cultivates which is the best incentive to industry, absolute security of tenure is necessary. This given, a man becomes virtually the owner of the land he holds from the State, and can deal with it like a freehold, only that it remains subject to such ground-rents and such general conditions as may from time to time be held to be for the good of the community. Another important principle is, that sub-letting must be absolutely prohibited; for if this were allowed the evils of landlordism would again arise, as middlemen would monopolize large quantities of land which they would let out at advanced rents and under onerous conditions, so that the actual cultivators might be no better off than under the present landlord system.

Recurring again to first principles we find, that although *land* itself cannot justly become private property, yet everything added to the land by human labour is truly and properly so; and this leads to the important subdivision of landed property into two parts (as is so common in Ireland), the one represented by the landlord's rent for the use of the bare land, the other the *tenant-right*, consisting of the houses, fences, gates, plantations, drains, and other permanent and tangible improvements which are there always made by the tenants. Now these improvements, which for purposes of sale or transfer may here also be conveniently termed *tenant-right*, should always be the absolute property of the occupier of any plot of land, the State or municipality being the owner of the bare land only; and by this simple and logical division it will be seen that all necessity for State *management*, with the long train of evils which Professor Fawcett so properly emphasizes, is absolutely done away with, and the cultivator may be left perfectly free to treat his estate as he pleases. For everything on the land which can be deteriorated by bad farming or wilful neglect is his private property, and its preservation may safely be left to the influence of self-interest; while the *land*, which is the property of the State, is practically incapable of deterioration; for its value depends on such natural causes as geological formation, arterial drainage, aspect, rainfall, and latitude, and on such social conditions as density of population, nearness to towns, to seaports, to railroads, or canals, the vicinity of manufactures, &c., none of which can be changed by the action of the tenant. The State, therefore, will have nothing whatever to *manage*, but need only collect its ground-rent as it now collects the land-tax or house-tax, leaving every land-holder perfectly free to do as he pleases, and only interfering with him by means of general enactments applicable to all holders alike. All arrangements that may be necessary for facilitating the acquisition of land by those who need it, should be in the hands of Local Land Courts, acting on principles determined by general enactment.

Having thus given the main outlines of a just and beneficial system of land tenure, let us consider briefly how to bring it into practical operation. And, first, we will explain how existing landlords may be equitably dealt with, as this is considered by many to be the real difficulty in the way of Land Nationalization.

In Mr. Gladstone's proposed scheme for buying out the Irish landlords the principle was laid down, and never controverted, that the landlords were entitled to compensation for the *net incomes* they derived from the land, and that the amount of government land-bonds given them in payment, at 20 or 25 years' purchase, was to be estimated on this *net income*, not on the *judicial rental* payable by the tenants. The difference between these consists of cost of agency, law expenses, and bad debts, and is estimated at from 20 to 30 per cent. of the judicial rents. It was this large margin which enabled Mr. Gladstone to offer the tenants the means of purchasing their farms by means of *terminable rentals* actually lower than the present judicial rents, which should yet leave a considerable surplus after paying the interest on the landlord's bonds, by means of which these bonds might be extinguished before the terminable rentals came to an end. This same principle is applicable in Great Britain, for the recent reports of the Agricultural Commission show that large English Estates cost from 15 to 30 per cent. for management, while as much more is often laid out for improvements. The net revenue will therefore be, in every case, very much below the gross rental of the farms.

But according to our proposals, all these costs of management will be saved when the State takes the lands, because we insist that the tenants must always become the owners of the buildings, fences, roads, drains, &c.—all in fact which can be destroyed, removed, or deteriorated, while he remains a tenant of the State for the bare land. This bare unimproved land will want no "management;" the tenant will be left perfectly free to do what he likes with it, so long as he pays the rent punctually into the proper office : and he will be sure to do this because, if he fails he will be liable to ejectment, and to have the house and improvements, which are his own property, sold. The improvements will therefore be a security for the payment of the rent, just as the house and premises in a town are a security for the payment of the ground rent. It is thus clear that, even in the case of agricultural land, the State could safely pay the landlord a fair price for his bare land (the tenants paying the landlord for any improvements made by him) and could still, with the surplus rent received from the tenant, redeem a number of the land-bonds every year. Taking the difference between the fair rental payable by the tenant and the net rental received by the landlord at 20 per cent., this difference will serve to pay off the whole purchase-money in 56 years, even if there is no increase whatever in the value of the land. But as large quantities of agricultural land are required every year for extension of towns and villages, for manufacturing purposes, for market gardens, &c., the much higher rents obtained for this portion of the land will render the transaction a still more secure one, and the period of repayment shorter.

With regard to town and building lands the conditions are much more advantageous, because they show a continuous rise in value, even in times of depression, owing to increase of population. It is calculated that this increase for the last forty years has been at the average rate of 2 per cent. per annum for all the land of the kingdom, agricultural and urban, and as we have allowed for no increase in the value of agricultural land we may fairly take the increase of the urban land, at 3 per cent. But an increase of only 2 per cent. per annum will suffice to pay off the whole of the ground rents in 40 years, while if the increase were 3 per cent. the repayment would be effected in a much shorter period.

When this was done, the whole of the revenues derived from the land of the country would be available in lieu of taxation, and would probably suffice for all the legitmate purposes of the community, both imperial and local; but there is no reason why this great boon should be reserved for our successors alone. The fair plan would be to devote one half of the surplus rents to relief of taxation, the other half to repayment of purchase money by extinguishing a portion of the land-bonds. This plan would have the double advantage of immediately reducing taxation, and effecting this by moderate steps annually so as not to cause too great a disturbance to trade.[1]

[1] This is an alternative proposal to that formulated in Chap. XVI. It is here suggested because the principle has been adopted in Ireland, but the present writer holds that the former proposal is the most sound, and would be the most beneficial.

How the Land is to be Distributed.

Having thus shown how the land may be acquired by the State for the use of the community without cost or risk, we may proceed to consider how it may best be used for the benefit of all; and here we shall be able to answer Professor Fawcett's questions (which he seemed to think unanswerable)—"What principles are to regulate the rents to be charged? Who is to decide the particular plots of land that should be allotted to those who apply for them?" The answer to both is easy. Rents will be fixed in the first place by official valuation, following the precedent of the Irish Land Act; afterwards, probably, by free competition. As to *who* is to have land, and what particular plots of land, it is essential that there should be the greatest freedom of choice compatible with the just rights of existing occupiers. How these two important matters may be settled I will now briefly indicate.

It will first be necessary to determine the value of the improvements on the land as distinguished from that of the land itself, and to facilitate subdivision or rearrangement of farms or holdings. This should be done for each separate inclosure shown on the large-scale ordnance maps. Some general principles being laid down for the guidance of the valuers there will be no real difficulty in making the separation. An old pasture field in which the hedges and gates have been constantly repaired by successive tenants may be considered to be, so far as the landlord is concerned, in an unimproved state. Here the whole value will be *land value*. From this as a datum the separation of *improvements* will take place where the landlord has recently put new gates or has drained, or has built sheds, bridges, or farm buildings. The valuation, when complete, will show the annual value of the *land* for the State *ground-rent*, the present *annual* value of the improvements, and the present *purchase* value of the improvements to a tenant calculated on a scale determined by their quality and probable duration.

This official valuation being made, it would be only fair that the existing occupier of any farm or other land should have the first offer of it under the new conditions; these being that he should become the owner of the improvements and agree to pay the State ground-rent. If he has capital he can purchase the improvements by a cash payment, but if not he should be entitled to purchase them by means of a terminable rental, as in the case of purchases under the Irish Church Act. This would be done through the Land Courts, which would decide on the annual payments to be made and the period for which they are to run, so as to meet the views of both parties. It is the opinion of good authorities that most farmers now hold too much land in proportion to their capital, and that with perfect security of tenure and absolute freedom of action they would reduce their holdings in order to farm more highly and to be able to effect permanent improvements. Some farms would therefore be divided, and remote fields be detached from others, and these would afford land for small holdings or for gardens and fields for labourers.

But much more than this is needed. The crofters and cottiers who have been ejected from their homes, the labourers who have been driven into towns, and all who have been robbed of their ancient rights by the inclosure of commons, require immediate redress. We have seen what beneficial results invariably follow the grant of small plots of land at fair rents and on a secure tenure, and Nationalization would not deserve the name did it not place this boon within the reach of all who desire it. There is no privilege so beneficial to all the members of a community as to have ample space of land on which to live. Surround the poorest cottage with a spacious vegetable garden, with fruit and shade trees, with room for keeping pigs and poultry, or cows, and the result invariably is untiring industry and thrift, which soon raise the occupiers above poverty, and diminish, if they do not abolish, drunkenness and crime. Every mechanic and tradesman should also be able to obtain this great benefit whenever he desired it; and this is far too important a matter for the whole community to be left to the chance of land being offered for sale when and where wanted. It is not sufficiently recognized that the use of land for the creation of healthy and happy homes is far higher than its use as a mere wealth-producing agent, in which latter aspect alone it has hitherto been chiefly viewed. To get the greatest benefit from the land of a country it is essential that every inhabitant should be, as far as possible, free to live where he pleases; and to attain this end the right to hold land for *profit* should always be subordinate to the right to occupy it as a *home*.

To carry these principles into effect, and to allow population to spread freely over the whole country, it is essential that all who desire a permanent home should have a right of free selection (once in their lives) of a plot of land for this purpose. A limit might be placed to the quantity so taken in proportion to the density of the population—near towns perhaps half an acre, in the country an acre or more. Such choice should of course be limited to agricultural or waste land, and, at first, to such land as borders public roads. Other limitations might be, that not more than a fixed proportion of any one farm should be so taken, and that a plot should never be chosen so near the farmer's house as to be an annoyance to him—questions to be decided by the Local Land Courts. Of course this land would be subject to the usual payment of ground-rent to the State according to the official valuation, which should always be a low one, while the improvements would have to be purchased from the farmer with some small addition as compensation for disturbance.

The effects of such freedom of choice in fixing upon a permanent residence would be gradually to check the increase of towns and to re-populate the country districts. Rural villages would begin a natural course of healthy growth, and if the minimum of land to be taken for one house were fixed at an acre (the maximum being four or five acres) these could never grow into crowded towns, but would always retain their rural character, picturesque surroundings, and sanitary advantages. The labourer would choose his acre of land near the farmer who gave him the most constant employment and treated him with most consideration; and besides those who would continue to work regularly at agricultural labour, there would be many with larger holdings or with other means of living, who would be ready to earn good wages during hay or harvest time. With a million of agricultural labourers, each holding an acre or more of land, and at least another million of mechanics doing the same thing, and all permanently attached to the soil by its secure possession, that scandal to our country, the scarcity of milk and the importation of poultry, eggs and butter from the Continent would come to an end, while the vast sums we now pay for this produce would go to increase the well-being, not only of the labourers themselves, but of all the retail and wholesale dealers who supply their wants. Our most important customers are, or should be, those at home, and there is no more certain cure for the almost chronic depression of trade than a system which would at once largely increase the purchasing power of the bulk of the community.

Before concluding this chapter, I would wish to point out how easily the principles of land tenure here advocated may be tried on a small scale without interfering with any private rights or interests; and so convinced am I of the soundness of these principles, that I would venture to stake the whole question of the practicability of land nationalization on the result of such a trial. I would suggest, then, that all Crown lands in any degree suitable for cultivation should be thrown freely open to applicants in small holdings for personal occupation, on the tenure which I have just explained; and I would earnestly press some Liberal member of Parliament to urge this trial on the Government by means of an annual motion. The result would certainly be a large increase of revenue from these lands, since all expenses of management would be saved; while it is equally certain that the localities would be benefited by the increased well-being of the inhabitants.

In view of such a trial being made and its further extension being desirable, a resolution should be passed declaring it inexpedient to sell any Crown lands or rights over commons; and the next step should be to stop entirely the further inclosure of common lands for the benefit of landlords, a proceeding which the Liberal portion of the community has long condemned as legalised robbery of the people. Many of the more extensive commons and heaths far removed from dense centres of populations offer the means for a further trial of this system of land-tenure, thus creating a considerable body of virtual peasant-proprietors of the best type. For this purpose all manorial rights of individuals should be declared to be (as they certainly are) injurious to the public, and should be at once acquired by the State. Their present owners might either be repaid the purchase-money if they had themselves bought them, or be compensated by means of terminable annuities of amounts equal to the actual average net incomes derived from the several manors. Thus would be offered ample means for a great social experiment, the result of which, if fairly tried, cannot be doubtful.

CHAPTER XVIII

HERBERT SPENCER ON THE LAND QUESTION:
A CRITICISM

ALL my readers know the name of our great philosophic thinker and writer, Herbert Spencer, but they are perhaps not aware that to him is primarily due the formation of the Land Nationalization Society. In 1853, soon after I returned from my travels in the Amazon Valley, I read his book on *Social Statics*, and from it first derived the conception of the radical injustice of private property in land. His irresistible logic convinced me once for all, and I have never since had the slightest doubt upon the subject. He taught me, that "to deprive others of their rights to the use of the earth is to commit a crime inferior only in wickedness to the crime of taking away their lives or their personal liberties;" and when he added, that however difficult it might be to find a practical means of restoring the land to the people, yet "justice sternly commands it to be done," a seed was sown in my mind which long afterwards developed into that principle of the separation of the inherent value of land from the improvements effected in or upon it, which was the foundation of the proposals in my article "How to Nationalize the Land" (see Chapter XVI.), and this article led to my association with Mr. Swinton, Dr. Clark, M.P., and other friends in the formation of the Land Nationalization Society. In one of his latest works, however, entitled *Justice*, and forming part of his *Principles of Ethics*, Mr. Spencer repudiates his legitimate offspring—Land Nationalization—

and for various stated reasons arrives at the conclusion that, though it may be, and is, right in principle, there are insuperable difficulties in putting it into practice, and that therefore "individual ownership subject to State-suzerainty should be maintained." This, of course, will be seized upon by our opponents as a great triumph for the cause of landlordism, though as yet they seem hardly to have realized that a Daniel has come to judgment in their behalf. But we must always remember that they mostly belong to what has been termed "the silly party," and that they cannot therefore be expected to read works on high philosophy. Land Nationalizers, however, who have long quoted, and will continue to quote, from *Social Statics*—not because the book was written by Herbert Spencer, but because it was among the earliest and the most forcible of the arguments against private property in land—are bound to show that the philosopher has not refuted his own work, and that it is his later and not his earlier writings that are illogical, and are even inconsistent with the main principles of his own philosophy.

And first let us see what he still admits. After showing how land-ownership has been derived from conquest or usurpation, and that all the land originally belonged to the Crown as representing the whole nation, he says:—

"If the representative body has practically inherited the governmental powers which in past times vested in the king, it has at the same time inherited that ultimate proprietorship of the soil which in past times vested in him. And since the representative body is but the agent of the community, this ultimate proprietorship now vests in the community."[1]

And he then remarks that even the Liberty and Property Defence League admit this, saying in their Report of 1889 that:

"The land can of course be resumed on payment of full compensation, and managed by the people if they so will it."

In another place Mr. Spencer states, as proving the

[1] *Justice*, p. 92.

spread of more correct ideas of justice, that the truth has now come to be recognized, that—

"private ownership of land is subject to the supreme ownership of the community, and that therefore each citizen has a latent claim to participate in the use of the earth."[1]

So far, then, Mr. Spencer and the Liberty and Property Defence League are perfectly in accord with us; but thenceforth we diverge. They believe and maintain that this latent claim of the people to the full and equal use of their native soil shall and will remain latent. We, on the other hand, believe and are determined that it shall now become an active claim, and very soon a realized possession.

Now let us see what are Mr. Spencer's grounds for believing that land must, at all events in the immediate future, remain private property. It is as follows:

"All which can be claimed for the community is the surface of the country in its original unsubdued state. To all that value given to it by clearing, fencing, draining, making roads, farm-buildings, &c., constituting nearly all its value the community has no claim. . . . All this value, artificially given, vests in existing owners and cannot without a gigantic robbery be taken from them. If, during the many transactions which have brought about existing landownership, there have been much violence and much fraud, these have been small compared with the violence and frauds which the community would be guilty of did it take possession without paying for it, of that artificial value which the labour of nearly two thousand years has given to the land."[2]

This is all that Mr. Spencer has to say on the question in the body of the work, and before going on to consider his further discussion of it in an appendix, we must just notice the gigantic and almost incredible misstatement, that the improvements such as he specifies, made upon the land by human labour, constitute "nearly all its value!" Setting aside houses, fences and things of like character, which we have always recognized as being personal property to be purchased at fair value by the new occupiers, how much is the soil of England on the whole better for agricultural purposes than it was one

[1] *Justice*, p. 152.   [2] *Justice*, p. 92.

hundred, or five hundred, or a thousand years ago? In all probability it is not ten per cent. better, because, though limited areas have been greatly improved, very large areas remain quite unimproved, and other large areas have been decidedly made worse. More than half the whole area of the country is permanent pasture which has been mown or grazed from time immemorial, and is probably no better and no worse than it was a hundred or five hundred years ago. But every one with an observant eye may notice all over the country poor weedy pastures bearing the ridge marks of former cultivation. These were once old pastures, broken up when wheat and rents were high, and afterwards left to return as they could to the poor weedy land we now see. This land has been positively deteriorated; and besides this, much of our farming is still so bad under yearly tenancies that a large part of the arable land is partially worn out, and is probably no better if it is not worse than five hundred years ago when we not only grew all the wheat we required but exported to the Continent.

But Mr. Spencer's chief error consists in the latent assumption that increased *value* of land implies *improvement* in the soil, ignoring altogether that this increase is almost wholly due to the growth of population and improved means of communication. Let us take as an illustration the land around London. The late J. C. Loudon, the celebrated gardener and agriculturist, came to London from Scotland about the year 1804, and found the land to the west of the city—now occupied by the suburbs or by market gardens—let in small farms at 10s. or 12s. an acre. Now, probably, the lowest rent is as many pounds as it was then shillings, while much that is built over brings a hundred times the rent it did then; but that is not owing to any improvement of the soil itself, but wholly, or almost wholly, to railroads and to the consequent growth of London. Again, portions of the New Forest have remained wholly unimproved since the time of the Norman kings, yet if any of this unimproved land were for sale it would probably fetch a higher price than the best agricultural land in the kingdom if situated in a worse or less attractive locality. But neither these well-known facts, nor the other fact that, whatever improvement there is in the land itself has mostly been effected, not by the landlords but by successive generations of tenants, are much to the purpose; because, as Mr. Spencer truly states, all the value, by whomsoever created, now vests in the landlords. This has been recognized, and is still recognized, by the law, and by a large preponderance of public opinion, and therefore, I admit, as I think do most land-nationalizers, that it must not be taken from existing owners without reasonable and equitable compensation; and as Mr. Spencer thinks such a vast transaction to be financially impossible, he concludes that "individual ownership must be maintained."

Before showing how superficial, illogical, and unjust is such a conclusion, we have to note the extraordinary argument set forth in his Appendix B. He there says:—

"Even supposing that the English as a race gained possession of the land equitably, which they did not; and even supposing that existing landowners are the posterity of those who spoiled their fellows, which in large part they are not; and even supposing that the existing landless are the posterity of the despoiled, which in large part they are not; there would still have to be recognized a transaction that goes far to prevent rectification of injustices."[1]

And what do you think this "transaction" is? I would give every one of my readers who is not familiar with the work we are discussing half-a-dozen guesses each, and probably not one person would hit upon this stupendous "transaction" which, apparently, in Mr. Spencer's opinion, settles the land question, and forbids all future generations of Englishmen from possessing their native land, the soil of which is to remain the absolute property of existing landlords—their heirs, administrators, or assigns, as the lawyers say—for ever. In no other way can I interpret the terrible dictum that it "goes far to prevent the rectification of injustices." This momentous transaction is nothing but our old and too familiar friend, the Poor-rate!

Now for the solution of the problem. Mr. Spencer

[1] *Justice*, p. 268.

gives elaborate figures to show that the amount of that portion of the poor-rate contributed by the land during the last two-and-a-half centuries amounts to about 500 millions, and this he thinks is more than the "prairie-value" of the whole of the land! He says:

"Thus, even if we ignore the fact that this amount, gradually contributed, would, if otherwise gradually invested, have yielded in returns of one kind or another a far larger sum, it is manifest that against the claim of the landless may be set off a large claim of the landed—perhaps a larger claim."

Here is a turning of the tables with a vengeance! If this is the true state of the case we had better at once present a humble petition to the landlords, praying that they will let us off whatever balance may be due to them, on our undertaking to pay all the poor-rates for the future. For, if Mr. Spencer's reasoning is sound, this is what, in his own words, "Equity sternly commands should be done." But, first, let us look a little closer at this "new way to pay old debts." Mr. Spencer says, that " if we are to go back upon the past at all, we must go back upon the past wholly." These are his own words. Let us then do so, and what shall we find ? We find that the landlords have, century by century, continuously evaded or thrown off the burdens and duties which appertained to their original tenure of the land. The whole costs of the maintenance of the crown, of the army and navy, of the church, and of the poor, were payable by the landlords or by the proceeds of land which they have stolen from the church and from the people. We find also that the tenants on their estates had originally rights of possession similar to their own, on performance of specified duties. We find that in the time of the Tudors, they themselves created the very pauperism that has been handed down to us, by over-riding those rights. They carried out wholesale evictions of the cultivators of the soil, because the high price of wool rendered the turning of arable land into pasture profitable to them.[1] Again we find

[1] See Mr. Joseph Fisher's *History of Landowning in England* for authentic records of these cruel evictions and their consequences. Also Greene's *Short History of the English People*, p. 320.

that the vast estates of the abbeys and monasteries whose inmates had educated the people and relieved the poor, were absorbed by them, often for no services at all, often for disgraceful services. We find a little later, in 1692, that the remnant of their feudal duties was, with their own consent, commuted into "a tax of 4s. in the pound on a rack-rent without abatement for any charges whatever;" and we find that the valuation made at that date, which we may be sure was a low one even then, has been fraudulently maintained by a landlord parliament to this day, notwithstanding the increase of land-values to manyfold its amount at that date. Yet again we find that even during the past century about three millions of acres of common lands have been most inequitably enclosed and divided among the landlords, thus robbing the people of the last remnant of their rights to their native soil, and creating more pauperism. And lastly, it must be remembered that pauperism itself has been a direct benefit to the landlords, inasmuch as the poor-rates were once openly, and are still actually, "relief in aid of low wages" paid by all classes of the community, and which enabled, and still enable, farmers to get cheap labour and landlords higher rents. Under these circumstances, and remembering all these iniquities of the past, even landlord assurance will probably recoil before making the claim Mr. Spencer suggests, that payment of poor-rates since 1630 is really a re-purchase of the land from the people!

But in all this discussion and in much more of a like kind that I have neither time nor inclination to notice, Mr. Spencer misses the real point at issue. It matters not to us, now, whether existing landlords or their ancestors got possession of the land equitably or fraudulently, or whether all landlords (as some have done) bought the land at full value with hard-earned money. It matters not whether the ancestors of the present landless class were serfs or nobles, whether they never had land, or whether they sold or gambled away their inheritance. All this has nothing whatever to do with the main question, which is, the essential wrong to the community of private property in land; whether to deprive others of the use of the earth, now and for future generations, is or is not, in Herbert Spencer's own words, "a crime inferior only in wickedness to the crime of taking away their lives or personal liberties." If Mr. Spencer had taken the trouble to study the Programme of the Land Nationalization Society, which it would have been a natural and proper thing for him to have done before arguing against the possibility of such nationalization—he would have found that, among the eight reasons we give for holding private ownership of land to be wrong, the mode in which it has been acquired either by past or present generations of landlords finds no place. We ground our claim on considerations of absolute justice as well as of practical expediency, and as against us, all the good or evil deeds of landlords or their ancestors are wholly beside the question. We maintain, that, when a great wrong has been done in the past, a wrong which still produces and must ever produce evil results, a wrong which is the fundamental cause of the wide-spread pauperism and misery that pervades our land—it is our primary duty to find a means of abolishing that wrong. And when the great philosopher who first taught us how enormous was this wrong, goes back on his own words, declares that he sees no way out of the difficulty, and that the huge injustice to the living and to the unborn must go on indefinitely—then we refuse to accept the teachings of such a helpless guide, who sets before us this most impotent conclusion under the holy name of JUSTICE.

Let us now turn for a while to consider the fundamental principles of Mr. Spencer's Social-philosophy—principles which are altogether excellent, and which, if he had boldly and logically followed them out, would have shown him how this great wrong—this wicked crime of land-monopoly may be easily and equitably abolished.

In the second paragraph of his Chapter entitled HUMAN JUSTICE (as distinguished from animal and sub-human justice previously discussed), Mr. Spencer thus lays down the ethical correlative of the law of survival of the fittest in the animal world:—" Each individual ought to receive the benefits and the evils of his own nature and consequent conduct: neither being prevented from having whatever good his actions normally bring to him, nor allowed to shoulder off on to other persons whatever ill is brought to him by his actions." This law is appealed to again and again throughout the book, as being a decisive test of the right or wrong, the usefulness or the hurtfulness of certain social or governmental agencies. It is generally given under a shorter form of words, such as—" Each adult shall receive the results of his own nature and consequent actions"—or still more briefly—" Each shall receive the benefits and evils due to his own nature and conduct." This is the fundamental principle of social development according to the Spencerian philosophy; and from it is derived the formula of JUSTICE—" Every man is free to do that which he wills, provided he infringes not the equal freedom of any other man," or briefly " The liberty of each limited only by the like liberty of all."

From these principles Mr. Spencer deduces many important results as to personal and social rights—among others the right of property, the right of free industry, and the right of gift and bequest. Under this latter heading he makes an important qualification, as follows:—" One who holds land subject to that supreme ownership of the Community which both ethics and law assert, cannot rightly have such power of willing the application of it as involves permanent alienation from the community."[1] With this rather vague statement he leaves the subject, and afterwards affirms those extraordinary propositions I have already quoted as to the permanence of private property in land, and the extinguishment of the people's right through payment of two centuries of poor-rates!

But if we logically follow out the Spencerian principles we shall find that the right of bequest has far more extensive limitations than the author gives it, limitations which render it easy for the State, that is the people, equitably to regain possession of their own land. For, if the law that—" each shall receive the good or evil results of his own nature and actions " be a true guide to social

[1] *Justice*, p. 124.

development, then the correlative of it must also be true, that no one shall receive throughout life, that which is *not* the result of his own nature and actions; and this will absolutely forbid such bequests to children or others as will render them independent of all personal exertion, enabling them to live idle lives on the labour of others, and thus neutralize the operation of that beneficent law which gives to each the results of "his own nature and acts." To permit unlimited bequest is, in fact, doubly injurious. It is a positive injury to the recipient whenever it enables him to live a life of idleness and pleasure—to be a mere drone in the human hive. And it is also a gross injustice to the rest of the community, for when such parasites abound they become a burden on the industrious who necessarily support these idlers by their labour. We must further consider, that, so long as landlordism continues, this idle and generally useless portion of the community may increase almost indefinitely, because savings can be made by all large landlords, on which savings a larger and ever larger number of succeeding generations may live without exertion, and thus render the lot of the workers harder in proportion.

It is strange that Mr. Spencer did not perceive that if this law of the connection between individual actions and their results is to be allowed free play, some social arrangement must be made by which all may start in life with an approach to equality of opportunities. While, as now, some are brought up from childhood among low and degrading surroundings—material, intellectual, and moral—and have to struggle amid fierce competition for the bare necessaries of life, it is absurd to maintain that they receive the legitimate results of their own nature and actions only; both of which may be and often are far superior to those of thousands whose early years are surrounded by all the refinements of a higher social life, and who find a place provided for them in which with little effort on their part they can provide for all their wants, their comforts, and their pleasures. Such a law as Mr. Spencer has formulated becomes a mockery and a delusion, unless each individual is given a fair start in life, and this can never be the case under a system of landlordism and unlimited bequest.

Mr. Spencer's fundamental principle of social justice, therefore, logically implies that the power of free gift and bequest should be placed under strict limitations, the State taking to itself all above the amount which may be judged necessary for providing each heir with such ample education and endowment as may give him or her a favourable start in life—after which they must be left to receive the results of their own nature and actions; while the surplus property thus accumulated will form a fund out of which to endow in like manner all those whose parents are not able to provide for them. This branch of the subject however does not directly concern us here except in so far as it leads us to consider the application of the Spencerian principles of JUSTICE to the land question.[1]

We are told distinctly that no landowner should be allowed to leave his land in such a way as to permanently alienate it from the community to which it rightly belongs. But to concede the right of unlimited gift or bequest *does* so alienate it; therefore no such right should exist. There is another principle, which was asserted by the great jurist Jeremy Bentham, and which is of some importance as a guide :—That laws should never be such as to disappoint "just expectations"—expectations which people had been brought up to consider both legal and equitable. Such an expectation, in our country, is that of succeeding to one's father's property. But no one can have such an expectation till he is born, nor even for some few years afterwards. We may therefore, from every point of view, equitably enact that, from the date of the law, no land shall descend to any person then unborn. We may also rightly add that it shall descend only in the direct line—that is to children and children's children living at the time of passing the Act, because no collateral relatives have any just or reasonable claim or expectation of succeeding to the land.

[1] This problem is more fully treated in chapter xxviii. of this volume.

Here then, by this simple and perfectly equitable principle, we have found a means of transferring the people's land back to the people, by a gradual process which would rob nobody and cost nothing; and thus the whole huge mountain of difficulty which has induced Mr. Spencer to look upon the restoration of the land as a moral and financial impossibility crumbles into dust. The mode of acquiring the land now suggested was advocated in my first article on Land Nationalization (see Chap. XVI.), and I myself, and many of my friends, still think it to be the best. Of course we may and do also advocate the power of compulsory purchase by local authorities to supply the immediate wants of the people. But while this was being done, wherever needed, land would be continually accruing to the State by the dying out of the direct heirs of landlords; and land thus acquired, always administered by the local authority and the proceeds of the rents equitably divided between the State and the locality, would continually reduce the weight of both imperial and local taxation.

This concludes all that needs now be said of Mr. Spencer's new work. There are some other points I should have liked to touch upon, but they are of less importance from our present point of view. I hope that I have shown with sufficient clearness the fallacies that underlie his recent utterances, and have thereby enabled all land reformers to continue with a good conscience to quote the burning and logical denunciation of landlordism to be found in *Social Statics*, notwithstanding the author's recent attempt to minimize the effect of them.

CHAPTER XIX

SOME OBJECTIONS TO LAND NATIONALIZATION ANSWERED.

IN the present chapter I deal with a few of the most frequent of the objections of those who oppose Land Nationalization, and also discuss a few of the difficulties which are often felt even by those who are fully in sympathy with the principle.

State-tenants versus Freeholders.

When Nationalization of the Land is advocated, a great many people reply, " I don't see the good of Nationalization, I prefer freeholders to State-tenants." Let us therefore see what are the comparative advantages of the two modes of tenure.

In order that the greatest number of people may become freeholders, many Liberals advocate the abolition of all restrictions on the sale and transfer of land. They say, make every man who owns land an absolute owner, with power to sell, or divide, or bequeath as he pleases, and plenty of land will come into the market. Then, every one who wants land can buy it, if able to do so; and if the mode of transfer is also made simple and cheap every thing will have been done that need be done. We shall then have free trade in land; there will be no limited or encumbered estates; and capital will flow to land and develop its resources.

But people who talk thus forget that we have already had two great experiments of this nature, both supported by these very arguments, and that both have utterly failed. About forty years ago the dreadful condition of the Irish peasantry was imputed to the prevalence of entailed and encumbered estates the owners of which had no money to spend on improvements, and a most radical measure was passed by which all these estates were brought into the market and sold to the highest bidder. But the result was not as expected. Capital flowed into the country, but with no benefit to any one but the capitalist. English manufacturers and speculators became owners of Irish land, and sometimes laid out money on it; but they were harder landlords than those whom they replaced; they looked upon the land they had bought merely as a means of making money, and utterly ignored the equitable or customary rights of the unhappy tenants. Irish distress was not in the least degree ameliorated by this drastic measure from which so much was expected, and it is now rarely spoken of, while legislation on totally different lines has been found necessary.

The second example of the utter uselessness of pouring capital into a country so long as the people are denied any *right* to the use of land is afforded by Scotland. In the early part of this century, the great demand for wool made sheep-farming profitable, and many of the highland landlords were persuaded that they could double their incomes by establishing great sheep-farms on their vast estates. They did so. Many thousands of valuable sheep were introduced; much money was spent in fencing and in building new farmhouses for the lowland farmers, while the rights of the hereditary dwellers on the soil were utterly ignored, and, by a series of barbarous evictions, these poor people were banished to the sea-shore, or forced to emigrate. The result was, for a time, beneficial to the landlords, who proclaimed the scheme a great success; but it was most disastrous to the people, who, ever since, have been kept in a state of serf-like subjection and pauperism. The present condition of the Highlands is a direct consequence of the application of capital to the land by landlords while the rights of the people were ignored; and the result of these two great experiments in Ireland and Scotland should teach us that any similar experiment in England cannot possibly lead to good results. It is true the conditions of society in England are different. There are here more capitalists ever competing for the possession of land; but "free trade" would simply enable those *wealthy* capitalists who desire land to obtain it more easily. What chance would the poor man have against such competitors? With population and wealth and manufactures ever increasing, as they are in England, the poor man will have less and less chance of getting land, so long as it is to be obtained solely by purchase, and there is neither compulsion to sell, nor right to buy at equitable prices.

Effects of Land Monopoly.

As land is ever getting scarcer in proportion to population, and in private hands must necessarily be a monopoly, it offers the greatest temptation to speculators, who even now frequently buy up estates offered for sale and resell them in small plots at competition prices which no poor man can afford to give; and this will continue to be the case so long as land is treated as a commodity to be bought and sold for profit. We maintain that this is a monstrous wrong and should never be permitted. Land is the first necessary of life, the source of food and of all kinds of wealth, and a sufficiency for health and enjoyment is absolutely needed by every one. It is a political crime to permit land to be monopolized by a few, to allow the wealthy to employ it for mere sport or aggrandisement, while thousands live in misery and have to suffer disease and want because they are denied the right to live and labour upon it.

In order that all may have equal rights to use and enjoy the land of their birth, it must become, not theoretically only, but actually, the property of the State or Local authority in trust for all; and for all to derive equal advantages from it, those who occupy it must pay a rental to the State for its use. This is the only way to equalise the advantages derived by the several occupiers of land of different qualities and in different situations

—the only way to enable the whole community to benefit by the increased value which the community itself gives to land.

The use of land is twofold. Its chief and primary use is to supply to every household in the kingdom, the conditions for healthy existence, and, whenever possible, some portion at least of their daily food. When all are thus supplied with the land necessary for a healthy home, the remainder should be devoted to cultivation in such a way as to produce the maximum of food, and at the same time to support and bring up the maximum number of healthy and happy food-producers. All experience shows that these two things go together, and that in any country the maximum of food is produced when the greatest possible population live upon and by the land. At one extreme we have the great farms of S. Australia and California, cultivated with the minimum of human labour and producing a net return of about ten bushels of wheat per acre, and at the other extreme the allotments of our farm labourers producing food to the value of £40 per acre.

But in order that our labourers and mechanics may each be enabled to have, say, an acre of land to live on, and an acre or two more to cultivate, if they require it, with the power of getting a small farm of, from ten to forty acres whenever they have obtained money enough to stock it, the land must be *let*, not *sold* to them. For at first a man wants all his little capital to enable him to cultivate even the smallest plot of land, and if he has to buy it, even by the easiest instalments, he is to that extent crippled. Moreover it is a bad thing for him to *own* the land absolutely, because he is then open to the temptations of the money-lender. Instead of economizing and pinching in bad seasons, he borrows money and mortgages his land, and thus falls under a tyranny as bad as that of the hardest landlord. In every part of the world the small freeholder falls a victim to the money-lender.

As a State-tenant the occupier would have all the essential rights and advantages of a freeholder. His tenure would be practically perpetual. He would have the right to sell or bequeath his holding, or any part of it, just as freely. His rent would never be raised on account of any improvement made by himself, but only on account of increased value of the ground-rent, due to the growth of population or other general causes, which would affect all the land around as well as his. He would therefore enjoy all the rights, all the privileges, and all the security which a freeholder enjoys. But he would have this great advantage over the freeholder, that he need not sink one penny of his capital in the purchase of the soil; and thus, for one man who could save money enough to acquire a farm or a homestead by purchase, two or three would be able to become State-tenants, with money in their pockets to stock their land or build their house, and to live upon till their first crops were gathered. Those who maintain the superiority of freeholds, therefore, speak without knowledge; the superiority is all the other way.

There is one more point to be considered, which is of great importance, that under a general system of small freeholders, one half of these would very soon be ruined by the other half—would be obliged to sell their farms to money-lenders or lawyers, and thus great estates would again monopolize the land. The way this would necessarily come about (as it always has come about) is as follows. Suppose there are a body of peasant proprietors all over the country. Their land necessarily varies in quality and position, and, therefore, in value from fifteen or twenty shillings an acre up to two, three, or four pounds an acre; and, all being freeholders, none of them pay rent. But the owner of the better land can afford to sell his produce of all kinds at a lower rate than the owner of the inferior land, because prices which will enable the former to live and save money will be starvation to the latter. Hence an unequal competition will arise between the two classes in which the one must necessarily starve out the other. The payment of rent in proportion to the *inherent value of the land* equalises the position of all. The occupier of poor land at a low rent can fairly compete with the occupier of rich land at a high rent; and thus while a system of *small proprietors* is sure to fail, a system of *small occupiers*, under the State, combines all the essential elements of stability.[1]

Thus far we have considered the question solely from the economical and practical point of view, but the great superiority of State tenants over freeholders is equally apparent when we treat it as a question of justice. Land necessarily increases in value as population and civilization increase, and that increase being the creation of the community at large is justly the property of the community. By a system of State tenants we shall obtain this increase for the benefit of all, by means of a periodical reassessment of the ground rents payable to the State; but if we create a body of small freeholders we shall perpetuate injustice and inequality. A. and B. may acquire two farms at the same cost and may bestow the same labour and skill in the cultivation of them. But in thirty or forty years the value of the two may be very different. Minerals may be discovered or some new industry may spring up causing the farm of A. to become the site of a populous town, while that of B. remains in a secluded agricultural district; so that, while the children of the one are earning their living by honest labour the children of the other may be all living in idleness by means of wealth which they have not created and to which they have no equitable claim, and to the same extent the community at large is robbed of its due. If on the other hand we establish a system of State-tenancy over the whole country, the natural increase of land-value by social development will produce an ever increasing revenue even if existing landlords continue to be paid the incomes they now receive from land, so that in addition to all the other advantages of the system we shall acquire the means of bringing about a steady diminution of taxation, by which all alike will benefit.

Briefly to sum up the argument: SMALL FREEHOLDS ARE BAD BECAUSE—

1. Money must be sunk in the *purchase* which can be better invested in the *cultivation* of the soil.

[1] This danger has been attempted to be obviated on the Continent by the farms consisting of scores or hundreds of scattered patches of land of different qualities. But this system renders economical cultivation impossible, and the remedy is worse than the disease.

2. The number of men who can advantageously acquire small farms is therefore greatly reduced.

3. The unearned increment of the land is taken from the community who create it and is given to individuals.

4. The inheritors of these small farms of different qualities of land will compete unequally with each other, and those holding the poorer land must sooner or later sell their farms or fall into the hands of the money-lender.

The system therefore contains within itself the elements of decay and failure.

In all these respects State-tenancy is greatly to be preferred to small-freeholding, and a general system of State-tenancy can only be secured by a complete Nationalization of the Land.

Land Taxation—Who will Pay it?

One of the most important points of difference between the followers of Henry George and Land-Nationalizers is on the question whether the landlord can throw the burden of a heavy land-tax upon the tenant. This question has been the subject of much discussion for some years past, and the proposal to tax land-values, especially in large cities and wherever land is held by the owners in expectation of a great rise in value for building purposes, has been generally adopted by the more advanced section of the Liberal party. It therefore becomes very important to ascertain whether such a tax will really fall upon the landlord, or whether it will not, in the course of a few years at furthest, be shifted to the tenant, who will thus be no better off than before.

The one strong point of the advocates of land-taxation is the appeal to authority. Adam Smith says:

"A tax upon ground rents would not raise the rents of houses. It would fall altogether upon the owner of the ground rent, who acts always as a monopolist, and exacts the greatest rent which can be got for his ground."

Ricardo says:

"A tax on rent would fall wholly on the landlords, and could not be shifted to any class of consumers. . . . It would leave unaltered the difference between the produce obtained from the least productive land in cultivation and that obtained from land of every other quality."

John Stuart Mill says:

"A tax on rent falls wholly on the landlords. There are no means by which he can shift the burden upon any one else."

Professor Thorold Rogers and most other writers on political economy have followed these great authorities, repeating their statement in slightly modified words; and in many recent articles as well as in discussion in the County Council, either these authorities are accepted as conclusive, or if any attempt is made to prove that the same results would follow the imposition of a 4s. or higher land tax, the logic is at fault, and the attempted proof utterly breaks down. This I shall hope to show in a very few words.

The whole essence of the question at issue is contained in the concluding words I have quoted from Adam Smith, that the landlord "always acts as a monopolist, and exacts the highest rent which can be got for his land." It follows, therefore, that however much you tax *him*, however much you impoverish *him*, you do not, by that act alone, enable him to extract more rent from the tenant, who is not benefited by the landlord's tax. Now this was the only kind of tax contemplated by the early writers. Governments were always in want of money, always seeking to impose new taxes; and having taxed the poor as much as they could bear, if they put a special tax on land the landlord must pay it, since the tenant was already rented and taxed up to the hilt. He could bear no more. Adam Smith, and Ricardo, and Mill never contemplated the case of a landlord-government taxing land, not to supply its own dire necessities, but to relieve the tenant. Such a thing was inconceivable to them, was beyond the range of their practical politics, and therefore they did not deal with it. But this is the very essence of the modern proposal. The 4s. tax is not to be a war-tax, or, what is much the same thing, the money obtained is not to be thrown into the sea. It is not proposed to tax the landlords in order to ruin them without benefiting the rest of the community, who are all tenants. That would be pure "cussedness," as our American friends would say. *But the tax on landlords is for the express purpose of relieving tenants.* Just as the landlords are made poorer by it, the tenants are to be made richer, taxes are to be transferred from tenants to landlords, and would be exactly equivalent to a reduction of rent. What then would happen—the landlord, as Adam Smith says, acting "always as a monopolist and exacting the greatest rent which can be got for his ground?" Is it not absolutely certain that, he being poorer and the tenant richer, he would raise the rent and thus get back the tax he had just paid? Mr. Fletcher Moulton says this is a fallacy. He urges, that.

"under the proposed system the ground value will represent the full rental value that the landowner can obtain for the use of his land, *unburdened by any rates*. This is a sum which will be determined by considerations relating to the land itself, and by these alone. In ascertaining the rental which he can afford to give for the *free* use of the land, the proposed tenant will not be affected by the question, what portion of that rent will be retained by the landlord, and what proportion will be paid over in rates?"

Of course he would not; nobody has ever suggested that he would. But Mr. Moulton has slipped in the words, "unburdened by any rates," and thereafter ignores them. Are tenants now "unburdened by any rates?" And if their rates are removed and put on the landlord, does not this affect the tenant's power of paying more rent, and the landlord's power, "acting always as a monopolist," to obtain more rent? Surely the "simple fallacy" is on Mr. Moulton's part, not on ours.

A writer in the *Democrat* also trusts mainly to the authorities, and in his arguments also misses the main point. He says:

"A tax upon land value, or economic rent, would not raise prices, but would simply transfer to the State a portion or the whole of that premium paid for the use of better land which now constitutes the unearned incomes of the landlords."

Here, "transferring to the State" is spoken of as if it

were a foreign State, or a private individual who would spend the money on himself. But the State in this case is the community, consisting almost wholly of tenants who are to benefit directly by reduction of rates and taxes. They would therefore be *able* to pay higher rents, and the landlord, "acting always as a monopolist," would exact those higher rents so as to bring back the respective position of landlord and tenant to what it was before the tax was imposed.

The only other argument used—that if it would have the effect we urge the landlords would not object to it, is hardly worth answering. Landlords are not a specially intellectual body of men, and why should they not be blinded by the alleged "authorities" as well as the followers of Mr. George? Besides, there would be *some* loss to the landlords, especially to those who had granted leases. All we urge is that in a very few years the landlords would necessarily recoup themselves under the inevitable laws of supply and demand, they being left in possession of a monopoly of what is essential to all men. In his "Scheme for the Abolition of Landlordism," in the *Westminster Review* for May, 1890, Mr. Charles Wicksteed hit the nail on the head by his suggestion that all receipts for taxes on land should be applied in the *purchase of land*, not in relieving tenants of rates and taxes, and thus prevent the landlords from recouping themselves by raising rents. This simple proposal has satisfied one member of the Land Restoration League that the imposition of the 4s. or any other tax, to be applied in relief of other taxation, would be useless, and would ultimately leave the landlords as much masters of the situation as they are now. It is to be hoped that others will be equally convinced and equally candid in acknowledging it, and thereafter retire from an untenable position.

The Conditions Essential to the Success of Small Holdings.

As there seems to be considerable difference of opinion on this question, and as success or failure in the first steps towards obtaining free access to land for all who desire it is a matter of vital importance, it will be useful to set forth for the consideration of Land-reformers what seem to be the principles which should guide us in this matter.

I begin with the proposition, which will probably be generally accepted, that the primary object of giving the people free access to land is, to enable as many as possible to obtain either the whole or a portion of their living by its cultivation, and, by thus withdrawing some of them altogether from competition with wage-earners; while by giving others a temporary alternative to wage labour, we shall raise the rate of wages of the less remunerative kinds of work all over the country. And as secondary benefits, we shall have, in the first place, an enormous increase in the production of food, displacing much that is now imported from abroad; secondly, an increased consumption of manufactured goods by these self-supporting workers; thirdly, a diminution in the poors' rate from the disappearance of paupers and out-of-works; and, lastly, a public revenue arising from the continuous increase in the value of land owing to the growth and increased well-being of the population. These various benefits I conceive to be important, somewhat in the order in which I have given them; at all events the first is undoubtedly the most important, while the last—the money profit—is the least important. It is a valuable incidental result, but is not to be sought after as one of the chief ends in itself in depreciation of the other kinds of benefit.

Now in order that the various good results above enumerated shall be, in their due order, most certainly obtained it is of the highest importance that the workers who gain access to the soil shall be able, not only to live upon it, but to live well and thrive upon it. We do not want them merely to earn a living as long as they can work, and go to the poor-house in their old age or during sickness; neither do we want them to be ruined by the first bad season, or by any of the chances and misfortunes to which agriculture is especially liable. On the contrary, it is of the highest importance to the whole community that the land-cultivators should be in such a position that even in bad years they should not lose, while in average years they should obtain such profits, that, either by increase of the size of their holding or the amount of their stock, their condition should steadily improve, and thus ensure them against poverty in old age. If land-nationalization is to be the success we hope it will be, some such result as this must be aimed at, and we must be sure that we do nothing to prevent its attainment.

In order to bring about this result two things are especially required—first, a certain amount of knowledge, of experience, of prudence, and of industry in those who obtain these small holdings; in the second place, such very moderate rents and such favourable conditions of tenure as to give them not only a chance but almost a certainty of success. Nothing, in my opinion, is more likely to bring about the failure of the first, and therefore the most important experiments in this direction than the methods usually suggested by politicians, and more or less implied in the various Acts of Parliament, whether already passed or proposed, dealing with this matter. It is, for example, almost always proposed or taken for granted, that before the men can have the land a large cost must be incurred in preparing it for them. We hear of road-making, fencing, draining, and house-building, as if these were absolute necessities; and in addition to all this, it is generally thought that it will be necessary to advance capital to the proposed tenants to enable them to stock and crop their holdings, and support themselves till they get a return from the land. Now I can hardly imagine a more certain way of bringing discredit on the whole system of small holdings—and with it of land-nationalization—than such a method. In the first place, this work of "laying out" and "improvement," when it is not done by the occupiers themselves, but for them, by persons who have no interest in doing the work economically but often the reverse, and who have besides no personal knowledge of what these small cultivators really require, will often be unnecessary work and will always be done in an unnecessarily costly way, and will thus add to the rent of the land, without proportionately increasing its value. Then again, the advance of money on loan to men who have never, perhaps, had ten pounds to spend at once in their lives, will in many cases lead to its injudicious expenditure, and be antagonistic to that prudence, thrift, and industry, which are vital to success.

There is yet another, and a very important objection to such methods as these. All this expenditure of public money by other people than those who are to directly benefit by it, will certainly lead to wastefulness and jobbery, since they constitute that very "management of land by public officials," the evils of which form one of the chief objections to land-nationalization, and which all our proposals and methods have been calculated to avoid. We must therefore never cease to urge that such management is not only unnecessary, but is calculated to defeat the very purpose for which free access to land is required.

If we consult the reports of the various Royal Commissions on Agriculture we shall find numerous cases of labourers, miners, mechanics, and others, who have become successful cultivators of small holdings and sometimes of considerable farms, often having begun with a lease of waste land which they enclosed, improved, and built houses on, entirely on their own resources and through their own industry; and whenever we find a successful small farmer he has usually worked his way up by some such method. It is an extraordinary thing that whenever it is proposed to allow men to obtain small holdings in England, there is always this talk of "improvements" and "housebuilding," in addition to giving the land at a fair rent and on a secure tenure; while over a large part of Scotland and Ireland all improvements and all buildings have been done by the tenants themselves, with no security of tenure whatever, so that whenever a misfortune prevents payment of rent—usually rent on the tenants' own improvements—the landlord ejects the poor tenant and confiscates his improvements. The Irish cottar and the Highland crofter ask nothing better than a sufficiency of land at a moderate rent and on a secure tenure. All the perennial misery and often-recurring famine of these two countries have arisen from a denial of this very moderate instalment of bare justice; and if we give this easy access to the land to our English workers, they too will ask for nothing else, and will be far more likely to succeed without any attempt to do for them what they will do much better and more economically themselves.

But we do undoubtedly require some process of selection of the best men for this great experiment in the regeneration of our country; and the natural, self-acting, and therefore best mode of selection, will arise from the fact that no man *can* take a holding unless he has saved money to stock and crop it, or has such a character for industry, sobriety, and capacity as to induce some friend to advance him the money; while the certainty that he is risking the loss of his own savings if he fail, will be the best guarantee that he will have some amount of intelligence and some agricultural experience. No artificial mode of selection will compare with this. A man may get testimonials to character, but no testimonials can show that he will spend borrowed money prudently, or be able to make a profit on land burdened with unnecessary and costly improvements, which, for his purpose, will often be no improvements at all.

How to fix the Rent.

We have now to consider the second great essential of success, which is, the rental to be paid for the land and the conditions of tenure. Many of our fellow-workers maintain that the competition-rent offered for land is the best, and in fact the only certain way, of determining its value, and therefore what should be paid for it. From a landlord's or speculator's point of view—considering the money income to be got from the land to be everything, the well-being of the tenants nothing—this is undoubtedly the case; but, from our point of view—looking at the cultivation of the land as leading primarily to the well-being and progressive advancement of the cultivator, and through him the similar advancement of all other manual labourers—it seems to me to be the very worst mode possible. Let us therefore consider it a little in detail.

At the present time, wherever there are allotments to let, there we find it to be the rule that agricultural labourers willingly hire them at a rental per acre, sometimes double, sometimes four or five times as much as is paid by farmers for the same quality of land; and there can be no doubt that if arable land were now offered for allotments and small holdings almost anywhere in England, and the quantity thus offered was not in excess of the demand at the time, it would, if let by auction, realize somewhat similar rentals. Many agricultural labourers, as well as village tradesmen, and mechanics, find it advantageous to them to have land even at these high rents, and there is sufficient land thus held all over the country to afford a guide to the prices at which such land would let by auction, even in quantities of from one to five acres. Now the reason such high rents have been, and are paid, is, simply, that the labourers' wages have been always so low and his condition so miserable, that anything by which he could add two or three shillings a week to his earnings by means of overtime work and the assistance of his wife and family, was eagerly accepted. It was his low standard of living that rendered him willing to pay a rent which left him a mere trifle of profit on his labour; and if he were now offered a small holding on which to live he would be willing (if he could not get it cheaper) to pay a rent which would enable him to live in about the same way as he had hitherto done, but with the chance of occasional better luck and with the satisfaction of being his own master. We see this result very prevalent on the Continent, especially in Belgium and in parts of France, where the price of land, and consequently its rental, is very much higher than with us, and as a consequence the small holders often work harder and live as near the starvation line as our poorly paid agricultural labourers or our rack-rented Irish cottars. It will be said, no doubt, that this arises from the demand for land being greater than the supply, and that if land were offered in larger quantities, competition would be less keen and prices lower. But is it at all likely that for a long time the supply of land will be greater than the demand, except quite locally, and temporarily? Is it not, on the contrary, almost certain that the demand will, at first and for a long time to come, perhaps always, be greater than the supply? Is it not our contention that the depression of agriculture and the deplorable condition of so many of our workers is due to the denial of access to land, and that when that access is freely given it will bring about the well-being, first of those who cultivate it, and afterwards of all other wage-earners? There will therefore be a constant and ever increasing demand for land; but unless we take care that those who apply for it have it on such terms that they can not only make a good living from it, but also provide for a comfortable old age, the benefits we anticipate will not arise. It is to avoid any such failure, to prevent the recurrence of the miserable spectacle of men being ejected from their holdings for non-payment of high rents; to secure for them something better than a struggle ending in dependence on charity in old age, that I urge the fixing of rents by valution, taking always the amount paid by prosperous farmers in the same district, rather than that of allotment-holders, as the standard of value. The land when purchased by the local authorities, will be purchased at the farm value, and it can be let at that value at first without loss to the community.

The above sketch of the reasons why I object to the system of competition-rents sufficiently exhibits the principle on which, in my opinion, our dealings with the land should be founded. But there is also a practical objection to that system—that it would be very unequal in its results, and also that it can hardly be carried out unless based on a preliminary valuation. I presume the advocates of competition-rents do not propose that land should *always* be let to the highest bidder, without any reserve whatever. For, if so, whenever the intending tenants were less numerous than the lots to be offered, these lots might be let at much less than agricultural rents. No doubt it will be said there must be a reserved price for each lot, or group of lots of the same kind of land; but such a reserve cannot be fixed without a careful valuation by an expert; so that we should require two processes both involving some expense, first the valuation then the auction. The result would be, that in some cases, where there happened to be little competition, the land would be let at the reserved rent, while in other cases—perhaps a few months later, and in the same place—similar land would be run up by competition to much higher rents; and this would inevitably lead to dissatisfaction and inequality in the prosperity of the tenants—a dissatisfaction which would compel the authorities to adopt the plan of letting all land at the reserved rent, that is at that fair but low rent which should have been adopted at first.

I maintain, therefore, that it is essential to adopt from the first the only just and equal method, which is, the valuation of the lots by an expert, founded on what would be fair rents of similar land to a large tenant farmer. These lots, with the rents thus determined, would then be open to selection, either on the system of "first come first served," or if thought fairer, of a ballot for the order of choice on certain fixed days. By either of these two methods, supposing the valuation to be fairly made, there would be no inequality of opportunities, no feeling that either by chance or through any other cause, some of the tenants were paying higher rents than others.

Another point of some importance is, that men should be allowed to have as much land as they wished, up to a certain limit—say five or ten acres according to circumstances; and also that the land first let should always be that abutting upon roads or lanes, the inner portion of the farms thus let being reserved for some years, so that any man wishing to add to his holding could have it extended by taking a plot or field behind it, thus avoiding the great inconvenience and loss arising from the separation of plots under one holding. In the meantime this central portion of the farm could be let by the year to any adjacent farmer.

One other point arises in connection with this question —the vital importance of *security* to the occupier and cultivator of land. We want men to be able to form and keep a HOME; to be practically as secure in that home, so long as they pay the moderate ground-rent for the land, as if they were the actual owners of the freehold, subject only to the payment of a tax. We want the new tenants under land-nationalization to be really *free-holders* in the old sense—free men holding land from the community, never to be interfered with so long as they continued to pay the moderate dues and to be law-abiding citizens. To give this full security all the rights of bequest or sale now appertaining to freehold land should appertain to these State tenancies.

It is, I believe, only by some such process as that which I have here indicated that we can possibly obtain the full benefit of land-nationalization or of the first steps which we may be able to make towards it. We must always remember that the community will be benefited just in proportion to the well-being of the cultivators and of those who obtain access to land. If we rack-rent them so that they just make a living out of the land, they will have little influence in raising the wages of other workers or in enabling them to make a successful bargain with capitalist employers. But if, on the other hand, we allow all occupiers of land to have it on such terms and conditions that they are able to make a good living, provide well for their families, and enjoy an old age of secure repose in their own homesteads, this state of well-being will serve to establish a standard of living which will react on the whole working population, and lead to a corresponding rise in the rate of wages throughout the whole industrial world. There appears to me to be no proposition in the domain of political and social science more certain than this. On the other hand, no mistake can be more fatal than to think that the community would be benefited by screwing from twenty to fifty per cent. more rental from the occupiers of land, thus reducing their profits, rendering their position less secure, lowering their standard of living, and with it that of the whole working population, and often creating a body of prospective paupers.

In conclusion, therefore, I would, urge most strongly, that in all arrangements or proposals with regard to the land, we should throw aside altogether the idea of getting the highest possible rents, but should always aim at the maximum of well-being for the cultivators. By thus acting we shall best secure the equal well-being of the whole of the industrial community, and shall initiate that progressive improvement, with the diminution and ultimate abolition both of enforced idleness and of undeserved poverty, which is the whole aim and object of Land Nationalization.

CHAPTER XX

A COUNSEL OF PERFECTION FOR SABBATARIANS [1]

ALMOST all the Christian Churches of Great Britain have adopted the Sabbath of the Jewish lawgiver as a divine institution, only changing the day from Saturday to Sunday, though many of the Nonconformists retain the Jewish term, Sabbath. Many, perhaps most, religious persons hold that to work on Sunday is an actual sin comparable in gravity with most other acts forbidden in the Ten Commandments; and the strong condemnation of Sabbath-breaking in religious tracts and Sunday-school teaching is a sufficient proof of the importance attached to a due observance of the day.

An impartial onlooker is, however, somewhat puzzled by the circumstance that, notwithstanding this general uniformity of precept, the practice, even of the teachers, is exceedingly lax, since there is hardly a Christian family in the whole country, not excluding those of the clergy of the various denominations, where the Sabbath is not broken fifty-two times in every year. Now the fourth commandment, as read every Sunday in our churches, is either binding on Christians or it is not. In the latter case breaking it is no sin, and any observance of a seventh day of rest is merely a matter of expediency or of human law. It is, however, nearly certain that the majority of Protestant clergy do not accept this latter view, and I therefore propose to discuss the question—how Sunday may be most consistently and

[1] This article appeared in the *Nineteenth Century* (October, 1894) under the editor's title "A Suggestion to Sabbath Keepers."

beneficially observed by those who believe it to be a divine institution; and my argument will apply equally to those who maintain that we are only bound by the spirit of the commandment, not by the letter, still less by the special interpretation of it adopted by the Jews.

Let us then first inquire what is the spirit and purport of the law; and in this there can be little difficulty, because it is more fully explained than any other of the commandments, so that its whole meaning and purpose cannot possibly be misunderstood. This command is not given briefly, as so many others are; not merely " thou shalt not work on the Sabbath," as in " thou shalt not kill," or " thou shalt not steal;" but with full and impressive reiteration and detail.

First, we are told, " Six days shalt thou labour and do all thy work;" then, on the Sabbath, " thou shalt not do any work;" and then, to show how wide and complete is the law, there is added, " thou, nor thy son, nor thy daughter, thy manservant, nor thy maidservant, nor thy cattle, nor the stranger that is within thy gates." If ever there were plain words with a plain meaning these are such. They mean, as clearly as words can convey meaning, that each one's work during the week, that work which is the duty of our lives, and by which we maintain ourselves, is to cease on the Sabbath; and that the law is especially to apply to all servants of every kind, and to all beasts of burden, which are included under the generic term " cattle."

This being the commandment, how is it obeyed by those who uphold the sanctity of the law; by those who are continually urging others to keep the Sabbath; by those who take every opportunity of putting in force human laws against Sabbath-breakers? Are not manservants and maidservants all at work on Sunday? Are not servants and horses employed by the thousand to take people to church on Sunday? Many persons, if asked why they go to church or chapel, will say that it is to save their souls or to please God, and yet they seem to think that they may break what they believe is God's own commandment week after week, without any chance of displeasing Him or of losing the souls they are so anxious to save.

What makes the matter worse is that, while they are

thus disobeying the scriptural commandment in the most flagrant manner, they are salving their consciences by abstaining, and trying to force others to abstain, from things which are *not* forbidden by the commandment, and which are not in any way opposed to its spirit. To walk for health or pleasure, to row in a boat, to play at cricket, or at chess, to whistle, or sing, to read amusing books, to look at great pictures in art galleries, or to admire the beauties and wonders of nature in museums or gardens—all these things have been, and many of them are still considered by the more strictly religious to be " breaking the Sabbath," and are denounced as such in many a tract and sermon. And the good people who hold these views seem quite unconscious that they themselves are far greater sinners than the people they denounce as "Sabbath breakers;" for to direct Sabbath-breaking they add the sin of pharisaism, inasmuch as they condemn in others what is, at the worst, a far less offence than their own, and are guilty of impious presumption in venturing to add to and improve upon the divine commandment, while constantly and knowingly disobeying the commandment itself. Do not the words of Christ exactly apply to such, when He rebuked the Pharisees from the mouth of Esaias?— "But in vain they do worship me, teaching for doctrines the commandments of men."

And when we inquire the reason for this strange and inconsistent conduct, we find only a series of excuses. They say, that the requirements of health and decency render a certain amount of work necessary on Sunday; that we keep a Christian and not a Jewish Sabbath; that we reduce the work of our labourers as much as possible; and that we only recognize works of necessity and of mercy as permissible on the holy day. It is true that Christ justified deeds of charity and of mercy to both man and beast on the Sabbath, but He nowhere abrogates the law of rest for each labourer, whether man or beast, from his six days' work. To tend the sick and supply the wants of the animals which serve us in various ways is not to break the Sabbath; but all these things and much more may be done without infringing even the letter of the Commandment, if we choose to seek out the right way of doing them. Christ clearly emphasized the spirit of the law when He declared that the Sabbath was made for man, not man for the Sabbath; by which we are taught, that the essential principle of rest on the seventh day for all who have laboured during six days is what we must seek to preserve. How we may preserve this, and yet have everything done that is necessary for health, comfort, and refreshment of mind and body, I now propose to show.

The whole essence of the Sabbath-question rests upon giving the proper meaning to the words "labour," "work," "thy work," as used in the fourth commandment. These words, as the context shows, do not refer to any particular acts, but to the work done by each one of us in the business or profession by which we live. To the summer tourist in the Alps the ascent of a mountain or the passage of a glacier is pleasure and health-giving recreation; to the guides who accompany him it is their work. A hired gardener works for his living in a garden; but though I do many of the same things as he does, to me they are not my work, but my recreation. So, a domestic servant's work is to cook or to prepare a meal, or to wait at table; but when a party go out for a picnic, light a fire, make tea, roast potatoes, arrange the meal, and help the guests, they are certainly not working but pleasuring. When a doctor attends the sick in a hospital, or the wounded on a battlefield, he is doing the work of his life; but if any one of us nurses a sick person or binds up a wound, we may be doing acts of mercy or of charity, but we are not doing "our work." Even if we take upon ourselves some of the work of others, carry a heavy load for a weary woman, or do an hour's stone-breaking to help an old rheumatic labourer, what we do ceases to be work in the true meaning of the term but is transformed into a deed of love or mercy; and such deeds are not only permissible, but even commendable, on whatever day they are done.

We have here the clue to a method by which all that needs doing for health, for enjoyment, or for charity, may be done on Sunday without any one breaking the fourth commandment. Almost all this necessary work is now done by various classes of hired servants who, are employed on similar work for six days every week, and who also have not much less to do on the seventh day. To keep the Sabbath, both in the letter and the spirit, these workers must be allowed full and complete rest; they must do none of their special work on that day. All that portion of their weekly duties which is necessary for the well-being of their employers, and for the rational enjoyment of their lives, must be done by those other members of the household who have spent the week largely in idleness or in pleasure, or if in work, in work of a quite different character from that of their servants. In doing this work; in helping each other; in sharing among themselves the various household occupations which during all the week have been undertaken by others; and in doing all this in order that those others may enjoy the full and unbroken rest which their six days' continuous labour requires and deserves, each member of the family will be doing deeds of self-sacrifice and of charity (in however small a degree), and such deeds do not constitute the "work" which is so strictly forbidden on the Sabbath day.

In the ordinary middle-class household, where there are six or eight in family and two or three servants, all that is necessary may be easily done, and allow every member of the family to go to church or chapel once or oftener. In other cases there will, no doubt, be difficulties, but none which may not be overcome by a little arrangement and mutual helpfulness. Where a household consists only of aged or elderly people to whom the needful operations of housework would be painful or even impossible, there are always younger relatives or friends, or even acquaintances, who could, either regularly or occasionally, spend the Sunday with such old people; and there is probably not a single difficulty of this kind which could not be overcome by two or more households combining for the Sunday in such a way as to divide the work and thus render it as little irksome as possible. If it were once really felt that the thing *must* be done, that on no account must the commandment be broken by servants doing any of their usual work on Sunday, and that the truest and most divine "service" would thus be "performed," all difficulties would vanish, and the day would become, not in name only but truly, a holy one, inasmuch as it would witness in every household deeds of true charity and mercy, because in every case they would involve some amount of personal effort and self-sacrifice.

In the larger establishments of the higher classes there would be no greater difficulty, since it would be easy to effect such a division of labour as to render the work light for each. The son or other relative who was fondest of horses and dogs would of course see after their wants on Sunday; another might undertake the fire-lighting; while the young ladies would prepare the meals and do all other really necessary domestic work. And as all visitors would be acquisitions, almost the whole of the lodging- and boarding-houses would be emptied, their occupants becoming guests at the houses of their friends and taking their share of the Sabbath day's duties. Of course the greater part of the servants thus released from their regular work would also visit their friends, and by giving some little voluntary assistance would take their part in the great altruistic movement that would characterize the day.

Among the more important of these deeds of mercy would be the relief of the nurses in hospitals and asylums, and of the attendants in workhouses and prisons. When the great principle of rest for each individual from the weary monotony of his or her daily work was once thoroughly accepted, volunteers by thousands would be found to take part in every duty of the kind; and it would probably not be necessary for any one to undertake the more repulsive duties more frequently than once a month, or perhaps three or four times a year. This would of course imply some general instruction of the young in the principles and practice of nursing, which is much to be desired on other grounds.

In the same way all the national treasures of art and nature in our galleries and museums, our libraries and gardens, might be thrown open to the great body of toilers who can enjoy them at no other time, the place of the

week-day guardians of these treasures being taken by
volunteers from among the more leisured classes, or from
the higher ranks of workmen. Thus would be remedied
the great injustice, that these grand institutions, for the
support of which all alike pay, are yet closed at the only
time when those who contribute most toward them would
be able to benefit by them. Of course the police would
also be relieved by a body of special constables who would
volunteer for the service. This occupation might be restricted to the Volunteer force, whose recognisable uniform
and military organization would render them admirably
fitted for the purpose. Further details on this part of the
subject are unnecessary, since it is evident that by an
extension of the same principle it would be possible to
relieve every one whose week-day labour is now extended
over some portion of Sunday also.

And now, having briefly set forth the arguments and
suggestions which seem to me needful for illustrating my
views as to the consistent observance of the day of rest
by all who look upon it as a divine institution, I will
state with equal brevity the good effects which such an
observance of it would produce. The substance of the
present chapter had been in my mind for twenty years
before it was written, and I made it public because many
circumstances seemed to render it less likely to give
offence and also more likely to do good than at an earlier
period, on account of the ever-growing strength of the
altruistic movement with the principles of which it
so well harmonises. For, the latter part of the nineteenth century will be characterized in history by the
awakening of the cultured classes to the terrible failure of
our civilization to provide even the barest necessaries and
decencies of life for thousands and tens of thousands of
those by means of whose work *they* live in luxury; and
also by their strenuous effort no longer to rely on mere
almsgiving, but to devote themselves to a sympathetic
study of the condition and needs of the poorest among the
workers, and to helping them with personal advice and
assistance. Toynbee Hall and Dr. Barnardo's homes,

missions innumerable and General Booth's slum-lasses, serve to indicate a few of the many ways in which this great movement is now making itself felt.

And it has begun none too soon if society is to be saved from a great catastrophe. Nearly sixty years ago Thomas Hood caused a spasmodic excitement among the well-to-do by the pictures of hopeless misery he set before them in his *Song of the Shirt* and *Bridge of Sighs*. Nearly half a century passed away; England's wealth had increased to an unprecedented extent, when society was again startled by the *Bitter Cry of Outcast London*, showing that the utter and hopeless misery of the earlier period was still with us, but increased and multiplied in quantity, just as the great city which produced it had increased and multiplied in size and riches. Then came official inquiries; and the "Commissions" on the Housing of the Poor and on the Sweating System, revealed horrors so terrible that it is simply impossible for men and women to live in a lower condition of want and misery and continue to exist. And during all this period there has been an ever-increasing growth of charitable institutions, trying in vain to cope with the ever-renewed crop of human misery; yet, notwithstanding all this effort, in each recurring winter the only difference of opinion seems to be whether the distress is worse than ever or only as great as it has been for years past. How bad it is may be inferred from the constant records in the daily press of suicide from hopeless misery, and death from want of food, fire, and clothing.

On the other hand, a change is now taking place in the attitude of the sufferers. They are no longer like the dumb beasts which perish uncomplainingly. They ask for work in order to live, and will no longer silently submit to be driven back to their cellars and slums by the police. They march by thousands into the churches, and listen to the platitudes of the preacher with murmurs of dissent. Many of them are now educated, and are quite as well able as their social superiors to reason on their condition. They begin to ask why it is that multitudes are enabled to live their whole lives idly and in luxury, while they themselves cannot obtain the poor privilege of constant work in order to provide the scantiest necessaries for their families. They now possess an amount of political power sufficient to overturn governments which do not satisfy them, and year by year they are becoming more able to make effectual use of that power; and it becomes more and more evident that, unless some real and great improvement in their condition is soon effected, very drastic, and perhaps dangerous, attempts at reform will be made.

To those who watch the growing enlightenment of the workers, it is clear that they will not much longer be satisfied with mere administrative reforms, or with petty palliatives which in no way touch the real causes of their unhappy condition. Many of them have learnt enough of political economy to know that the whole of the wealth annually consumed by the nation is the annual product of the labour, physical and mental, of the working classes; and that, just in proportion to the number of the non-producers and to the extent that labour is expended on the useless luxuries of pleasure, pomp, and fashion, to that extent are they deprived of the product of their labour and have to live in comparative penury. They begin to see clearly that hereditary wealth of all kinds, and especially the possession of land, enabling millions to live luxurious and idle lives, is the fundamental cause of the poverty of the workers, and the time will soon come when they will determine that this state of things must cease. They do not wish to rob any one of what he has been allowed by law and custom to consider his own, but they will not consent to the indefinite continuance of hereditary idlers any more than of hereditary legislators. They will probably say, as they will be perfectly justified in saying, "We recognize no rights in any portion of the next generation to live upon the labour of others. No child born after the passing of this Act shall inherit land, nor any greater amount of wealth than is necessary for a thorough education and such an endowment as to give him a fair start in life."

Such radical opinions as these are common among the workers, but they are also spreading beyond them, owing to the efforts of many talented and energetic thinkers, who expound analogous views with eloquence in the lecture hall, and with argumentative power and literary skill in numerous books and periodicals. The effect of this teaching is manifested in the growing opinion among the more thoughtful even of the wealthy and leisured classes, that a life spent in ease and idleness and the pursuit of pleasure is not the admirable and desirable thing it was once thought to be. The vices and frivolity, the extravagance and the barrenness of modern society are now felt, and are being fully exposed by its own members; and one of these modern prophets, Lady Lyttelton Gell, ably urged, in the *Nineteenth Century* of November, 1892, "that definite work of some sort should be the law, not merely the accessory of every girl's life," and that it should be the means of bringing about more union between the classes, and a real friendship between the highest and the lowest.

Now, I venture to think that nothing would tend more to bring about these desirable results than a method of observing Sunday in some way resembling that here advocated, while the beneficial effect on all concerned would be very great. The upper classes would learn, many of them for the first time, how great and how fatiguing is the labour daily expended in securing them the unvarying comfort and æsthetic enjoyment of their surroundings, and how often they cause unnecessary work by their thoughtlessness or extravagance. The need they would have, at first, of learning the duties of the particular department they were going to undertake, would bring them into friendly and intimate relations with their servants; and, in seeing how much care was often required to secure the comfort of the family, they might begin to appreciate that "dignity of labour" which is so often preached to the poor but so seldom practised by the rich. To many this "Sunday service" in their own families, or in that of some of their friends, would be the introduction to some serious occupation for their weekday lives, and thus inaugurate the great reform which the more thoughtful leaders of society see to be of imperative necessity.

On the whole body of the workers the effect would be

great indeed, since it would at once bring about better relations with the wealthy classes, and especially with those who teach or profess religion. They would see, what they had hitherto doubted or denied, that the religion of the upper classes had some real influence on their lives, by leading them, not merely to give away a portion of their surplus wealth in charity, or to take part in the public proceedings of charitable institutions, but really to sacrifice something which they have hitherto considered necessary to their comfort in order to obey the laws of that religion. They would further see, everywhere, men and women of culture voluntarily undertaking various public and private duties, in order to allow all kinds of workers to enjoy repose and recreation on one day in seven; and this great object-lesson in brotherhood and sympathy would lead to a general good feeling between all classes. The harmonious relations which would be thus produced might be of inestimable value when the time comes for those radical reforms in our social organization which are more and more clearly seen to be inevitable in the not distant future.

It is, perhaps, too much to expect that the "counsel of perfection" here set forth for the consideration of the religious world by an outsider, will have much effect on conduct. But even if it should influence a few here and there to alter their mode of life on the day they hold to be divinely instituted as a period of complete rest for all servants and beasts of burden, and if it should render others less severe in their judgment of those they term "Sabbath-breakers," but who often less deserve that name than do their accusers—and if it thus helps, in however small a degree, to lower the barriers which now divide class from class, and to remove one of the causes which lead many of the workers to look upon the religion of the rich as little better than hypocrisy, the object with which it was written will have been fulfilled.

CHAPTER XXI

WHY LIVE A MORAL LIFE?[1]

TAKING morality in its ordinary meaning, as including all actions for personal ends which are not knowingly injurious or painful to others, the question asked is, What are the sanctions of morality to the pure Rationalist—to the person who does not actively believe in a future state of existence? Can such a person give clear and logical reasons of sufficient cogency to induce him, even under the stress of temptation, and when any detection or evil results to himself appear out of the question, yet to act with strict conformity to moral principles?

. In existing society the abstention from immoral actions by individuals is usually due to one or more of the following causes :—(1) A natural upright and sympathetic disposition, to which any act hurtful or disagreeable to others is repugnant, and is, therefore avoided. (2) The fear of punishment, or of the condemnation of public opinion, leading to ostracism by the society in which they live. (3) The influence of religious belief, which declares certain acts to be offensive to the Deity, and to lead to punishment in a future life. (4) The belief expressed in the saying, " Honesty is the best policy," which may be expanded into the general principle that the moral life is, emphatically, the happiest life.

With the first cause, on which, probably, the largest proportion of moral action depends, we have here nothing to do, since it does not involve any process of reason—of *why* we should act in one way rather than in another—

---

[1] This article formed part of a Symposium on the above question which appeared in the *Agnostic Annual*, 1895.

but rests entirely on feeling, due to natural disposition. It is, however, the greater or less proportion of such persons in any community that determines the action of the next most powerful incentive to morality—public opinion; since dread of the criminal law is not so much dread of the punishment itself as of the disgrace attending it. To the great majority of educated people this is undoubtedly the most powerful incentive to abstain from immoral conduct; while the correlative approval of society has a large share in producing actively moral conduct, especially under conditions when such conduct is more or less open to public notice.

The other two causes enumerated above have, comparatively, very little influence on conduct. Innumerable examples show that the firmest belief in the doctrine of future rewards and punishments has hardly any influence on conduct in cases where it is not enforced by the approval or disapproval of public opinion. It is now generally admitted that the believer in religious dogma is, on the average, neither more honest nor more moral than the Agnostic or the Atheist. No doubt, in exceptional cases, religious enthusiasm acts upon character and conduct in a very powerful degree. We are, however, concerned here, not with exceptional cases, but with the average individual, and it has not been shown by any statistical inquiry that belief in the system of future rewards and punishments leads to exceptionally moral conduct. The same may be said of the believers in the essential reasonableness of a moral life as the best guarantee of permanent happiness. It is doubtful whether such a belief, however firmly held, really influences any one in time of temptation, or leads to any change of conduct which society does not condemn, but which is yet fundamentally immoral. It was, and is held by great numbers of persons, both religious and sceptical, that slavery was absolutely immoral; yet, probably, not one in a thousand followed the Quakers in refusing to purchase slave-grown sugar. Neither will it be maintained that any belief in the abstract principle of the beneficial results of morality would restrain a poor, selfish, and naturally unsympathetic man from pressing the electric button which would at once destroy some unknown millionaire and make the agent of his destruction the honoured inheritor of his wealth.

It is under circumstances analogous to the last-mentioned case that we can alone have a real test of the efficiency of any alleged sanction for morality. When a man can greatly benefit himself by an act which he believes can never be known, and which will, perhaps, only slightly injure others—as by destroying a will of whose existence no other person is aware—no belief in the general principle that honesty is the best policy can be depended on to secure a strictly moral line of conduct. Why, in fact, should a man give up what he knows will ensure freedom from anxiety, and from a constant and laborious struggle for bare existence, and afford him the means of living a pleasurable and luxurious life—the only life in which he has any belief—and all for the sake of a general principle which the society around him does not, as a rule, act upon? Why should he thus injure himself and his own family in order to benefit strangers of whom he knows nothing? Of course there are many men, without either religion or any formulated ethical principles, who would not hesitate a moment in such a case, because their natural sentiments of right and justice, enforced by constant association with men of honour and morality, would render the strict line of moral action natural and easy to them; but with such men we have, so far as the present discussion is concerned, nothing to do.

For these reasons, it seems to me that the Rationalist or Agnostic has no adequate motive for living a moral life, except so far as he is influenced by public opinion and by a belief that, generally, it pays best to do so. But neither of these influences is of the least value, either in exceptional cases of temptation, or in those very common circumstances when the usual actions of the society in which a man lives are not justified by morality; as in the innumerable adulterations, falsehoods, and deceptions so common in trade that it has been even asserted that no thoroughly honest manufacturer or tradesman can make a living.

Religious belief would, on the other hand, furnish an adequate incentive to morality, if it were so firmly held and fully realized as to be constantly present to the mind in all its dread reality. But, as a matter of fact, it produces little effect of the kind, and we must impute this, not to any shadow of doubt as to the reality of future rewards and punishments, but rather to the undue importance attached to belief, to prayer, to church-going, and to repentance, which are often held to be sufficient to ensure salvation, notwithstanding repeated lapses from morality during an otherwise religious life. The existence of such a possible escape from the consequences of immoral acts is quite sufficient to explain why the most sincere religious belief of the ordinary kind is no adequate guarantee against vice or crime under the stress of temptation.

There is, however, one form of religious belief which, if it were to become general, would, I believe, afford a better sanction for a moral life than can now be found either in rationalism or in religion. It is to be found in the teachings of Modern Spiritualism, which, though they were to some extent anticipated by a few spiritual and poetical natures, have never been so fully and authoritatively set forth as through those exceptionally gifted individuals termed mediums. We have here nothing to do with the evidence for the truth of Spiritualistic phenomena, which the present writer has discussed elsewhere,[1] but only with the question whether its teachings do really afford the required sanction for a moral life. Let us then see what these teachings are.

The uniform and consistent statements, obtained through various forms of alleged spiritual communications during the last fifty years, declare that we are, all of us, in every act and thought of our lives, helping to build up a mental fabric which will be and constitute ourselves in the future life, even more completely than now. Just in proportion as we have developed our higher intellectual and moral nature, or starved it by disuse, shall we be well or ill fitted

[1] See *Miracles and Modern Spiritualism* (Trübner and Co.); and the article "Spiritualism," in the new edition of *Chambers's Encyclopædia*.

for the new life we shall enter on. The Spiritualist who, by repeated experiences, becomes convinced of the absolute reality and the complete reasonableness of these facts regarding the future state—who knows that, just in proportion as he indulges in passion, or selfishness or the reckless pursuit of wealth, and neglects to cultivate his moral and intellectual nature, so does he inevitably prepare for himself misery in a world in which there are no physical wants to be provided for, no struggle to maintain mere existence, no sensual enjoyments except those directly associated with sympathy and affection, no occupations but those having for their object social, moral, and intellectual progress—is impelled towards a pure and moral life by motives far stronger than any which either philosophy or religion can supply. He dreads to give way to passion or to falsehood, to selfishness, or to a life of mere luxurious physical enjoyment, because he knows that the natural and inevitable consequences of such a life are future misery. He will be deterred from crime by the knowledge that its unforeseen consequences may cause him ages of remorse; while the bad passions which it encourages will be a long-enduring torment to himself in a state of being in which mental emotions cannot be put aside and forgotten amid the fierce struggles and sensual excitements of a physical existence.

Again, the Spiritualist not only believes, but often obtains direct evidence of the fact, that his dearest friends and relations, who have gone to the higher life, are anxiously watching his career, and themselves suffer whenever he gives way to temptation. An American Spiritualist writes :

"To the son or daughter that has been deprived of parents' care, and perhaps has strayed from the paths of rectitude and purity, will not the knowledge that loving hearts are cognisant of every departure from the right way be an incentive for them to retrace their steps, to strive to so live as to deserve the approval of the angelic ministers ?. . . The knowledge that the loving eyes of a mother or father, a beloved child or companion, are watching us with tender solicitude will be a restraining influence from evil courses, and an incentive to a higher and purer life, when all other influences fail."

Some of the highest teachings of Modern Spiritualism have been given through the automatic writings of the late Mr. Stainton Moses, and are to be found in his work entitled *Spirit Teachings*. His perfect integrity is guaranteed by Mr. F. W. H. Myers, and there is the very strongest internal evidence that the substance of the writings emanated from some intelligence other than his own. But, however this may be, these teachings are perfectly consistent with those of Spiritualism generally, and the following short extracts will illustrate their bearing on the question we are here discussing:

"As the soul lives in the earth-life, so it goes to the spirit-life. . . The soul's character has been a daily, hourly growth. It has not been an overlaying of the soul with that which can be thrown off; rather it has been a weaving into the nature of the spirit that which becomes part of itself, identified with its nature, inseparable from its character."

And again :

"We know of no hell save that within the soul—a hell which is fed by the flame of unpurified and untamed lust and passion, which is kept alive by remorse and agony of sorrow, which is fraught with the pangs that spring unbidden from the results of past misdeeds, and from which the only escape lies in retracing the steps and cultivating the qualities which bear fruit in love and knowledge of God."

And, as a final epitome of this spiritual teaching, we have the following:

"We may sum up man's highest duty as a spiritual entity in the word 'Progress'—in knowledge of himself, and of all that makes for spiritual development. The duty of man, considered as an intellectual being possessed of mind and intelligence, is summed up in the word 'Culture' in all its infinite ramifications, not in one direction only, but in all; not for earthly aims alone, but for the grand purpose of developing the faculties which are to be perpetuated in endless development. Man's duty to himself, as a spirit incarnated in a body of flesh, is purity in thought, word, and act. In these three words, 'Progress,' 'Culture,' 'Purity,' we roughly sum up man's duty to himself as a spiritual, an intellectual, and a corporeal being."

The same teaching is embodied in the following lines, forming part of a long poem purporting to come from the late Edgar Allen Poe:

"Sons of earth, where'er ye dwell,
Shun temptation's magic spell,
TRUTH is HEAVEN, and falsehood, HELL,
Lawless lust a demon fell."

The general answer I would now give to the question, "Why live a moral life?" is—from the purely Rationalistic point of view—first, that we shall thereby generally secure the good opinion of the world at large, and more especially of the society among which we live; and that this good opinion counts for much, both as a factor in our happiness and in our material success. Secondly, that, in the long run, morality pays best; that it conduces to health, to peace of mind, to social advancement; and at the same time, avoids all those risks to which immoral conduct, especially if it goes so far as criminality, renders us liable.

It must be conceded that both these reasons, which are really but one, are of a somewhat low character; yet it seems to me they are all which the Agnostic can, logically, rely upon. It will also be evident that they will be of little value in cases of great temptation, or in those more frequent cases in every-day life where the standard of morality is already low. To raise this standard, and thus increase the force of public opinion as an incentive to morality, we require to increase the proportionate number of the naturally moral, and we have at present no way of doing that.

There remains only one other answer, which, at present, is only applicable among that section of the community which has obtained conviction of the reality of a future life through Modern Spiritualism, and is, therefore, influenced by the teachings as to the nature of that life of which I have sketched the barest outlines. Some of my readers may object that Modern Spiritualism is not Rationalism, and is, therefore, outside this discussion; to which I reply —Why not? It is founded on a personal and critical observation of *facts*. Is not that rational? Is it more rational to refuse to investigate these facts, or to deny them without investigation? I myself had been for nearly thirty years an Agnostic when I began to investigate these phenomena, and found them, against all my prepossessions, to be realities. Is it rational to ignore or deny phenomena which have been demonstrated to the satisfaction of such men as Robert Chambers, Professor De Morgan, Dr. Lockhart Robertson, William Crookes, and scores of other great thinkers, and has drawn from the ranks of English Secularists Robert Owen, George Sexton, and Annie Besant? But, it may be said, admitting the facts, the theory is irrational. Here, again, I ask, who can judge better of the correctness of the theory—those who have personally investigated the facts, or those who have not? But really, it is not a question of theory, since, when the whole facts are known to be realities, no other conclusion is possible or rational than that of the Spiritualists.

It has been shown, and will, I am sure, be admitted by all unprejudiced readers, that we have derived from Spiritualism a conception of a future state and of its connection with our life here very different from, and far superior to, the ordinary religious teaching which formerly prevailed. That teaching has now been partly modified through the influence of Spiritualistic ideas; but by the religious preacher it is taught dogmatically, not as it comes to the Spiritualist with all the force of personal communication with those called dead, but who, again and again, tell us they are far more alive than ever they were here. This Spiritualistic teaching as to another life enforces upon us, that our condition and happiness in the future life depends, by the action of strictly natural law, on our life and conduct here. There is no reward or punishment meted out to us by superior beings; but, just as surely as cleanliness and exercise and wholesome food and air produce health of body, so surely does a moral life here produce health and happiness in the spirit-world. Every well-informed Spiritualist realises that, by every thought and word and deed of his daily earth-life, he is actually and inevitably determining his own happiness or misery in a future life which is continuous with this—that he has the power of creating for himself his own heaven or hell. The Spiritualists alone, therefore, or those who accept with equal confidence the Spiritualistic teachings in this respect, can give fully adequate reasons why they should live a moral life. These reasons are in no way dependent on public opinion, or on any relation to success or happiness here, and are, therefore, calculated to influence conduct under the most extreme conditions of temptation or secrecy. Hence the only Rationalistic and adequate incentive to morality—the only full and complete affirmative answer to the question, " Why live a moral life ? "—is that which is based upon the conception of a future state of existence systematically taught by Modern Spiritualism.

CHAPTER XXII

THE CAUSES OF WAR, AND THE REMEDIES

IN response to a request from the Editor of *L'Humanité Nouvelle*, I give my views, briefly, on the questions submitted to me.[1]

(1) Under the existing conditions of society in all civilized communities, and as a consequence of the principles and methods of government which prevail in them, war cannot cease to be more or less prevalent among them.

The conditions which almost inevitably lead to war are the existence of specialized ruling and military classes, to whom the possession of power and the excitements and rewards of successful war are the great interests of life. So long as the people permit these distinct and independent classes to exist, and—more than this—continue to look up to them as superiors and as necessary for the proper government of the country and for the effective protection of individual and national freedom, so long will these rulers continue to make wars.

---

[1] These questions were as follows :—
1. Is war among civilized nations still necessary on the grounds of history, right, and progress?
2. What are the effects of militarism—intellectual, moral, physical, economic, and political?
3. What is the best solution of the problems of war and militarism in the interests of the future civilization of the world?
4. What is the most rapid means of arriving at this solution?

A French translation of the larger part of this article appeared in the special supplementary number of *L'Humanité Nouvelle* for May, 1899, which contains replies by more than 130 writers, including Tolstoy and seven English authors.

All civilized governments, whatever may be their professions, *act* on the principle that extension of territory and the absorption of adjacent or remote lands, so as to increase both the extent of country and the population over which they have sway, is a good in itself, quite irrespective of the consent of the peoples so absorbed and governed, and even when the peoples are alien in race, in language, and in religion. Although they may not openly avow their acceptance of this doctrine, yet they invariably *act* upon it, though in some cases they think it necessary to make excuses for their action. They declare that such conquest and absorption is necessary for the national safety, for the increase of trade, and for many other reasons. The majority of the workers, and of educated people who do not belong to the ruling or the military classes, however, do not accept this principle. They more or less decisively hold the opinion that governments can only justly derive their power from the consent of the governed, and that all wars for territory and all conquests of alien peoples are wrong.

The reason of this difference of opinion is very simple. Every addition of territory, every fresh conquest even of barbarous nations or of savages, provides outlets and additional places of power and profit for the ever-increasing numbers of the ruling classes, while it also provides employment and advancement for an increased military class, in first subduing and then coercing the subject populations, and in preparing for the inevitable frontier disputes and the resulting further extensions of territory. Wars and conquests and ever-expanding territories are thus found to be essential to their existence and continued power as superior classes. But the people outside these classes derive little, if any, benefit from such extensions, while they invariably suffer from increased taxation, either temporarily or permanently, due to increased armaments which the protection of the enlarged territory requires. Almost without exception every war of modern times has been a dynastic war—a war conceived and carried out in the interests of the two great governing classes, but having no relation whatever to the well-being of the peoples who have been forced to fight each other. In every case the people suffer by the loss or disablement of sons, husbands, and fathers, by the destruction of crops, houses, and other property, and by increased taxation, due to the increase of armaments that always follows such wars even in the case of the victors. Hence the material and moral interests of the mass of the people of every country are wholly opposed to war, except in the one case of defending their country against invasion and conquest. They are therefore more open to the influence of moral and humane considerations, while they alone feel the full force of the numberless evils which war brings upon them. Except in very rare cases, a plébiscite fairly taken would decide against any other than a defensive war.

(2) To discuss the effects of militarism under the various heads suggested in the question would require much space and some special knowledge which I do not possess. That these effects have both good and evil aspects may be admitted. The evil effects have been often set forth and are sufficiently known, both in their vast extent and far-reaching consequences, while the greatest of them—the perpetuation of war and the desire for military glory—has already been alluded to. I will, therefore, confine my remarks to the partial good that undoubtedly exists in this fundamentally evil thing, chiefly for the purpose of showing that whatever good there is in it may be obtained in other ways which are as essentially humane, moral, and beneficial as war is essentially cruel, immoral, and hurtful.

The good that results from militarism arises wholly from the perfection of its organisation, of its training, of the habits of order, cleanliness, and obedience which the soldier soon learns are essentials to efficiency; from the social and brotherly life of the soldier, whether in camp or in the field; from the *esprit de corps* which grows out of its systematic organisation and companionship, leading to generous rivalry and to those deeds of heroism and self-sacrifice which are universally admired. And, further, every soldier learns by experience the marvellous power of organized labour under skilled direction to overcome what to the ordinary man seem insurmountable difficulties. He sees how foaming torrents or broad rivers can be rapidly bridged; how roads can be made over morasses or across mountains; how the most formidable and apparently impregnable defences are attacked and taken; and how a few bold men in a "forlorn hope," by the sacrifice of their lives, often ensure the success of the army to which they belong. Many of the finest qualities of our nature are thus called into action by the soldier's training and during his struggle against the enemy; and so greatly has humanity developed among us that it may be fairly argued that these good effects more than balance the evil passions of cruelty, lust, and plunder which even now are to some extent manifested in every great war, though to a far less degree than even fifty years back.

But every one of these good results of militarism could certainly be obtained by any equally extensive and equally skilful organization for wholly beneficial purposes. If labour, where organized for military ends, is so effective in results and so beneficial as a training, it would be equally effective and equally beneficial when devoted to overcoming the obstacles to man's progress presented by nature; to the production of the necessaries of civil life; to sanitary works for the preservation of health; and to everything that facilitates communication and benefits humanity. If the same amount of knowledge, the same amount of energy, and the same lavish expenditure where absolutely required, were devoted to the training of great industrial armies, to their maintenance in the most perfect health and efficiency, and to their employment in that great war which man is ever waging against Nature, subduing her myriad forces to his service, guarding against those sudden attacks by storm and flood, by avalanche and earthquake, which he cannot altogether avoid, and in the production of all the essentials of human life and of a true and beneficent civilization, the good effects on character would surely be as much greater than those produced by mere military training, as the objects aimed at and the results achieved would be more beneficial and more calculated to promote the higher interests of man. And if these industrial armies were allowed to reap the full advantages, material as well as moral, which they created, the results would be so striking that almost the entire population, male and female alike, would claim to be so trained and organized for their own physical, moral, and economic benefit. And the enjoyment of life under such a system of voluntary organized labour would be so enhanced that few indeed would wish to escape from it. Labour in companionship for the common good almost ceases to be labour at all. Friendly emulation takes the place of unfriendly competition, and *esprit de corps* urges each local organisation to surpass other local organizations in efficiency. In such a grand industrial organization, with equal opportunities of education and training for all, there would necessarily be numbers of inventors and students whose aim and delight would be to so improve the machinery and the methods of work as to continually diminish all the less pleasant forms of labour, and thus proportionately increase the amount of leisure and the higher enjoyments of social life.

It has been objected to all such proposals for the organization of industry that it would deteriorate character by destroying individuality; but no such objection is made to the military organization, while under its best forms the reverse is found to occur. In point of fact, *all* organization is beneficial to character just in proportion as it rises above slavery. And when it shall have reached the point of being the organization of social equals for the equal benefit of all, it will attain to its most beneficial influence. Then, character and merit will alone give authority, and the highest and best will inevitably rise to the highest positions. And, just in proportion as the rank and file became educated, and felt the inspiring influences of comradeship and emulation, they could be left more and more to their own initiative; each one's individuality would have the fullest play, controlled only by the influence and opinion of his immediate fellow-workers, and the whole great organization would become almost automatic in its harmonious operation.

Such is found to be the case in the best military organizations, in which the intelligence and individual action of both commissioned and non-commissioned officers, and even of privates, is cultivated, and becomes of the greatest value, giving to the army in which it most generally exists an undoubted superiority. In any army thus intelligently and sympathetically trained and organized none of the results so dreaded in industrial organization are found to occur. Men are *not* brought to a dull level of mediocrity; interest in the work they have to do is *not* lost; skulkers, malingerers, and deserters do *not* abound in any appreciable or hurtful proportion ; nor is there any indication that men of superior abilities refuse to exercise their talents for the common good because the money rewards of such ability are small as compared with those often obtained in civil life; and, lastly, the fact that all are provided with food and clothing, and are thus removed from the influence of economic competition, is *not* found to have any injurious effect on their effectiveness as workers, fighters, or organizers. And that these effects are not caused by compulsion and the severe penalties of military law is shown by the fact that during the civil war in America, where compulsion and punishment were rarely used, the whole of the opposing armies being practically volunteers cheerfully submitting to military drill and organization for the common good, these high qualities were equally manifested.

Yet objections of this class are held to be fatal to any proposal for national industrial organization for the benefit of all, and the very system of training and co-operation which in the one case is admitted to have beneficial effects on character, and is undoubtedly, even under very unfavourable conditions, attractive in its comradeship and freedom from care, is condemned as being injurious and unworkable when applied industrially. Oh! that some great ruler of men would arise to benefit humanity by organizing industrial armies, leading to the elevation and happiness of a whole people, and thus proving that peace may have its victories far greater and more glorious than those of war!

(3-4) The two last questions—as to the solution of the problems of war and militarism, and the means of arriving as rapidly as possible at such a solution—have already been partly answered in the preceding discussion of the problem itself, but a few words may here be added.

It is, I think, clear that no hope of a complete solution, hardly even of amelioration—is to be expected from the ruling classes, urged on as they are on the one hand by those who are ever seeking for place and power or for official appointments in newly-acquired territories, and on the other hand by the military classes, who ever seek to justify their existence and the enormous burden they are to the nation by obtaining for it extensions of territory or military glory, and with either of these an extension of their own influence. It is, therefore, the *people*, and the people alone, that must be relied upon to banish militarism and war, and for this end every possible effort must be made to educate and enlighten them, not only as to the horrors and iniquity of war, but as to the utter inadequacy and worthlessness of almost all the causes for which wars are waged. They must be shown that all modern wars are dynastic; that they are caused by the ambition, the interests, the jealousies, and the insatiable greed of power of their rulers, or of the wealthy mercantile and financial classes which have the greatest influence over their rulers; and that the results of war are *never* good for the people who yet bear all its burthens.

In the course of this education of the people there are certain points that should be specially advocated. For example, nothing is more inconsistent, more foolish, and more wicked than the universal practice of civilized and Christian nations in selling all the most improved weapons and instruments of destruction to semi-civilized, barbarous, or savage rulers, thereby rendering it more difficult—more costly in blood and treasure—to deal with such rulers when their crimes against their own peoples or against humanity becomes too great to be borne. This practice also renders it ever more and more difficult for advanced nations to disarm, and thus gives to militarism an additional reason for its existence. From every point of view, whether of Christianity, humanity, or human progress, the supply of modern instruments of war to barbarous rulers, for the coercion of their own subjects, and as a standing menace to civilization, should be absolutely forbidden. For this purpose, and in order that legal enactments to this end may be effective, we must try and create a sentiment of horror against those who continue thus to betray the cause of civilization, as being not only traitors to their country but enemies to the human race. In my opinion, men who, after due notice and in spite of its declared illegality, continue to supply these weapons to the possible enemies of their country should be declared outlaws in every Christian or civilized community.

Hardly less foolish and wicked is the free trade in these instruments and armaments of war, so that directly one or more of the civilized nations are preparing for war the workshops of all the other civilized nations are at once engaged in supplying every kind of destructive appliance, even though they may in a year or two be used against themselves. The time will surely soon come when this conduct will be looked upon as the very culminating point of combined folly and wickedness that the world has seen. The only rational mode of procedure would be to forbid altogether the private manufacture or sale of war material. War is a national act, and so long as it exists all preparation for it should be kept strictly in the hands of national governments.

This supply of the implements of war is the work of capitalists in their own interests; but even worse, if that be possible, is the action of the great civilized governments themselves in allowing their trained officers to engage in the organization of the armies of semi-barbarous rulers, thus rendering it more difficult to coerce these rulers in the interests of civilization, and indirectly, yet most certainly, leading to a vast extension of the horrors of war. The entire absence of ethical principle created by militarism is especially shown in the fact that no effective protest has been raised against this most pernicious and suicidal practice. Here, again, the people alone can take effective action, and the people want educating. Common justice, common humanity, even common sense, alike demand that this practice be absolutely forbidden, and that any officer engaging in the organization of the armies of semi-barbarous or alien rulers should be declared an outlaw by the Government in whose army he was trained, be demanded from the employing government as a traitor to his country, and the refusal to give him up be followed by an instant declaration of war from *all* the civilized governments.

Yet another point on which the people should be educated is, that they should claim and exercise the right to refuse, as soldiers, to act against their fellow-countrymen or against other countries with whose people they have no quarrel. Accepting the principle that the only just rights of governments rest upon the consent of the governed, what is termed rebellion is not a crime, but is usually the just demand of a community for self-government, a demand which, instead of being repressed by force, should be tested by a plébiscite. And smaller disturbances, termed riots, always arise from some injustice or supposed injustice, and are not proper subjects for massacre by armed soldiers. To use fire-arms against a crowd, and kill or maim innocent persons, women and children, as almost always happens, is to authorize murder. Whenever it may be necessary to prevent violence by a mob, and the available force of police is not sufficient, special constables should be enrolled. But a far better plan would be to organize the fire-brigades as coadjutors of the police, since it is certain that no unarmed (or even armed) mob can stand against the jet of a fire-engine or of several fire-engines. The mob would instantly disperse, and be rendered ridiculous without endangering life.

Of course, any proposed system of arbitration to settle

disputes between nations should be strongly supported; but the existing condition of all the great civilized governments renders it certain that, so long as the ruling and military classes exist, and are allowed to possess the almost absolute powers they now exercise, war, as the ultimate mode of settling national disputes, will not cease.

CHAPTER XXIII

THE SOCIAL QUAGMIRE AND THE WAY OUT OF IT[1]

I. THE FARMERS

IN the early years of the century, English readers enjoyed the perusal of many American works of fiction dealing with the rural life of the Eastern States in those almost forgotten days when railways and telegraphs were unknown, when all beyond the Mississippi was "the far west," when California and Texas were foreign countries, and when millionaires, tramps, and paupers were alike unknown. They introduced us to an almost idyllic life, so far as rude abundance, varied occupations, and mutual help and friendliness among neighbours constitute such a state of existence. Almost all the necessaries and many of the comforts of life were obtained by the farmer from his own land. He had abundance of bread, meat, and poultry, with occasional game. Of butter, cheese, fruit, and vegetables there was no lack. He made his own sugar from his maple trees, and soap from refuse fat and wood ashes; while his clothes were the produce of his own flocks, spun, and often dyed, woven, and made at home. His land contained timber, not only for firing, but for fencing and house-building materials, as well as for making many of his farm implements; and he easily sold in the nearest town enough of his surplus products to provide the few foreign

---

[1] This chapter appeared in the *Arena* (Boston U.S.A.), of March and April, 1893. It deals with the problem mainly from an American point of view; but for that very reason it is specially instructive to us, because similar evils have arisen there under conditions which are often alleged, by English writers, to be the remedy for them.

luxuries that the family required. The farmer of that day worked hard, no doubt, but he had also variety and recreation, and there was none of that continuous grinding, hopeless toil, that appears to characterize the life of the Western farmer to-day. As a rule, his farm was his own, unburdened by either rent or mortgage. Year by year it increased in value, and if he did not get rich he was at least able to live in comfort and to give his sons and daughters a suitable start in life. In those days wages of all kinds were high; food was cheap and abundant; and the strange phenomenon—yet so familiar and so sad a phenomenon now—of men seeking for work in order to live, and seeking it in vain, was absolutely unknown.

The impression of general well-being and contentment given by these tales was confirmed by the narratives of travellers and the more solid works of students of society. All agreed in telling us that not only the pauperism of Europe, but even ordinary poverty or want, were quite unknown. The absence of beggars was a noticeable fact; and except in cases of illness, accident, or old age, occasions for the exercise of charity could hardly arise. The extraordinary contrast between this state of things and that which prevailed in Europe had to be accounted for, and several different causes were suggested. A favourite explanation on both sides of the Atlantic was, that it was a matter of political institutions. On the one hand, it was said, you have a Republican government, in which all men have equal rights and no privileged classes can oppress or rob the people; on the other, there is a luxurious court, a bloated aristocracy, and an established church, quite sufficient to render a people poor and miserable; and this was long the opinion of the English radicals, who thought that the cost of the throne and of the church was the chief cause of the poverty of the working classes. Others maintained that it was entirely a matter of density of population. Europe, it was said, was overpeopled; and it was prophesied that, as time went on, poverty would surely arise in America and become intensified in Europe. More philosophical thinkers imputed the difference to the fact that there was an inexhaustible supply of unoccupied and fertile land in America, on which all who desired to work could easily support themselves; and that, all surplus labour being thus continually drawn off, wages were necessarily high, as the only means of inducing men to work for others instead of for themselves. When the accessible land was all occupied, it was anticipated that America would reproduce the phenomena of poverty in the midst of wealth which are prevalent throughout Europe.

It is needless to point out that these anticipations have been realized far sooner and far more completely than were ever thought possible. The periodical literature of America teems with facts which show that the workers of almost every class are now very little, if any, better off than those of the corresponding classes in England. For though their wages are nominally higher, the working hours are longer; many necessaries, especially clothing, tools, and house rent, are dearer; while employment is, on the whole, less continuous. The identity of conditions as regards the poverty and misery of the lower grades of workers is well shown by the condition of the great cities on both sides of the Atlantic. The description of the dwellers in the tenement houses of New York, Boston, and Chicago exactly parallels that of the poorer London workers, as revealed in the " Bitter Cry of Outcast London," in the " Report of the Sweating Commission," and in cases of misery and starvation recorded almost daily in the newspapers. In both we find the same horrible and almost incredible destitution, the same murderous hours of labour, the same starvation wages; and the official statistical outcome of this misery is almost the same also. The English registrar-general records that considerably over one tenth of all the deaths in London occur in the workhouses, while nearly the same proportions receive pauper burial in New York.[1]

Henry George, in his great work *Progress and Poverty*, declares in his title page that there is, in modern civilization, "increase of want with increase of wealth;" and in Book V., Chapter II., he traces out the causes of "the persistence of poverty amid advancing wealth."

[1] See James B. Weaver's *Call to Action*, p. 369.

The truth of this latter statement stares us in the face in every country, and especially in every great city, of the civilized world; no one can have the hardihood to deny it. But people are so dazzled by the palpable signs of wealth and luxury which everywhere surround them; so many comforts are now obtainable by the middle classes, which were formerly unknown; so many and so wonderful have been the gifts of science in labour-saving machinery, in the means of locomotion and of distant communication, and in a hundred arts and processes which add to the innocent pleasures and refinements of life; and again, so jubilant are our legislators and our political writers over our ever-increasing trade and the vast bulk of yearly growing wealth, that they cannot and will not believe in the increase, or even in the persistence of an equal amount of poverty as in former years. That there is far too much cruel and grinding poverty in the midst of our civilization, they admit; but they comfort themselves with the belief that it is decreasing; that bad as it is, it is far better than at any previous time during the present century, at all events; and they scout the very notion that it is even proportionally as great as ever, as too absurd to be seriously discussed.

These good people, however, believe what they wish to believe, and persistently shut their eyes to facts. Even in Great Britain it can, I believe, be demonstrated that there is actually a greater bulk of poverty and starvation than one hundred or even fifty years ago; probably even a larger proportion of the population suffering the cruel pangs of cold and hunger. I need not here go into the evidence for this statement, beyond referring to two facts. There has, during the last thirty or forty years, been an enormous extension of the sphere of private charity, together with a judicious organization calculated to minimize its pauperizing effects.

Besides the marvellous work of Dr. Barnardo and General Booth, there are in London, and in all our great cities, scores of general and hundreds of local charities; while the numbers of earnest men and women who devote their lives to alleviating the sorrows and sufferings of the poor, have been steadily increasing, and may now be counted by thousands. The fact of a slight diminution in the amount of State relief under the poor law is, therefore, quite consistent with a great increase of real poverty; yet this slight diminution is again and again cited to show that the people are really better off. This decrease is, however, wholly due to the growing system, favoured by the authorities, of refusing all outdoor relief, the place of which is fully taken by the increase of private and systematized charity. And there is good proof that this vast growth of charitable relief has not overtaken the still greater increase of real pauperism. This proof is to be found in the steadily increasing proportion of the population of London which dies in the workhouses. The registrar-general gives this number as *6,743* in 1872; in 1881 it had risen to *10,692*, and in 1891 to *12,473*. Thus the deaths of paupers in workhouses had increased 85 per cent. from 1872 to 1891, while the total deaths in London during the same period had increased from 70,893 to 90,216, or 27 per cent. It may be thought that this has been caused by the influx of the poor into the towns; but it is mainly the young that thus emigrate; and the registrar-general shows that the same increase of deaths in workhouses has occurred, though in a less degree, in the whole country. In his report for 1888, the only one I have at hand, he says:

"The proportion of deaths recorded in workhouses, which steadily increased from 5·6 per cent. in 1875 to 6·7 per cent. in 1885, further rose, after a slight decline in 1886 and 1887, to 6·9 per cent. in 1888."

The same continuous increase of aged pauperism is thus proved to occur in all England, but to be especially great in the larger cities; and this fact appears to me to demonstrate the increase of poverty during the last twenty years of rapidly increasing wealth and ever-growing luxury. And at the same time, notwithstanding all the efforts of all the charitable institutions and philanthropic associations, we see every week in the papers, though only a few of these cases get noticed, such headings as

"Shocking Destitution," "Destitution and Death," proving that the official records, terrible though they are, only show us a portion, perhaps only a small portion, of the wretchedness and poverty culminating in actual death from want of food, fire, and clothing, in the midst of the wealthiest city the world has ever seen.[1]

But if any real doubt can exist as to the actual increase of poverty in England, we have in America an object lesson in which the fact is demonstrated with a clearness and fulness that admits of no dispute. Fifty years ago there was, practically, no poverty, as we now understand the word, in the sense of men willing to work being unable to earn enough to support their families. Now these exist by tens of thousands, culminating in all the great cities, in actual death caused or accelerated by want of the barest necessaries of life. That the wealth of the community has increased enormously in this period, there is also no doubt. According to Mr. Mulhall, the great English statistician, the total wealth of the United States increased nearly seven-fold from 1850 to 1888, while the population had increased less than two and three-fourths fold. Here, then, we have a clear and palpable "increase of want with increase of wealth;" and as the causes which have been at work in the production of this increased wealth are of exactly the same nature in America and in England, only that they have acted with more intensity in America, we are supported in the conclusion that the coincident increase of want has occurred also, though with less intensity, in England. The causes of the enormous wealth-increase are simple and indisputable. First, steam power has increased in America seven-fold (and probably as much in England), and its application to ever-improving labour-saving machinery has given it an effective productive power of perhaps twenty-fold or even more; secondly, railways have spread over the country, enabling the varied products of the whole land to be more and

---

[1] Fuller and much later details on these points are given in my *Wonderful Century*, Chapter XX., proving that poverty and its attendant evils have gone on increasing at an increasing ratio, to close on the end of the century.

more utilized. The result of these two great factors has been the corresponding increase of agriculture, mining, manufactures, and commerce, by means of which the increased wealth has been directly produced. If, then, fifty years ago there was practically no want in the United States, and there is now, say, ten times the wealth, with about four times the population, not only ought there to be no want of any kind, but all those who had mere necessaries before should be able to have comforts and even luxuries now; hours of labour should be shorter, and the struggle for existence less severe. But the facts are the very opposite of these; and there has evidently been an increasing inequality in the distribution of the wealth produced. The result of this inequality is seen, broadly, in the increase of wealth and luxury on the one hand, and of the most grinding poverty on the other; and more particularly in the growth and increase of millionaires. Fifty years ago a millionaire was a rarity in England; now they are so common as to excite no special attention. In America in 1840, there was probably no one worth one million pounds (five million dollars). Now there are certainly hundreds, perhaps over a thousand, who own as much; and it has been estimated that two hundred and fifty thousand persons own three fourths of the whole wealth of the country.

The paradox of increasing want with increasing wealth is thus clearly explained. If we take the two hundred and fifty thousand persons above referred to to be heads of families, four to a family, we have a million persons absorbing three fourths of the wealth created by the whole community, the remaining fifty-nine millions having the remaining one fourth amongst them; and as probably half of these are comfortably off, the other half have to exist in various grades of destitution from genteel poverty down to absolute starvation.

The Problem to be Solved.

The problem we have now to solve is, to discover what are the special legal and social conditions that have enabled a small proportion of the community to possess themselves of so much of the wealth which the whole of the community have helped to produce. That much of this wealth has been obtained dishonestly, is quite certain, yet it has for the most part been obtained legally; and it is probable that if the whole of the transactions of some of the chief of American millionaires were made public, few of them would be found to be contrary to law, or even contrary to what public opinion holds to be quite justifiable modes of getting rich. Yet there is probably a very large majority of voters who see the evil results of the system, and would be glad to alter it if they knew how. They have a vague feeling that something is wrong in the social organization which renders such results possible. They begin to see that the old explanations of the poverty and starvation in Europe were all wrong; since, though America still possesses its republican constitution, though it is still free from hereditary aristocracy, state church, or the relics of a feudal system, though its population is less than twenty-five per square mile, while Great Britain has over three hundred, it has, nevertheless, reached an almost identical condition of great extremes of wealth and poverty, of fierce struggles between capitalists and labourers, of crowded cities where women are often compelled to work sixteen hours a day in order to sustain life, and where thousands of little children cry in vain for food. The causes that have led to such identical results, slowly in the one case, more rapidly in the other, must in all probability be identical in their fundamental nature.

The present writer has long since arrived at very definite conclusions as to what these causes are, and what are the measures which alone will remedy the evil. In America there has hitherto been a great prejudice against these measures because they run counter to one of the institutions which has profoundly influenced society, and which, till quite recently, has been considered to be almost perfect and to be of inestimable value—I allude, of course, to the land system of the United States. It is because the present generation has been taught to look upon this land system as almost perfect, that we now behold the curious phenomenon of a large and most important class of the community, the Western farmers, while almost on the brink of ruin, yet quite unable to discover the real cause of their suffering, and frantically asking help of the government through action which might, perhaps, alleviate their immediate distress, but could have no effect in permanently benefiting them. As this question of the farmers is one calculated to throw light on the whole problem of "increasing want with increasing wealth," it will be well to devote a little space to its discussion.

The farmers in the great food-producing States in the West are admitted to be very badly off. A large proportion of them are crushed down by heavy mortgages, others are tenants at high rents, and almost all have a hard struggle for a bare livelihood.[1] Their friends and representatives consider that their misfortunes depend primarily on financial and fiscal legislation, and advocate reforms of this nature. Mr. S. S. King, of Kansas City, says:

"The first step in legislation is for the people to undo, so far as they can, the things done by the hired tools of the monopolists, repeal the National Banking Act, pay off the bonds, stop the interest, call in the National Bank notes, and replace them with full legal-tender paper money issued by the government ... Then let

[1] Mr. Atkinson, the optimist statistician of Boston, in his paper read before the British Association in August, 1892, summarizes the special Census Report on this subject as follows: Dealing with Illinois, Alabama, Tennessee, Iowa, Kansas, and Nebraska, he states that "more than one half of the farms are free from any mortgage," and that "those which are under mortgage are encumbered for less than half their value." This is the optimist way of stating the case, as if it were something gratifying, something that indicated a successful agriculture and a contented body of farmers! Nearly half the farms in six great agricultural States mortgaged! And these mortgaged to nearly half their value, which, at the high rates of interest usually paid, is equivalent to a heavy, sometimes to a crushing rent!! I could scarcely have imagined a more terrible state of things, short of absolute ruin; and had the facts been stated by any less trustworthy authority, I should have thought there was certainly error or exaggeration. It must be remembered, also, that during past years many mortgages have been foreclosed, and the mortgagees are now the landlords. We are not told how many of the farmers in these States are tenants.

the government reclaim from the railroads all the land still held by them beyond what is necessary for the operation of the roads ... take absolute control of the roads ...then level to the ground the tariff-tax abomination."

Hon. James H. Kyle, U.S. Senator from South Dakota, says:

"To pass the income tax; to sweep away national banks; to restore free coinage of gold and silver; to have money issued directly to the people in sufficient volume to meet the needs of legitimate business—these are the reforms which are entirely within the reach of earnest and persistent agitation ... Land loans and produce loans would surely follow ... The nationalization of the great highways of commerce would inevitably follow."

These same reforms are advocated by General J. B. Weaver in his powerful work *A Call to Action*; and he imputes all the evils of the present land system—the increase of large land-owners, the rapidly increasing army of tenants, the numerous mortgages at high interest, and the universal distress of the agriculturists—to causes connected with the banking system and with the tariff.

Now, so far as I can understand these difficult questions, all the evils pointed out by these writers are real and very great evils, and the remedies they suggest may to some extent remove these evils; but I feel convinced that these are *not* the fundamental remedies as regards the farmers. The suggested remedies might benefit them slightly along with the rest of the community, but would not remove the troubles that specially affect the tillers of the soil. It would, no doubt, be an advantage to be able to pay off existing mortgages with money advanced by the government at very low interest; but an agriculture that rests on mortgages, whether at high or at low interest, is not a successful agriculture. General Weaver truly says:

"The cultivation of the soil should be, and in fact is, under natural conditions, the surest road to opulence known among men. Under just relations it would be impossible to impoverish this calling, for it feeds, clothes, and shelters the human family."

And again:

"What the farmer most wants is a good price for the products of his farm, rather than an advance in the value of the farm itself."

But he does not pursue this point, and does not show how any of the remedial measures suggested can possibly raise the price of farm-produce; and unless this is done, the farmer's condition, though it may be somewhat ameliorated, will never be raised to the degree of comfort and security which ought to be enjoyed by those whose labour provides the food of the community.

Let us then try and get at the root of this question. Why is it that the degree of comfort and safety of the American farmer has, during the last fifty years or less, so greatly diminished? What is the cause of the strange phenomenon of food being sold by its producers at such low prices as to be unremunerative to them? It is evident that these prices are determined by competition. How is it that in this particular business competition has forced prices down to such a point as to be permanently unprofitable? The causes that have brought this about are clearly twofold: the absence of the equalizing power of *rent*, and the competition of capitalist or *bonanza* farms. Why this is so will now be explained.

Owing to the almost universal custom in America (until recently) of purchase rather than rental of land, and the wide-spread interests involved in real-estate speculation, the true nature of *rent*, as thoroughly worked out by the political economists of Europe, is quite unknown except to the comparatively few who have made a special study of the subject. It is therefore necessary to show, in as clear a manner as possible, its economic importance, and that it is really the key to the whole problem of American agricultural distress.

The Social Importance of Rent.

Rent is the *equalizer of opportunities*, the means of giving fair play to all cultivators of the soil in the struggle for existence. Farms differ greatly in value, from two quite distinct causes: the fertility of the land itself, as dependent on soil and climate, is one cause; situation, as regards distance from a railroad or from a market, is the other. Let us suppose one farm to produce thirty bushels of wheat an acre, another only twenty, with the same labour and outlay, and that the first farm is only a mile from a railroad, while the other is ten miles over a bad and hilly track. The owner of the first farm will evidently have a double advantage over the owner of the other, both in the amount of his crops and the economy in getting them to market; and prices which will enable the first to live comfortably and lay by money, will mean poverty or ruin to the second. It is just the same as with shops or stores. The business done, other things being equal, will depend upon situation. If one store is situated in a main street, with five hundred people passing the door every hour, and another store just like it is in a bye-street where not more than fifty people pass per hour, and both sell exactly the same goods, of the same quality, and neither have any special connection or reputation, but depend mainly on chance customers, then it is quite certain that the one will make a living where the other will starve. Now prices are fixed by the competition of the whole of the stores of these two classes, and the more favoured class will run down prices just so low that the less favoured class can hardly live; and the inevitable result will be that many of them will be starved out, and the whole of the business be absorbed by the other class. But if all these shops belong to landlords, whether private individuals or the municipalities, then rents will be so much higher in the one class than in the other as to approximately equalize the opportunities of both. Both will then be able to earn a living for a time, and the ultimate superior success of either will be a matter of business capacity. The competition between them will be fair and equal.

The same thing happens with rival manufacturers. Facilities for getting raw material, cheapness of water power or fuel, and above all the possession of the best and most improved machinery, enable one to undersell another, and ultimately to drive him out of the market, unless the latter can improve his conditions, or the former is subject to an increased rent, to compensate for his advantages of position.

Now, in the case of the farmer there is no possibility of removing the disadvantages of some as compared with others. Land which is naturally poor can never be made equal to that which is naturally fertile; neither can a farm be moved bodily near to a market or to a railway. The competition between different farmers is, therefore, not a fair one. As more land is cultivated and more surplus grain produced, those having the advantage in land and situation will get their produce earliest to market; and those who come later, when the market is already well supplied, must take a lower price. Year by year, as the output of grain increases, the price becomes lower still, till it reaches a point at which those worst situated cannot afford to grow it at all. Then either the worst farms go out of cultivation, or some other crops are grown, or the owner, burthened with mortgages, is sold out, and his farm is perhaps joined to another, and goes to form one of the great capitalist farms, which are another means of driving down prices below the level at which the less favoured farmers can make a living.

Many people argue, that, if large farming pays where small farming will not pay, that large farming therefore produces more food and is better for the country. But this is a great mistake. The farms which are measured by thousands of acres rarely produce so much per acre as the small farms of fifty or a hundred acres. In the former the object is to reduce the cost of labour to a minimum, and so leave a larger profit to the owner. Whether in Australia, Dakota, or California, the great machine worked farms only produce from about eight to twelve or twenty bushels of wheat an acre; but on ten thousand acres a very small profit per bushel will give a large income, while the same profit on a much larger produce per acre will starve the small farmer. In 1879 the wheat produce of the United States varied in the several States from an average of seven bushels an acre in North Carolina and Mississippi, to nineteen and twenty bushels in Michigan and Indiana; and in the bad year, 1884, the range was from five bushels to twenty bushels. But as these are the averages of whole States, the produce of the several farms must have a very much wider range ; and the profit made will vary still more than the produce, owing to much greater cost of carriage to market in some cases than in others. It thus happens that the variations in the cost of producing and selling a bushel of wheat are, in the United States, extremely large, perhaps larger than in any other part of the world, because, in the first place, that cost is not equalized by any general payment of rent for the land in proportion to its better or worse quality ; and in the second place, because capitalists have been allowed to acquire enormous areas of land from which, by means of machinery and a very little hired labour, they can make large profits from a very small produce per acre.

Some people will say that this result is a good one. Bread is made cheap, and that benefits the whole community. This, however, is one of those utterly narrow views by which capitalist writers delude the people. All other things being equal, cheap bread is doubtless better than dear ; but if cheap bread is only obtained through the poverty or ruin of the bulk of those who grow it, and if its value to most other workers is discounted by lower wages or smaller earnings, both of which propositions are in the present state of society demonstrably true, then cheap bread is altogether evil.

There are few better definitions of good government than that it renders possible for all, and actually produces in the great majority of cases, happy homes and a contented people. Unless a number of the best writers of American fiction, and a considerable proportion of those who contribute to the most serious periodicals of the day, are deluding their readers, the present system of cheap bread-production is founded on privation, misery, or ruin in the homes of thousands of farmers, and on the unnatural growth of great cities, with a corresponding increase of millionaires, of pauperism, and of crime.

The Remedy.

If the exposition now given of the causes of the sufferings of the Western farmers is correct—and I have the greatest confidence that it is so—the only thorough remedy will be to bring the land back into the possession of the people, to be administered, locally, for the benefit of the men who actually use it, never for those who want it only for speculation; and by means of a carefully-adjusted system of rents or land-taxes, to equalize the benefits to be derived from the land (as regards quality and situation), so that none will be able to undersell others to their ruin. Prices will then be adjusted by fair competition, and will fall to the lowest level compatible with the usual standard of living of the time and place, and will be such as to leave a clear margin of profit for the support of a family and for provision for old age.

It will of course be understood that under such a system the farmers would be really as much the owners of their land as if they possessed the fee simple and were free of mortgage. So long as the very moderate differential rent or land-tax was paid, the farmer would have perpetual, undisturbed possession, with the right to bequeath or sell, just as he has now. Rents would never be raised on the farmer's improvements, but only on any increase of value of the land itself due to the action of the community, as when increase of population or new railroads so raised prices or cheapened production as to increase the inherent value of land in that locality in proportion to its value in other localities. But it should be always recognized that the creation of "happy-homes," so far as material well-being affects them, should be the first object of land legislation; and thus rents should in every case be assessed low enough to secure that end, always supposing reasonable care and industry in the farmer, which would be sufficiently indicated by the average result.

Under such a system of land-tenure as is here suggested, the farmer's life would become a peaceable and happy one, more like that of the early days, when he supplied most of his own wants, and only needed to sell a portion of his surplus products. Every benefit which the community at large may derive by the abolition of import duties, and the operation of the railroads by the State for the good of all, would be fully enjoyed by the farmer also, and his standard of comfort would gradually rise. If, however, these last mentioned reforms are made without any alteration of land-tenure, he will not be permanently benefited, because the competition of the better, rent-free land, and also that of the great capitalist farmers, will still drive prices down to the lowest point at which he can just exist. This competition will act quite as surely and quite as cruelly as the competition of labourers in the towns and cities, which always drives down the earnings of unskilled labour to the very lowest point, a point which is kept stationary by the presence of a large body of the unemployed on the verge of starvation and always ready to work at a little above starvation wages.

It will no doubt be objected that, even admitting such a land-system to be desirable, there is now no equitable means of getting the land back, except the impossible one of purchase from existing owners. But this is a mistake, and several practical methods have been or can be suggested. We have, first, the "single tax" of Mr. George, which has already obtained many adherents. At first sight farmers may think this would increase their burthens; but it would, on the contrary, relieve them, because all land would be taxed on its inherent, not on its improved, value. Now the inherent value of land in and around cities is enormous, and is not now fairly assessed. This city land would bear a much larger share of taxation than now; farm land proportionally less; and as this single tax would be accompanied by the removal of all duties on imported goods and produce, the farmer's tools, machinery, and clothing would be greatly cheapened.

But notwithstanding this single tax on land values, it might still be worth the while of great capitalists, companies, and trusts to hold large areas of land, because they could derive both profit and power from it in various indirect ways. The people will never be free from the countless evils of land-monopoly and land-speculation until it is declared contrary to public policy for any one to hold land except for personal use and occupation. A date might then be fixed before which all land not personally occupied must be sold; and that it should be really sold might be insured by enacting that afterwards, no rental or other charge on land payable to individuals or companies would be recoverable at law.

All municipalities, townships, or other local authorities should, however, have a prior and also a continuous right to purchase all such land at a moderate but fair valuation, paying for it with bonds bearing a low interest and redeemable at fixed dates. In this way the public would be able to acquire most of the land for some miles around all towns and cities; and as much of this would certainly increase rapidly in value, through growth of population and municipal improvements, the bonds could in a few years be redeemed out of the increased rents.

There is, however, another quite distinct method of reclaiming the land for the community, which has many advantages. This may be effected by carrying into practice two great ethical principles. These are, first, that the unborn have no individual rights to succeed to property; and, second, that there is no equitable principle involved in *collateral* succession to property, whatever there may be in *direct* succession. By the application of these two principles the people may, if they so will, in the course of some eighty years gradually regain possession of the whole national domain, without either confiscation or purchase. The law should declare that, after a certain date, land would cease to be transferable except to direct descendants—children or grandchildren; and, that, when all the children of these direct descendants, who were living at the time of passing the law, had died out, the land should revert to the State. As people owning land, but having no children, are dying daily, while even whole families often die off in a few years, land would be continually falling in, to be let out to applicants on a secure and permanent tenure, as already explained, so as best to subserve the wants of the community.

Here, then, are two very distinct methods of obtaining the land, both thoroughly justifiable when the welfare of a whole nation is at stake. The last named is that which seems best to the present writer, since it would at once abolish the greatest evils of the American social system—

those founded on land-speculation and land-monopoly; while the land itself would be acquired by means involving the minimum of interference with the property or welfare of any living persons. But, unless by these or some analogous measures farmers are relieved from the competition of great capitalists, while competition among themselves is rendered fair and equal by a differential rent or land-tax, no other kind of legislation can possibly relieve the majority of them from the state of poverty and continuous labour in which they now exist. In an unfair and unequal competition the less favoured must always be beaten.

II. WAGE-WORKERS.

The once familiar term "republican simplicity" is now an unmeaning one, since both in France and in America there is an amount of wealth and luxury not surpassed in any of the old monarchies. Yet it serves to show us the ideal which the founders of republics fondly hoped to attain. They aimed at abolishing for ever, not only the rank and titles of hereditary nobility, but also those vast differences of wealth and social grade which were supposed to depend upon monarchical government. Their objects were to secure, not only political and religious freedom, but also an approximate equality of social conditions; or, at the very least, an adequate share of the comforts and enjoyments of life for every industrious citizen. Yet after a century of unprecedented growth, and the utilization of the natural riches of a great continent, we find to-day, in all the great cities of the United States, thousands and tens of thousands who by constant toil cannot secure necessaries and comforts for their children or make any provision for an old age of peaceful repose. One great object of republican institutions has, it is clear, not been attained. Let us now endeavour to form some idea of the extent and nature of the disease of the social organism, so as more effectually to provide the true remedy.

In his *Social Problems* (written in 1883) Henry

George thus refers to the condition of one of the richest states of the Union, Illinois:

"In their last report the Illinois Commissioners of Labour Statistics say, that, their tables of wages and cost of living are representative only of intelligent working men who make the most of their advantages, and do not reach 'the confines of that world of helpless ignorance and destitution in which multitudes in all large cities continually live, and whose only statistics are those of epidemics, pauperism, and crime.' Nevertheless, they go on to say, an examination of these tables will demonstrate that one half of these intelligent working men of Illinois 'are not even able to earn enough for their daily bread, and have to depend upon the labour of women and children to eke out their miserable existence.'"

Dr. Edward Aveling in his book on the *Working Class Movement in America*, quotes from the same reports for other States as follows:

"In Massachusetts a physician gives evidence as to the condition of Fall River: 'Every mill in the city is making money, . . . but the operatives travel in the same old path—sickness, suffering, and small pay.'"

In Pennsylvania the Commissioners say:

"The rich and poor are further apart than ever."

In New Jersey:

"The struggle for existence is daily becoming keener, and the average wage-labourer must practise the strictest economy, or he will find himself behind at the end of the season."

In Kansas:

"The condition of the labouring classes is too bad for utterance. . . . It is useless to disguise the fact that out of this . . . enforced idleness grows much of the discontent amd dissatisfaction now pervading the country, and which has obtained a strong footing now upon the soil of Kansas, where only the other day her pioneers were staking out homesteads almost within sight of her capital city."

In Michigan:

"Labour to-day is poorer paid than ever before; more discontent exists, more men in despair; and if a change is not soon devised, trouble must come."

In the pages of *The Arena*, within the last two years, I find the following statements :—

"In the city of New York there are over one hundred and fifty thousand people who earn less than sixty cents a day. Thousands of this number are poor girls who work from eleven to sixteen hours a day. Last year there were over twenty-three thousand families forcibly evicted in that city owing to their inability to pay their rent." (*Arena*, February, 1891, p. 375.)

"During the ten years which ended in 1889, the great metropolis of the western continent added to the assessed value of its taxable property almost half a billion dollars. In all other essential respects save one, the decade was a period of retrogression for New York City. Crime, pauperism, insanity, and suicide increased; repression by brute force personified in an armed police was fostered, while the education of the children of the masses ebbed lower and lower. The standing army of the homeless swelled to twelve thousand nightly lodgers in a single precinct, and forty thousand children were forced to toil for scanty bread." (*Arena*, August, 1891, p. 365.)

"When the compulsory education law went into effect (in Chicago), the inspectors found in the squalid region a great number of children so destitute that they were absolutely unfit to attend school on account of their far more than semi-nude condition; and although a number of noble-hearted ladies banded together and decently clothed three hundred of these almost naked boys and girls, they were compelled to admit the humiliating fact that they had only reached the outskirts, while the great mass of poverty had not been touched... On one night last February, one hundred and twenty four destitute men begged for shelter at the cells of one of the city police stations." (*Arena*, November, 1891, p. 761.)

"Within cannon shot of Beacon Hill, where proudly rises the golden dome of the Capitol, are hundreds of families slowly starving and stifling; families who are bravely battling for life's barest necessities, while year by year the conditions are becoming more hopeless, the struggle for bread fiercer, the outlook more dismal." (*Arena*, March, 1892, p. 524.)

The above extracts may serve to give an imperfect indication of the condition of those whose labour produces much of America's phenomenal wealth. Volumes would not suffice to picture a tithe of the misery, starvation, and degradation that pervades all the great cities, and to a less extent the smaller manufacturing towns and rural districts ; and one of the latest writers on the subject gives it as his conclusion,

"that there is in the heart of America's money-centre a poverty as appalling, as hopeless, as degrading, as exists in any civilized community on earth." (*Arena*, December, 1892, p. 49.)

Let it be clearly understood that I do not in any way imply that republicanism is itself the cause of this state of things. It simply exists in spite of republicanism, and serves to demonstrate the great truth that systems of government are in themselves powerless to abolish poverty. The startling, and at first sight depressing, fact that grinding poverty dogs the footsteps of civilization under all forms of government alike, is really, from one point of view, a hopeful circumstance, since it assures us that the source of the evil is one that is common alike to republic, constitutional monarchy, and despotism, and we are thus taught where *not* to look for the remedy. We find it prevailing where militarism is at a maximum, as in France, Italy, and Germany, and where it is at a minimum, as in the United States. It is quite as bad in thinly as in thickly populated countries; but the one thing that it always accompanies is CAPITALISM. Wherever wealth accumulates most rapidly in the hands of private capitalists, there—notwithstanding the most favourable conditions, such as general education, free institutions, a fertile soil, and the fullest use of labour-saving machinery—poverty not only persists but increases. We must therefore look for the source of the evil in something that favours the accumulation of individual wealth.

Capitalism the Cause of Poverty.

Now, great wealth is obtained by individuals in two ways: either by speculation, which is but a form of gambling and perhaps the very worst form, since it impoverishes, not a few fellow-gamblers only but the whole community; or by large industrial enterprises, and these depend for their success on the existence of great bodies of labourers who have no means of living except by wage-labour, and are thus absolutely dependent on employment by capitalists in order to sustain life, and are compelled in the last resort to accept such wages as the capitalists choose to give. The result of these conditions is very low wages, or if nominally higher wages, then intermittent work; and thus we find in all great cities—in New York, Chicago, London, Vienna, for example, in each recurring winter—many thousands of men out of work, and either partially supported by charity or undergoing slow starvation. Now I propose to show that these terrible phenomena pervading all modern civilizations alike—speculation, capitalism, compulsory idleness of those willing to labour, women and children starving or killed by overwork—all arise as the natural consequences and direct results of private property, and consequent monopoly, of land. This is the *one* identical feature in the social economy of modern civilizations, and it alone is an adequate cause for an identity of results when so many of the social and political conditions in the great civilized communities of to-day are not identical but altogether diverse.

We must always remember that the existence of large numbers of surplus labourers, which is at once the indication and the measure of poverty, is a purely artificial phenomenon. There is no surplus, as regards land and natural products waiting to be transmuted by labour into various forms of wealth; there is no surplus, as regards demand for this wealth by those in want of all the comforts and many of the barest necessaries of life, and who only ask to be allowed to call those necessaries into existence by their labour. The only surplus is a surplus as regards demand for labourers by capitalists, a surplus which owes its existence wholly to artificial conditions which are fundamentally wrong. It is not a natural but a man-created surplus, and all the want and misery and crime that spring from it is equally man-created, and altogether unnatural and unnecessary.[1]

[1] In his most admirable and thoughtful work, *Poverty and the State*, Mr. Herbert V. Mills relates how he was led to study the subject by finding in three adjoining houses in Liverpool a baker, a tailor, and shoemaker, all out of work. They all wanted bread and clothes and shoes; all were anxious to work to supply their own and the others wants. But the social system of which they formed a part did not permit of their so working for each other, the alleged reason being that there was already an overstocked market of all these commodities;

In those early times to which I referred in the first portion of this chapter, wages were higher, food cheaper, and there were practically none unemployed. Why was this? The country was then far less rich; there was almost no labour-saving machinery; yet no one wanted food, clothing, or fire. The reason simply was, that immediately around most of the smaller towns there was land which could be had for little or nothing; and farther off was everywhere the forest or the prairie, where any one might build his log hut, grow his corn, or even hunt or fish to support life. Every one could then easily obtain land from which he could, by his own labour, support himself and his family. There was a charm in this free life, and men were continually drifting away from civilization to enjoy it. Therefore it was that wage-labour was scarce and wages high, for no one would work for low wages when he had the alternative of working for himself. The labourer could then really make a "free contract" with the capitalist who required his services, because he had always an alternative; or, at all events, a sufficient number had this alternative and would avail themselves of it, to prevent there being any surplus labour vainly seeking employment.

Now, the great majority of the unemployed have no such alternative. It is either work for the capitalist or starve. Hence "free contract" is a mockery; the wages of unskilled labour have sunk to the minimum that will support life in a working condition—it cannot permanently be less—and skilled labour obtains somewhat better terms just in proportion as it is plentiful or scarce. If, then, we really desire that labourers shall all be better paid, and none be unemployed (and the two things necessarily go together), we *must* enable a large proportion of all wage workers to have a sufficiency of land, by the cultivation of which they can obtain food for

therefore they must either remain starved and naked or be supported in idleness by their fellows. It is hardly possible to imagine a more complete failure of civilization than such a fact as this; and the failure is rendered more grotesque and horrible by the additional fact that no politician or legislator has any effectual remedy to suggest, while the majority maintain that no remedy is needed or is possible!

themselves and their families, for at least part of the year, and thus have an alternative to starvation wages. There is absolutely no other way, because it is from land alone that a man can, by his own labour, obtain food and clothing, without the intervention of a capitalist employer. But in order to ensure his doing this, he must have the land on a permanent tenure; he must be able to live on it, and must never be taxed on the improvements he himself makes on it; and though he may be allowed to sell or bequeath it, he must not be allowed to mortgage it, since what we want is to create as many secure and permanent *homes* as possible, as the only safe foundation for a prosperous and happy community.

The Remedy—Free Access to Land for All.

But in order to do this—not here and there in certain localities, but everywhere throughout the length and breadth of the Union—the people must resume the land, which should never have been parted with, to be administered locally for the benefit of all, and to be held always for *use*, never for speculation. People are now beginning to see that land speculation is the curse of the country. Millionaires have, in almost every case, grown by what is, fundamentally, land speculation. It is this which has enabled the few to acquire the bulk of the wealth created by the many toilers; and it is by the monopoly of land, whether in city lots, in railroads, mines, bonanza farms, vast forests, or vaster cattle ranches, that the rich are ever growing richer, and the poor more numerous.

In an American town or city to-day, it is practically impossible for the worker to obtain land for cultivation, except at town-lot prices; while beyond the municipal limits the land is usually held in farms of 160 acres or more, the owners of which are all holding for a rise in value when the town limits are extended so as to include some of their property. But so soon as the land becomes all municipal or township property, and it becomes recognized that on its proper use depends the well-being of the workers, these workers, being everywhere in the majority, will determine that beyond the central business part of the town, the land shall be let in lots of from one to five or ten acres, at fair agricultural rents and on a permanent tenure. Such small lots would be a twofold benefit to the community. In the first place they would constitute homesteads for workers, where they and their families could utilize every portion of spare time in the production of vegetables, fruit, poultry, eggs, pork, and other foodstuffs, which would supply their families with a considerable portion of their daily food. In the second place they would supply the town itself with fresh and wholesome vegetables, fruit, eggs, &c., and also, from the larger plots of five or ten acres, abundance of fresh milk and butter and other farm and garden produce. A little farther off, the regular farms, held in the same way, would provide the town with wheat, corn, hay, beef, mutton, and other necessaries; and thus each town and the surrounding district would be to a large extent self-contained and self-supporting.

But at the present time the very reverse of this is the case; most of the towns and cities drawing their supplies mainly from great distances, the country immediately around them being often but half cultivated. Certain districts grow cattle, others wheat, others vegetables, others again fruit, each kind having its special district where it is raised on an enormous scale and sent by rail for hundreds or even thousands of miles to where it is to be consumed. This is thought to be economy, but it is really waste from every point of view except that of the capitalist farmer. He chooses a place where land can be had cheaply (though probably not more suitable for the special purpose than plenty of land within a few miles of every city), where communication is easy, and labour abundant, and therefore cheap; and by growing on a large scale, and employing the greatest amount of machinery and the least amount of labour, he obtains large profits. But this profit is derived, not from superior cultivation, but from the practical monopoly of a large area of land, and by the labour of hundreds of men and women who work hard and live poorly to make him rich. The same land, if cultivated *for themselves* by an equal or a larger number of workers, would produce far more per acre, and would keep them all in comfort, instead of making *one* man exceptionally well off while all the rest live in uncertainty and poverty. And besides this material difference, there is the moral effect of work on a man's own homestead, where every hour's extra labour increases the value of his property or the comfort of his home, as compared with wage-work for a master who will discharge him as soon as he ceases to want him, and in whose work, therefore, he can take no interest. Experience in every part of the world shows that this moral effect is one of the greatest advantages of securing to the mass of the people homesteads of their very own. As this aspect of the question is hardly ever discussed in America, a few illustrative examples must be given.

And first as to the profits of small farms as compared with large ones. Lord Carrington has eight hundred tenants of small plots of land around the town of High Wycombe, Bucks., and he has recently stated that these tenants get a net produce of forty pounds an acre, while the most that the farmers can obtain from the same land by plough cultivation is seven pounds an acre. Here is a gain to the country of thirty-three pounds an acre by peasant cultivation; and it is all clear gain, for these men are wage labourers, and their little plots of land are cultivated by themselves and their families in time that would otherwise be wasted.

Another case is that of the Rev. Mr. Tuckwell, of Stockton, Warwickshire, who has let two hundred acres of land to labourers in plots of from one to four acres, at fair rents, and with security for fourteen years. Most of the men with two acres grow enough wheat and potatoes to supply their families for the whole year, besides providing food for a pig, and all this by utilizing the spare time of the family. These men grow forty bushels of wheat to the acre, the farmer's average being less than thirty; and their other crops are good in proportion.

Still more interesting is the Wellingborough Allotment

Association in Northamptonshire, where two hundred and twenty-three men rent and occupy a farm of one hundred and eighty-four acres at three hundred pounds a year rent, though the land is rather poor. It is divided into plots from one eighth of an acre to six acres, the occupiers being various wage workers, small tradesmen, mechanics, and comparatively few farm labourers. The farm was visited by Mr. Impey, who states that it was excellently cultivated, and that the wheat averaged forty-eight bushels an acre—nearly twice the average of Great Britain—while one man got fifty-six bushels an acre from two and one-fourth acres. When this farm was let to a farmer, four men on the average were employed on it; now an amount of work equal to that of forty men is expended on it, and a considerable portion of the work is done during time that would otherwise be wasted.

The reports issued by the last Royal Commission on Agriculture in 1882 give numerous similar illustrations, showing that in periods of agricultural distress, when large farmers were being ruined, the small farmers who cultivated the land themselves were prosperous. Thus Mr. F. Winn Knight, M.P., of Exmoor, Devonshire, had sixteen tenants paying rents from thirteen pounds up to two hundred pounds a year, all being paid regularly to the last shilling, and every one of these men had been agricultural labourers. More remarkable is the case of Penstrasse Moor in Cornwall, a barren, sandy waste, which neither landlord nor tenant-farmer thought worth cultivating. Yet five hundred acres of this waste have been enclosed and reclaimed by miners, mechanics, and other labourers, on the security of leases for three lives at a low rent. This land now carries more stock than any of the surrounding farms, and the total produce is estimated by the assistant commissioner, Mr. Little, at nearly twice the average of the county.

Pages could be filled with similar cases, but these are sufficient to show the importance of land in improving the condition of the workers.

Moral Effects of the secure Homestead.

In the same reports remarks are made on the material and moral effects of this experiment. Mr. Little says:

"The family have a much more comfortable home, and many advantages, such as milk, butter, eggs, which they would not otherwise enjoy. The man has a motive for saving his money and employing his spare time; he enjoys a position of independence; he is elevated in the social scale; his self-respect is awakened and stimulated, and he acquires a stake and an interest in the country."

And the same reporter again recurs to the subject in the following weighty remarks:

"Interesting as this subject is in its relation to agriculture, as showing the capacity for improvement which some barren spots possess, and as a triumph of patience and industry, it is most valuable as an instance where the opportunity of investing surplus wages and spare hours in the acquirement of a home for the family, an independent position for the labourer, a provision for wife and children in the future, has been a great encouragement to thrift and providence. It is not only that the estate represents so much land reclaimed from the waste and put to a good use, it represents also so much time well spent, which would, without this incentive, have most probably been wasted, and wages, which would otherwise probably have been squandered, employed in securing a homestead and some support for the widow and family when the workman dies."

The men who reclaimed this waste, it must be remembered, are all miners, hence the references to their "wages"; and all these good results are secured on an uncertain tenure dependent on the duration of the longest of three lives, after which it all reverts to the landlord, who has not spent a penny on it, but has, on the contrary, received rent the whole time for giving the tenants permission to reclaim it! Under an equitable system of permanent tenure, the interest of the labourer in improving the land would be greater, his position more secure, and the benefit to the nation in the creation of happy homes more certain to be brought about.

Another illustration of the moral effects of even a moderately good land-system is given by the Honourable George C. Brodrick, in his interesting work *English*

Land and English Landlords. It occurred on the Annandale Estate in Dumfriesshire, where farm labourers were given leases for twenty-five years, at ordinary farm rents, of from two to six acres of land each, on which they built their own cottages with stone and timber supplied by the landlord.

"All the work on these little farms was done at by hours, and by members of the family; the cottager buying roots of the farmer, and producing milk, butter, and pork, besides rearing calves. Among such peasant farmers pauperism soon ceased to exist, and many of them became comparatively well off. It was particularly observed that habits of marketing and the constant demands on thrift and forethought, brought out new virtues and powers in the wives. In fact, the moral effects of the system, in fostering industry, sobriety, and contentment, were described as no less satisfactory than its economical success."

These moral effects of the secure tenure of land in small farms or cottage homesteads have been observed by politicians, travellers, and moralists wherever the system prevails. Thus, William Howitt, in his *Rural and Domestic Life of Germany*, says:

"The German peasants work hard, but they have no actual want. Every man has his house, his orchard, his roadside-trees, commonly so heavy with fruit that he is obliged to prop and secure them or they would be torn to pieces. He has his corn plot, his plots for mangold-wurzel, for hemp, and so forth. He is his own master, and he and every member of his family have the strongest motives to labour. You see the effects of this in that unremitting diligence which is beyond that of the whole world besides, and his economy, which is still greater. . . . The German bauer looks on the country as made for him and his fellow-men. He feels himself a man; he has a stake in the country as good as that of the bulk of his neighbours; no man can threaten him with ejection or the workhouse so long as he is active and economical. He walks, therefore, with a bold step; he looks you in the face with the air of a free man, but a respectful air."

That admirable historian and novelist, Mr. Baring Gould, confirms this. Writing at a much later period, he says in his *Germany Past and Present:*

"The artisan is restless and dissatisfied. He is mechanized. He finds no interest in his work, and his soul frets at the routine. He is miserable and he knows not why. But the man who toils on his own plot of ground is morally and physically healthy. He is a freeman; the sense he has of independence gives him his upright carriage, his fearless brow, and his joyous laugh." [1]

Results of Large and Small Holdings.

We see, then, that the statements continually made by economical writers as to the advantages of large farms—and repeated by press-writers as if they were demonstrated facts—are either partially or wholly untrue. Large farms, as compared with smaller farms—one thousand acres with two huudred acres, for instance—both being capitalists—may be more profitable, but partly because the larger farmer usually has more capital, and employs more machinery. His individual profits may also be much larger, even if he gets a smaller profit per acre, on account of his larger acreage; and for this reason landlords like large farmers because they can afford to pay a higher rent. But this has nothing whatever to do with the question as between peasant or cottage farmers who do their own work, and capitalist farmers employing wage labour. In every case known, and in all parts of the world, the former raise a much larger produce from the land, and it is this question of the amount of *produce* that is the important question for the community.

It is often the case, perhaps even generally the case with capitalist farmers, that a larger profit is obtained from a small than from a large production. This is the reason that, during the last twenty years, about two million acres of English arable land have been converted into pasture. But the average produce of arable land in Great Britain has been estimated by the best authorities as worth about ten pounds, while the average produce of pasture land does not exceed one pound ten shillings. Here is an enormous difference, yet the profit to the farmer is often larger per acre from the small than from the large produce. This is because the cost of raising

[1] Fuller details of the results of permanent land tenure are given in Mill's *Political Economy*, Book II., Chapter VI., and also in the present writer's *Land Nationalization*, Chapter VI.

the larger produce is much greater, labour of men and horses being the most important item of this great cost. When prices of wheat and other arable crops are low, it therefore pays both landlord and farmer to discharge their labourers, sow grass, and keep cattle or sheep, which require the minimum amount of labour per hundred acres. We have already seen in the case of the Wellingborough Allotment Association, that men working for themselves can profitably put ten times as much labour on the land as a tenant farmer usually employs; and this last number is again reduced to one-fifth when the land is turned into grass. It follows that the two millions of acres recently thrown out of cultivation in Great Britain would support in comfort, at the lowest computation, more than a hundred thousand families in excess of those who are now employed there.

The *reductio ad absurdum* of this method of confounding *profit* with *produce* was seen when, in reply to the demand of the Highland crofters to be allowed to occupy and cultivate the valleys formerly cultivated by their ancestors, but from which these were expelled to make room for deer, the late Duke of Argyll replied that there was no unoccupied land available, since all the land in Scotland was applied to its best "economic use." By this he meant that the rental received by the landlords for their deer and grouse shootings was greater than they could obtain from the Highland cultivators! The difference in produce of food might be a hundred to one in favour of the Highlanders; but that, in the duke's opinion, had nothing to do with the matter. Political economists, as a rule, never allude to this most important point, of the essential difference between *production and profit*. Mill just mentions it while showing that peasant farms are the most productive; but he does not reason the thing out, and few other writers mention it at all. Hence political writers, in the face of the clearest and most abundant evidence, again and again deny that labourers can possibly grow wheat and other crops at a profit, because capitalist farmers cannot do so. But the peasant gives that daily, minute, and loving attention to his small plot which the capitalist farmer cannot possibly give to his hundreds of acres; he works early and late at critical periods of the growth of each crop; and as a result he often obtains double the produce at less than half the money cost.

There is yet another objection made to peasant cultivators, and repeated again and again with the greatest confidence, but which is shown to be equally unfounded by the inexorable logic of facts. Peasants and small farmers, it is said, cannot afford to have the best machinery, neither can they make those great improvements which require large expenditure of capital; therefore they should not be encouraged. Yet fifty years ago Mr. S. Laing, in his *Journal of a Residence in Norway*, showed how far advanced were the peasant farmers of Scandinavia in co-operative works. The droughts of summer are very severe; and to prevent their evil consequences, the peasants have combined to carry out extensive irrigation works. The water is brought in wooden troughs from high up the valleys and then distributed to the several plots. In one case the main troughs extended along a valley for forty miles. Another writer, Mr. Kay, in his work on the *Social Condition of the People in Europe*, shows that the countries where the most extensive irrigation works are carried on are always those where small proprietors prevail, such as Vaucluse and the Bouches-du-Rhone in France, Sienna, Lucca, and other portions of Italy, and also in parts of Germany.

Again, in the French Jura and in Switzerland, the peasants of each parish combine together for co-operative cheese-making, each receiving his share of the product when sold, in proportion to the quantity of milk he has contributed. This system is also at work in Australia, where in the districts suited to dairying, co-operative butter and cheese factories are established, where the best machinery and the newest methods are used, the result being that some of the butter is so good that, after supplying the great cities, the surplus is exported to England. Of course it would be easy to apply the same principle to mowing machines, harvesters, or even flour mills, all of which might be obtained and worked by the co-operation of peasant farmers, each paying in proportion to the days or hours he made use of the machine. Neither is there anything in the superior education and intelligence often claimed for the large farmer. Mr. Kay tells us that in Saxony

> "the peasants endeavour to outstrip one another in the quantity and quality of the produce, in the preparation of the ground, and in the general cultivation of their respective portions. All the little proprietors are eager to find out how to farm so as to produce the greatest results; they diligently seek after improvements; they send their children to agricultural schools, in order to fit them to assist their fathers; and each proprietor soon adopts a new improvement introduced by any of his neighbours."

Finally, under this system of small peasant cultivators, who reap all the fruits of their own labours, the land is improved in an almost incredible manner. The bare sands of Belgium and Flanders have been gradually converted into gardens, and Mr. Kay sums up his observations by saying:

> "The peasant farming of Prussia, Holland, Saxony, and Switzerland is the most perfect and economical farming I have ever witnessed in any country;"

thus illustrating the famous axiom of Arthur Young a century ago:

> "Give a man secure possession of a bleak rock, and he will turn it into a garden."

It will hardly be said that the workers of America and of England to-day are *less* industrious, *less* intelligent, *less* influenced by the desire for an independent life and a home which shall be indeed each man's castle, than were the peasants of various parts of Europe half a century ago. Give them, therefore, equal or even superior opportunities, and you will obtain at least equal, probably far superior results.

The reason why we may expect better results is, that the system of peasant-proprietors, from which most of our illustrations have necessarily been drawn, had in it the seeds of decay and failure from the very same causes as those which have led to the failure of the homestead system of the United States: unequal competition, owing to differences in quality and situation of farms, as well as to capitalist farmers, the influence of both having been greatly increased by railroads and other means of rapid communication, and by the growth of great cities offering a practically unlimited market. Added to this there has been the twofold influence of the millionaire and the speculator, ever seeking to buy land, and o the money lender, ever seeking to lend money on land mortgages. These combined influences have led to the almost complete extermination of the statesmen and other small land-owning farmers of England, and have greatly diminished the number and the prosperity of the peasant farmers in France, Belgium, Germany, and Austria. The lawyers and money lenders have now absorbed many of the peasant properties of France and Belgium, whose former owners are now tenants, subject to the grinding pressure of rack-rents; while many others are struggling in the meshes of the mortgagees, as are so many of the farmers in the Western States of America.

Conclusions from the Inquiry.

The present inquiry has, I venture to think, established some definite and almost unassailable conclusions as to the fundamental causes which have led all civilized nations into the Social Quagmire in which they find themselves to-day; and in doing so it has furnished us with an answer to the vital question—What should be our next step towards better social conditions, such as will not render the term " civilization " the mockery it is now?

In the first place, we have demonstrated that a permanently successful agriculture, in which the food producer shall be sure of an adequate reward for his labour, is absolutely impossible without national or state ownership of the soil, so as to ensure the farmer undisturbed occupation at a low but equitably graded rent.

It is equally clear, in the second place, that the condition of the great body of industrial workers can only be improved, permanently, by giving them free access to land—the primary source of all food and all wealth—in the form of cottage homesteads around all cities, towns, and villages, by which they may be enabled to provide food for their families and to carry on such home industries as they may find convenient. Thus only will it be possible for them to enter into really " free contracts " with capitalists ; thus only can we get rid of the great army of the unemployed, and ensure to the worker a much larger proportion of the product of his labour.

When these two great radical reforms have been effected in every part of the country, the industrial classes of every kind will have before them a vista of permanent well-being and progressive prosperity. Many industries now carried on in factories, for the benefit of the individual capitalist, can be just as well prosecuted in the home of the worker, if "power" to work his machine is supplied to him. And so soon as there is a demand for such power it will be supplied, either by compressed air or water, or by electricity. A hundred looms, or knitting frames, or spinning mules, can be worked quite as well in separate houses as in one large building, with the enormous advantage to the worker that he could work at them during winter or in wet weather or at times when he would be otherwise idle, while carrying on another occupation out of doors in summer or in fine weather. At such machines different members of the family could work alternately, thus giving them all the relaxation derived from diversity of occupation, and the interest due to the fact that the whole product of their labour would be their own. In the case of those processes that require to be carried on in special buildings and on a larger scale, the workers could combine to erect such a building in their midst, and carry on the work themselves, just as they carry on co-operative cheese and butter making in so many parts of the world already. And, gradually, as men came to enjoy the health and the profit derived from varied work, and especially the pleasure which every cultivator of the soil feels in the products he sees grow daily as the result of his own labour, there seems no reason to doubt that co-operation of the kind suggested would spread, till the greater part or the whole of the manufacturers of the country were in the hands of associated workers.

At the present time all the boasted division of labour and economy of manufacture on a great scale, tends solely to the benefit of the capitalist. It is advantageous for him to have a thousand men working in one huge building, or agglomeration of buildings, and all the workers are made to come there, though their homes may be a long way off. The gain is his, the loss theirs. It is better for him that each man should do one kind of work only all day long, and from year's end to year's end, because he thus does it quicker and with less supervision. But the man suffers in the monotony of his work and in the injury to his health; while, doing one thing only, he is helpless when out of work. But in the future the arrangements will all be made in the interest of the worker. When possible, he will do his work at home, neither tied to special hours nor compelled to walk long distances, and thus lose precious time, besides adding so much unpaid labour to his daily toil. He will then be able to work some hours a day in his garden or farm, or in some occupation possessing more variety and interest than being a mere intelligent part of a vast machine. And when he works at his own machine he will not need to keep at it more than four or six hours a day. Being thus able to work, even as a manufacturer, on his own account, in association with his fellow-townsmen, he will not be induced to work for a capitalist except for very high wages and for very moderate hours of labour. He will then soon compete successfully with the capitalist, and ultimately drive him out of the field altogether. For it must always be remembered that, once the workers get homes of their very own, with the means of obtaining a considerable portion of their food direct from the soil, they can save their own capital, and thereafter employ their own labour; whereas the capitalists, though possessing abundance of money and machinery, cannot make a single piece of calico or an ingot of steel, cannot raise a ton of coal or turn out a single watch, unless they can induce men to work for them.

The workers of America, like those of Great Britain, have their future in their own hands. They have the majority of votes, and can return representatives to do their bidding. Let them turn their whole attention to the one point—of rescuing the land from the hands of monopolists and speculators. In this direction only lies the way out of the terrible Social Quagmire in which they are now floundering; this is the next step forward towards a happier social condition and a truer civilization.

In conclusion, I should not have ventured to make these suggestions to Americans on matters which, it may be supposed, they are quite able to deal with themselves, were it not that the principles on which my proposals are founded are fundamental in their nature and of universal application. For many years I have advocated similar remedies for my own country, and these are at length being very widely accepted by the chief organizations of our workers. These remedies are equally applicable and equally needed in Australia and New Zealand; while every country in Europe, from Spain to Russia, is at this moment suffering the evils which necessarily result from a vicious land system. Americans received this system from us, as they received slavery from us. To abolish the latter they incurred a fearful cost and made heroic sacrifices. The system which permits and even encourages land monopoly and land speculation inevitably brings about another form of slavery, more far-reaching, more terrible in its results, than the chattel slavery they have abolished. Let the tenement houses of New York and Chicago, with their thousands of families in hopeless misery, their crowds of half naked and famishing children, bear witness! These white slaves of our modern civilization everywhere cry out against the system of private ownership and monopoly of land, which is, from its very nature, the robbery of the poor and landless. This system needs no gigantic war to overthrow it; it can be destroyed without really injuring a single human being. Only we must not waste our time and strength in the advocacy of half-measures and petty palliatives, which will leave the system itself to produce ever a fresh crop of evil. The voice of the working and suffering millions must give out no uncertain sound, but must declare unmistakably to those who claim to represent them—Our land-system is the fundamental cause of the persistent misery and poverty of the workers; root and branch it is wholly evil; its fruits are deadly poison; cut it down; why cumbereth it the ground?

CHAPTER XXIV

ECONOMIC AND SOCIAL JUSTICE

During many past centuries of oppression and wrong there has been an ever-present but rarely expressed cry for redress, for some small instalment of Justice to the downtrodden workers. It has been the aspiration alike of the peasant and the philosopher, of the poet and the saint. But the rule of the lords of the soil has ever been so hard, and supported by power so overwhelming and punishment so severe, that the born thralls or serfs have rarely dared to do more than humbly petition for some partial relief; or, if roused to rebel by unbearable misery and wrongs, they have soon been crushed by the power of mailed knights and armed retainers. The peasant revolt at the end of the fourteenth century was to gain relief from the oppressive serfdom that was enforced after the black death had diminished the number of workers. John Ball then preached Socialism for the first time.

"By what right," he said, "are they whom we call lords greater folk than we? Why do they hold us in serfage?... They are clothed in velvet, while we are covered with rags. They have wine and spices and fair bread; and we oat-cake and straw, and water to drink. They have leisure and fine houses; we have pain and labour, the rain and the wind in the fields. And yet it is of us and our toil that these men hold their state."

John Ball and Wat Tyler lived five hundred years too soon. To-day the very same claims are made by men who, having got political power, cannot be so easily suppressed.

A century passed, and the great martyr of freedom, Sir Thomas More, powerfully set forth the wrongs of the workers and the crimes of their rulers in his ever-memorable *Utopia*. Near the end of this work he thus summarizes the governments of his time in words that will apply almost, if not quite, as accurately to-day:

"Is not that government both unjust and ungrateful that is so prodigal of its favours to those that are called gentlemen, or such others who are idle, or live either by flattery or by contriving the arts of vain pleasure, and, on the other hand, takes no care of those of a meaner sort, such as ploughmen, colliers, and smiths, without whom we could not subsist? But after the public has reaped all the advantage of their service, and they come to be oppressed with age, sickness, and want, all their labours and the good they have done is forgotten, and all the recompense given them is that they are left to die in great misery. The richer sort are often endeavouring to bring the hire of labourers lower—not only by their fraudulent practices, but by the laws which they procure to be made to that effect; so that though it is a thing most unjust in itself to give such small rewards to those who deserve so well of the public, yet they have given those hardships the name and colour of justice, by procuring laws to be made for regulating them.

"Therefore I must say that, as I hope for mercy, I can have no other notion of all the governments that I see or know than that they are a conspiracy of the rich, who, on pretence of managing the public, only pursue their private ends, and devise all the ways and arts they can find out; first, that they may, without danger, preserve all that they have so ill acquired, and then that they may engage the poor to toil and labour for them at as low rates as possible, and oppress them as much as they please."[1]

Here we have a stern demand for justice to the workers who produce all the wealth of the rich, as clearly and as forcibly expressed as by any of our modern socialists. Sir Thomas More might, in fact, be well taken as the hero and patron-saint of Socialism.

A century passed away before Bacon in England, and Campanelli in Italy, again set forth schemes of social regeneration. Bacon's *New Atalantis* supposed that the desired improvement would come from man's increased command over the powers of nature, which would give wealth enough for all. We have, however, obtained this command to a far greater extent than Bacon could

[1] Cassell's National Library—*Utopia*, p. 17.

possibly have anticipated; yet its chief social effect has been the increase of luxury and the widening of the gulf between rich and poor. Although material wealth, reckoned not in money but in things, has increased perhaps twenty or thirty fold in the last century, while the population has little more than doubled, yet millions of our people still live in the most wretched penury, the whole vast increase of wealth having gone to increase the luxury and waste of the rich and the comfort of the middle classes.

Campanelli, more far-sighted than Bacon, saw the need of social justice as well as increased knowledge, and proposed a system of refined communism. But all these ideas were but as dreams of a golden age, and had no influence whatever in ameliorating the condition of the workers, which, with minor fluctuations, and having due regard to the progress of material civilization, may be said to have remained practically unchanged for the last three centuries. When one-fourth of all the deaths in London occur in workhouses and hospitals notwithstanding that four millions are spent there annually in public charity, while untold thousands die in their wretched cellars and attics from the direct or indirect effects of starvation, cold, and unhealthy surroundings; and while all these terrible facts are repeated proportionately in all our great manufacturing towns, it is simply impossible that, within the time I have mentioned, the condition of the workers as a whole can have been much, if any, worse than it is now.

At the end of the seventeenth, and during the eighteenth century, a new school of reformers arose, of whom Locke, Rousseau, and Turgot were representatives. They saw the necessity of a fundamental justice, especially as regards land, the source of all wealth. Locke declared that labour gave the only just title to land; while Rousseau was the author of the maxim, that the produce of the land belongs to all men, the land itself to no one. The first Englishman, however, who saw clearly the vast importance of the land question, and who laid down those principles with regard to it which are now becoming widely accepted, was an obscure Newcastle schoolmaster, Thomas Spence, who in 1775 gave a lecture before the Philosophical Society of that town, which was so much in advance of the age that when he printed his lecture the society expelled him, and he was soon afterwards obliged to leave the town. He maintained the sound doctrine that the land of any country or district justly belongs to those who live upon it, not to any individuals to the exclusion of the rest; and he points out, as did Herbert Spencer at a later period, the logical result of admitting private property in land. He says:

"And any one of them (the landlords) still can, by laws of their own making, oblige every living creature to remove off his property (which, to the great distress of mankind, is too often put in execution); so, of consequence, were all the landholders to be of one mind, and determined to take their properties into their own hands, all the rest of mankind might go to heaven if they would, for there would be no place found for them here. Thus men may not live in any part of this world, not even where they are born, but as strangers, and by the permission of the pretender to the property thereof."

He maintained that every parish should have possession of its own land, to be let out to the inhabitants, and that each parish should govern itself and be interfered with as little as possible by the central government, thus anticipating the views as to local self-government which we are now beginning to put into practice.

A few years later, in 1782, Professor Ogilvie published anonymously, *An Essay on the Right of Property in Land, with respect to its foundation in the Law of Nature, its present establishment by the Municipal Laws of Europe, and the Regulations by which it might be rendered more beneficial to the Lower Ranks of Mankind.* This small work contains an elaborate and well-reasoned exposition of the whole land question, anticipating the arguments of Herbert Spencer in *Social Statics*, of Mill, and of the most advanced modern land-reformers. But all these ideas were before their time, and produced little or no effect on public opinion. The workers were too ignorant, too much oppressed by the struggle for bare existence, while the middle classes were too short-sighted to be influenced by theoretical views which even to this day many of the most liberal thinkers seem unable fully to appreciate. But the chief cause that prevented the development of sound views on the vital problems of the land and of social justice, was, undoubtedly, that men's minds were forcibly directed towards the great struggles for political freedom then in progress. The success of the American revolutionists and the establishment of a republic founded on a Declaration of the Rights of Man, followed by the great French Revolution and the Napoleonic wars, entirely obscured all lesser questions, and also led to a temporary and fictitious prosperity, founded on a gigantic debt the burden of which still oppresses us. These great events irresistibly led to the discussion of questions of political and personal freedom rather than to those deeper problems of social justice of which we are now only beginning to perceive the full importance. The rapid growth of the use of steam power, the vast extension of our manufactures, and the rise of our factory system with its attendant horrors of woman and infant labour, crowded populations, spread of disease, and increase of mortality, loudly cried for palliation and restrictive legislation, and thus occupied much of the attention of philanthropists and politicians.

Character of Nineteenth Century Legislation.

Owing to this combination of events, the nineteenth century has been almost wholly devoted to two classes of legislation—the one directed to reform and popularize the machinery of government itself, the other to neutralize or palliate the evils arising from the unchecked powers of landlords and capitalists in their continual efforts to increase their wealth while almost wholly regardless of the life-shortening labour, the insanitary surroundings, and the hopeless misery of the great body of the workers. To the first class belong the successive Reform Bills, the adoption of the ballot in elections for members of Parliament, household and lodger suffrage, improved registration, and the repression of bribery. To the second, restriction of children's and women's labour in factories and mines, government inspection of these industries; attempts to diminish the dangers of unhealthy employments, and to check the ever-increasing pollution of rivers; the new poor law, casual wards, and other attempts to cope with pauperism; while various fiscal reforms, such as the abolition of the corn-laws and the extension of free trade, though advocated in the supposed interest of the wage-earners, were really carried by the efforts of great capitalists and manufacturers as a means of extending their foreign trade. Later on came the Elementary Education Act of 1870, which was thought by many to be the crowning of the edifice, and to complete all that could be done by legislation to bring about the well-being of the workers, and, through them, of the whole community.

Its Outcome.

Now that we have had nearly a century of the two classes of legislation here referred to, it may be well briefly to take stock of its general outcome, and see how far it has secured—what all such legislation aims at securing—a fair share to all the workers of the mass of wealth they annually produce; a sufficiency of food, clothing, house-room, and fuel; healthy surroundings; and some amount of leisure and surplus means for the lesser enjoyments of life. And it must be remembered that never in the whole course of human history has there been a century which has added so much to man's command over the forces of nature, and which has so enormously extended his power of creating and distributing all forms of wealth. Steam, gas, photography, and electricity, in all their endless applications, have given us almost unlimited power to obtain all necessaries, comforts, and luxuries that the world can supply us with. It has been calculated that the labour-saving machinery of all kinds now in use produces about a hundred times the result that could be produced if our workers had only the tools and appliances available in the last century. But even in the last century, not only was there produced a sufficiency of food, clothing, and houses for all workers, but an enormous surplus, which was appropriated by the landlords and other capitalists for their own consumption, while large numbers, then as now, were unprofitably employed in ministering to the luxury of the rich, or wastefully and wickedly employed in destroying life and property in civil or foreign wars.

Taking first the anti-capitalistic or social legislation, we find that, though the horrible destruction of the health, the happiness, and the very lives of factory children has been largely reduced, there has grown up in our great cities a system of child-labour as cruel and destructive, if not quite so extensive. Infants of four years and upwards are employed at matchbox-making and similar employments to assist in supporting the family. A widow and two children, working all day and much of the night, can only earn a shilling or eighteenpence from which to pay rent and support life. Children of school age have thus often to work till midnight after having had five hours' schooling; and till quite recently a poor mother in this state of penury was fined if she did not send the children to school and pay a penny daily for each, meaning so much less bread for herself or for the children. Of course for the children this is physical and mental destruction. The number of women thus struggling for a most miserable living—often a mere prolonged starvation—is certainly greater than at any previous period of our history, and even if the proportion of the population thus employed is somewhat less—and this is doubtful—the fact that the actual mass of human misery and degradation of this kind is absolutely greater, is a horrible result of a century's remedial legislation, together with an increase of national wealth altogether unprecedented.

Again, if we turn to the amount of poverty and pauperism as a measure of the success of remedial legislation combined with a vast extension of private and systematised charity, we shall have cause for still more serious reflection. In 1888 the Registrar-General called attention to the fact that, both throughout the country and to a still greater extent in London, deaths in workhouses, hospitals, and other public charitable institutions had been steadily increasing since 1875. A reference to the Annual Summaries of deaths in London shows the increase to have been continuous from 1860 to 1890, the five-year periods giving the following results :—

In 1860–65 of total deaths in London, 16·2 per cent. occurred in charitable institutions.
1866–70 (no material at hand).
1871–75 of total deaths in London, 17·4 ,, ,,
1876–80 ,, ,, 18·6 ,, ,,
1881–85 ,, ,, 21·1 ,, ,,
1886–90 ,, ,, 23·4 ,, ,,
1891–95 ,, ,, 26·7 ,, ,,

When we add to this the admitted facts, that organized charity has greatly increased during the same period, while the press still teems with records of the most terrible destitution, of suicides from the dread of starvation, and deaths directly caused or indirectly due to want, we are brought face to face with a mass of human wretchedness that is absolutely appalling in its magnitude. And all this time Royal and Parliamentary Commissions have been inquiring and reporting, Mansion House and other Committees have been collecting funds and relieving distress at every exceptional period of trouble, emigration has been actively at work, improved dwellings have been provided, and education has been systematically urged on, with the final result that one-fourth of all the deaths in the richest city in the world occur in workhouses, hospitals, &c., and, in addition, unknown thousands die in their miserable garrets and cellars from various forms of slow or rapid starvation.

Can a state of society which leads to this result be called civilization? Can a government which, after a century of continuous reforms and gigantic labours and struggles, is unable to organize society so that every willing worker may earn a decent living, be called a successful government? Is it beyond the wit of man to save a large proportion of one of the most industrious people in the world, inhabiting a rich and fertile country, from grinding poverty or absolute starvation? Is it impossible so to arrange matters that a sufficient portion of the wealth they create may be retained by the workers, even if the idle rich have a little less of profuse and wasteful luxury?

The Impotence of our Legislators.

Our legislators, our economists, our religious teachers, almost with one voice tell the people that any better organization of society than that which we now possess is impossible. That we must go on as we have been going on, patching here, altering there, now mitigating the severity of a distressing symptom, now slightly clipping the wings of the landlord, the capitalist, or the sweater; but never going down to the root of the evil; never interfering with vested interests in ancestral wrong; never daring to do anything which shall diminish rent and interest and profit, and to the same extent increase the reward of labour; never seek out the fundamental injustice which deprives men of their birthright in their native land, and enables a small number of landlords to tax the rest of the community to the amount of hundreds of millions for permission to cultivate and live upon the soil in the country of their birth. Can we, then, wonder that both workers and thinkers are getting tired of all this hopeless incapacity in their rulers? That, possessing education which has made them acquainted with the works of great writers on these matters, from Sir Thomas More to Robert Owen, from Henry George to Edward Bellamy, from Karl Marx to Carlyle and Ruskin; and possessing as they do ability, and honesty, and determination, fully equal to that of the coterie of landlords and capitalists which has hitherto governed them, they are determined, as soon as may be, to govern themselves.

The Work of the Twentieth Century.

Now, I believe that the great work of this century, that which is the true preparation for the work to be done in the coming twentieth century, is not its well-meant and temporarily useful but petty and tentative social legislation, but rather that gradual reform of the political machine—to be completed, it is to be hoped, within the next few years—which will enable the most thoughtful and able and honest among the manual workers to at once turn the balance of political power, and, at no distant period, to become the real and permanent rulers of the country. The very idea of such a government will excite a smile of derision or a groan of horror among the classes who have hitherto blundered and plundered at their will, and have thought they were heaven-inspired rulers. But I feel sure that the workers will do very much better; and, forming as they do the great majority of the people, it is only bare justice that, after centuries of misgovernment by the idle and wealthy, they should have their turn. The larger part of the invention that has enriched the country has come from the workers; much of scientific discovery has also come from their ranks; and it is certain that, given equality of opportunity, they would fully equal, in every high mental and moral characteristic, the bluest blood in the nation. In the organization of their trades-unions and co-operative societies, no less than in their choice of the small body of their fellow-workers who represent them in Parliament, they show that they are in no way inferior in judgment and in organizing power to the commercial, the literary, or the wealthy classes. The way in which, during the past few years, they have forced their very moderate claims upon the notice of the public, have secured advocates in the press and in Parliament, and have led both political economists and politicians to accept measures which were not long before scouted as utterly beyond the sphere of practical politics, shows that they have already become a power in the state. Looking forward, then, to a government by workers, and largely in the interest of workers, at a not distant date, I propose to set forth a few principles and suggestions as to the course of legislation calculated to abolish pauperism, poverty, and enforced idleness, and thus lay the foundation for a true civilization which will be beneficial to all.

Suggestions for Real Reforms.

That the ownership of large estates in land by private individuals is an injustice to the workers and the source of much of their poverty and misery, is held by all the great writers I have alluded to, and has been fully demonstrated in many volumes as well as in the four chapters on the Land-question in the present work. It has led directly to the depopulation of the rural districts, the abnormal growth of great cities, the diminished cultivation of the soil and reduced food-supply, and is thus at once a social evil and a national danger. Some petty attempts are now making to restore the people to the land, but in a very imperfect manner. The first and highest use of our land is to provide healthy and happy homes, where all who desire it may live in permanent security and produce a considerable portion of the food required by their families. Every other consideration must give way to this one, and all restrictions on its realization must be abolished. Hence, the first work of the people's Parliament should be, to give to the Parish and District Councils unrestricted power to take all land necessary for this purpose, so as to afford every citizen the freest possible choice of a home in which he can live absolutely secure (so long as he pays the very moderate ground rent) and reap the full reward of his labour. Every man, in his turn, should be able to choose both where he will live and how much land he desires to have, since each one is the best judge of how much he can enjoy and make profitable. Our object is that all working men should succeed in life, should be able to live well and happily, and provide for an old age of comfort and repose. Every such landholder is a gain and a safety instead of a loss and a danger to the community, and no outcry, either of existing landlords or of tenants of large farms, must be allowed to stand in their way. The well-being of the community is the highest law, and no private interests can be permitted to prevent its realization. When land can be thus obtained, co-operative communities, on the plan so clearly laid down by Mr. Herbert V. Mills in his work on *Poverty and the State* (and sufficiently explained in Chap. XXVI. of this volume) may also be established, and various forms of co-operative manufacture can be tried.

The Inviolability of the Home.

But until this great reform can be effected there is a smaller and less radical measure of relief to all tenants, which should at once be advocated and adopted by the Liberal party. It is an old boast that the Englishman's house is his castle, but never was a boast less justified by facts. In a large number of cases a working man's house might be better described as an instrument of torture, by means of which he can be forced to comply with his landlord's demands, and both in religion and politics submit himself entirely to the landlord's will. So long as the agricultural labourer, the village mechanic, and the village shopkeeper are the tenants of the landowner, the parson, or the farmer, religious freedom or political independence is impossible. And when those employed in factories or workshops are obliged to live, as they so often are, in houses which are the property of their employers, that employer can force his will upon them by the double threat of loss of employment and loss of a home. Under such conditions a man possesses neither freedom nor safety, nor the possibility of happiness, except so far as his landlord and employer thinks proper. A secure HOME is the very first essential alike of political freedom, of personal security, and of social well-being.

Now that every worker, even to the hitherto despised and down-trodden agricultural labourer, has been given a share in local self-government, it is time that, so far as affects the inviolability of the home, the landlord's power should be at once taken away from him. This is the logical sequence of the creation of Parish Councils. For, to declare that it is for the public benefit that every inhabitant of a parish shall be free to vote and to be chosen as a representative by his fellow parishioners, and at the same time to leave him at the mercy of the individual who owns his house to punish him in a most cruel manner for using the privileges thus granted him, is surely the height of unreason and injustice. It is giving a stone in place of bread; the shadow rather than the substance of political enfranchisement.

There is, however, a very simple and effectual way of rendering tenants secure, and that is, by a short Act of Parliament declaring all evictions, or seizure of household goods, other than for non-payment of rent, to be illegal. And to prevent the landlord from driving away a tenant by raising his rent to an impossible amount, all alterations of rent must be approved of as reasonable by a committee of the Parish or District Council, and be determined on the application of either the tenant or the landlord. Of course, at the first letting of a house the landlord could ask what rent he pleased, and if it was exorbitant he would get no tenant. But having once let it, the tenant should be secure as long as he wished to occupy it, and the rent should not be raised except as allowed by some competent tribunal. No doubt a claim will be made on behalf of the landlords for a compulsory, not voluntary, tenancy on the part of the tenant; that is, that if the tenant has security of occupation, the landlord should have equal security of having a tenant. But the two cases are totally different. Eviction from his home may be, and often is, ruinous loss and misery to the tenant, who is therefore, to avoid such loss, often compelled to submit to the landlord's will. But who ever heard of a tenant, by the threat of giving notice to quit, compelling his landlord to vote against his conscience, or to go to chapel instead of to church? The tenant needs protection, the landlord does not.

The same result might perhaps be gained by giving the Parish and District Councils power to take over all houses whose tenants are threatened with eviction, or with an unfair increase of rent; but this would involve so many complications and would so burthen these Councils with new and responsible work, that there is no chance of its being enacted for many years. But the plan of giving a legal permanent tenure to every tenant is so simple, so obviously reasonable, and so free from all interference with the fair money-value of the landlord's property, that, with a little energy and persistent agitation, it might possibly be carried in two or three years. Such an Act might be more or less in the following form :—

"Whereas the security and inviolability of the HOME is an essential condition of political freedom and social well-being, it is hereby enacted, that no tenant shall hereafter be evicted from his house or homestead, or have his household goods seized, for any other cause than non-payment of rent, and every heir or successor of such tenant shall be equally secure so long as the rent is paid."

A second clause would provide for a permanently fair rent.

Now, will not some advanced Liberal bring in such a Bill annually till it is carried? It is, I think, one that would receive the support of a large number of reformers, because it is absolutely essential to the free and fair operation of the Parish and District Councils, and is equally necessary for the well-being of the farmer and the tradesman, as well as for the mechanic and labourer. The annual discussion of the subject in Parliament would be of inestimable value, since it would afford the opportunity of bringing prominently before the voters the numerous cases of gross tyranny and cruel injustice which are yearly occurring, but which now receive little consideration.

The Unborn not to Inherit Property.

The next great guiding principle, and one that will enable us to carry out the resumption of the land without real injury to any individual, is, that we should recognize no rights to property in the unborn, or even in persons under legal age, except so far as to provide for their education and give them a suitable but moderate provision against want. This may be justified on two grounds. Firstly, the law allows to individuals the right to will away their property as they please, so that not even the eldest son has any vested interest, as against the power of the actual owner of the property to leave it to whom or for what purpose he likes. Now, what an individual is permitted to do for individual reasons which may be good or bad, the State may do if it considers it necessary for the good of the community. If an individual may justly disinherit other individuals who have not already a vested interest in property, however just may be their expectations of succeeding to it, *a fortiori* the State may, partially, disinherit them for good and important reasons. In the second place, it is almost universally admitted by moralists and advanced thinkers, that to be the heir to a great estate from birth is generally injurious to the individual, and is necessarily unjust to the community. It enables the individual to live a life of idleness and pleasure, which often becomes one of luxury and vice; while the community suffers from the bad example, and by the vicious standard of happiness which is set up by the spectacle of so much idleness and luxury. The working part of the community, on the other hand, suffers directly in having to provide the whole of the wealth thus injuriously wasted. Many people think that if such a rich man *pays* for everything he purchases and wastes, the workers do not suffer because they receive an equivalent for their labour; but such persons overlook the fact that every pound spent by the idle is first provided *by* the workers. If the income thus spent is derived from land, it is *they* who really pay the rents to the landlord, inasmuch as if the landlord did not receive them they would go in reduction of taxation. If it comes from the funds or from railway shares, *they* equally provide it, in the taxes, in high railway fares, and increased price of goods due to exorbitant railway charges. Even if *all* taxes were raised by an income tax paid by rich men only, the workers would be the real payers, because there is no other possible source of annual income in the country but productive labour. If any one doubts this, let him consider what would happen were the people to resume the land as their right, and thenceforth apply the rents, locally, to establish the various factories and other machinery needful to supply all the wants of the community. Gradually all workers would be employed on the land, or in the various co-operative or municipal industries, and would themselves receive the full product of their labours. To facilitate their exchanges they might establish a token or paper currency, and they would then have little use for gold or silver. How, then, could idlers live, if these workers, in the Parliament of the country, simply declined to pay the interest on debts contracted before they were born? What good would be their much-vaunted "capital," consisting as it mostly does of mere legal power to take from the workers a portion of the product of their labours, which power would then have ceased; while their real capital—buildings, machinery, &c.—would bring them not one penny, since the workers would all possess their own, purchased by their own labour and the rents of their own land? Let but the workers resume possession of the soil, which was first obtained by private holders by force or fraud, or by the gift of successive kings who had no right to give it, and capitalists as a distinct class from workers must soon cease to exist.

No Right to Tax Future Generations.

Another principle of equal importance is to refuse to recognize the right of any bygone rulers to tax future generations. Thus all grants of land by kings or nobles, all "perpetual" pensions, and all war-debts of the past, should be declared to be legally and equitably invalid, and henceforth dealt with in such a way as to relieve the workers of the burden of their payment as speedily as is consistent with due consideration for those whose chief support is derived from such sources. Just as we are now coming to recognize that a "living wage" is due to all workers, so we should recognize a "maximum income" determined by the standard of comfort of the various classes of fund-holders and State or family pensioners. As a rule, these persons might be left to enjoy whatever income they now possess during their lives, and when they had relatives dependent on them the income might be continued to these, either for their lives or for a limited period according to the circumstances of each case. There would be no necessity, and I trust no inclination, to cause the slightest real privation, or even inconvenience, to those who are but the product of a vicious system; but on every principle of justice and equity it is impossible to recognize the rights of deceased kings—most of them the worst and most contemptible of men—to burthen the workers for all time in order to keep large bodies of their fellow-citizens in idleness and luxury.

How to deal with Accumulated Wealth.

By means of the principles now laid down, we can see how to deal fairly with the present possessors of great estates, and with millionaires, whose vast wealth confers no real benefit on themselves while it necessarily robs the workers, since, as we have seen, it has all to be provided by the workers. It will, I think, be admitted that, if a man has an income, say, of ten thousand a year, that is sufficient to supply him with every possible necessary, comfort and rational luxury, and that the possession of one or more additional ten thousands of income would not really add to his enjoyment. But all such excessive incomes necessarily produce evil results, in the large number of idle dependants they support, and in keeping up habits of gambling and excessive luxury. Further, in the case of landed estates the management of which is necessarily left to agents and bailiffs, it leads to injurious interference with agriculture and with the political and religious freedom of tenants, to oppression of labourers, to the depopulation of villages, and other well-known evils. It will therefore be for the public benefit to fix on a maximum income to be owned by any citizen; and, thereupon, to arrange a progressive income-tax, beginning with a very small tax on a minimum income from land or realized property of, say, £500, the tax progressively rising, at first slowly, afterwards more rapidly, so as to absorb all above the fixed maximum.

When a landed estate was taken over for the use of the community, the net income which had been derived from it would be paid the late holder for his life, and might be continued for the lives of such of his direct heirs as were of age at the time of passing the Act, or it might even be extended to all direct heirs living at that time. In the case of a person owning many landed estates in different counties, he might be given the option of retaining any one or more of them up to the maximum income, and that income would be secured to him (and his direct heirs as above stated) in case any of the land were taken for public use. In the case of fundholders, all above the maximum income would be extinguished, and thus reduce taxation.

The process here sketched out—by which the continuous robbery of the people through the systems of land and fundholding, may be at first greatly reduced and in the course of one or two generations completely stopped, without, as I maintain, real injury to any living person and for the great benefit both of existing workers and of the whole nation in the future—will, of course, be denounced as confiscation and robbery. That is the point of view of those who now benefit by the acts of former robbers and confiscators. From another, and I maintain a truer point of view, it may be described as an act of just and merciful restitution. Let us, therefore, consider the case a little more closely.

Origin of Great Estates.

Taking the inherited estates of the great landed proprietors of England, almost all can be traced back to some act of confiscation of former owners or to gifts from kings, often as the reward for what we now consider to be disgraceful services or great crimes. The whole of the property of the abbeys and monasteries, stolen by Henry VIII. and mostly given to the worst characters among the nobles of his court, was really a robbery of the people, who obtained relief and protection from the former owners. The successive steps by which the landlords got rid of the duties attached to landholding under the feudal system, and threw the main burden of defence and of the cost of government on non-landholders, was another direct robbery of the people. Then in later times, and down to the present century, we have that barefaced robbery by form of law, the enclosure of the commons, leading, perhaps more than anything else, to the misery and destruction of the rural population. Much of this enclosure was made by means of false pretences. The general Enclosure Acts declare that the purpose of enclosure is to facilitate 'the productive employment of labour' in the improvement of the land. Yet hundreds of thousands of acres in all parts of the country, especially in Surrey, Hampshire, Dorsetshire, and other southern counties, were simply taken from the people and divided among the surrounding landlords, and then only used for sport, not a single pound being spent in cultivating it. Now, however, during the last twenty years, much of this land is being sold for building at high building prices, a purpose never contemplated when the Enclosure Acts were obtained. During the last two centuries more than seven millions of acres have been thus taken from the poor by men who were already rich, and the more land they already possessed the larger share of the commons was allotted to them. Even a Royal Commission, in 1869, declared that these enclosures were often made "without any compensation to the smaller commoners, deprived agricultural labourers of ancient rights over the waste, and disabled the occupants of new cottages from acquiring new rights."

Now, in this long series of acts of plunder of the people's land, we have every circumstance tending to aggravate the crime. It was robbery of the poor by the rich. It was robbery of the weak and helpless by the strong. And it had that worst feature which distinguishes robbery from mere confiscation—the plunder was divided among the individual robbers. Yet, again, it was a form of robbery specially forbidden by the religion of the robbers, a religion for which they professed the deepest reverence, and of which they considered themselves the special defenders. They read in what they call *the Word of God*, " Woe unto them that join house to house, that lay field to field, till there be no place, that they may be placed alone in the midst of the earth," yet this is what they are constantly striving for, not by purchase only, but by robbery. Again they are told, "The land shall not be sold for ever, for the land is Mine;" and at every fiftieth year all land was to return to the family that had sold it, so that no one could keep land beyond the year of jubilee; and the reason was that no man or family should remain permanently impoverished.

Both in law and morality the receiver of stolen goods is as bad as the thief; and even if he has purchased a stolen article unknowingly, an honourable man will, when he discovers the fact, restore it to the rightful owner. Now, our great hereditary landlords know very well that they are the legal possessors of much stolen property, and, moreover, property which their religion forbids them to hold in great quantities. Yet we have never heard of a single landlord making restitution to the robbed nation. On the contrary, they take every opportunity of adding to their vast possessions, not only by purchase, but by that meanest form of robbery—the enclosing of every scrap of roadside grass they can lay their hands on, so that the wayfarer or the tourist may have nothing but dust or gravel to walk upon, and the last bit of food for the cottager's donkey or goose is taken away from him.

This all-embracing system of land-robbery, for which nothing is too great and nothing too little; which has absorbed meadow and forest, moor and mountain; which has secured most of our rivers and lakes, and the fish which inhabit them; which often claims the very seashore and rocky coast-line of our island home, making the peasant pay for his seaweed-manure and the fisherman for his bait of shellfish; which has desolated whole counties to replace men by sheep or cattle, and has destroyed fields and cottages to make a wilderness for deer; which has stolen the commons and filched the roadside wastes; which has driven the labouring poor into the cities, and has thus been the primary and chief cause of the lifelong misery, disease, and early death of thousands who might have lived lives of honest toil and comparative comfort had they been permitted free access to land in their native villages;—it is the advocates and beneficiaries of this inhuman system, the members of this "cruel organization," who, when a partial restitution of their unholy gains is proposed, are the loudest in their cries of "robbery!" But all the robbery, all the spoliation, all the legal and illegal filching has been on *their* side, and they still hold the stolen property. *They* made laws to justify their actions, and *we* propose equally to make laws which will really justify ours, because, unlike their laws which always took from the poor to give to the rich, ours will take only from the superfluity of the rich, not to give to the poor individually, but to enable the poor to live by honest work, to restore to the whole people their birthright in their native soil, and to relieve all alike from a heavy burden of unnecessary taxation. This will be the true statesmanship of the future, and will be justified alike by equity, by ethics, and by religion.

The Teaching of the Priests.

And now, what has been the conduct and teaching of those priests and bishops who profess to be followers of Him who declared that a rich man shall hardly enter into the kingdom of heaven, and who gave this rule as being above all the Commandments, "If thou wilt be perfect, go and sell that thou hast, and give to the poor, and thou shalt have treasure in heaven." Have they ever preached to the squires and nobles restitution of some portion of the land so unjustly obtained by their ancestors? Have they even insisted on the duty of those who hold the land to allow free use of it to all their fellow-citizens on fair terms? Have they even set before these men the inevitable and now well-known results of land-monopoly, and the deadly sin of using their power to oppress the poor and needy? It is notorious that, with some few noble exceptions, they have done none of these things, but have ever taken the side of the landed against the landless, and too often, whether in the character of landlords or magistrates, have so acted as to lose the confidence and even gain the hatred of the poor. We look in vain among priests and bishops of the Established Church for any real comprehension of what this land-question is to the poor; but we find it in the following words of a dignitary of the older Church, that good man and true follower of Christ, the late Cardinal Manning:

"The land-question means hunger, thirst, nakedness, notice to quit, labour spent in vain, the toil of years seized upon, the breaking up of houses; the misery, sicknesses, deaths of parents, children, wives; the despair and wildness which spring up in the hearts of the poor, where legal force, like a sharp harrow, goes over the most sensitive and vital rights of mankind. All this is contained in the land-question."

But our archbishops and bishops know or care nothing whatever of all this! They are truly blind guides; and, as pastors of a Church which should be pre-eminently the Church of the poor, how applicable are the words of Isaiah:

" They are all dumb dogs, they cannot bark; sleeping, lying down, loving to slumber. Yea, they are greedy dogs which can never have enough, and they are shepherds that cannot understand: they all look to their own way, every one for his gain!"

And now, in conclusion, I will give one or two extracts from a book written by a self-taught worker for workers, to show how workers feel on the questions we have touched upon.

"At present the working people of this country live under conditions altogether monstrous. Their labour is much too heavy, their pleasures are too few; and in their close streets and crowded houses, decency and health and cleanliness are well-nigh impossible. It is not only the wrong of this that I resent, it is the *waste*. Look through the slums, and see what childhood, girlhood, womanhood, and manhood have there become. Think what a waste of beauty, of virtue, of strength, and of all the power and goodness that go to make a nation great, is being consummated there by ignorance and by injustice. For, depend upon it, every one of our brothers or sisters ruined or slain by poverty or vice, is a loss to the nation of so much bone and sinew, of so much courage and skill, of so much glory and delight. Cast your eyes, then, over the Registrar-General's returns, and imagine, if you can, how many gentle nurses, good mothers, sweet singers, brave soldiers, clever artists, inventors and thinkers, are swallowed up every year in that ocean of crime and sorrow which is known to the official mind as 'the high death-rate of the wage-earning classes.' Alas! the pity of it."

And again, from the same writer:

"A short time ago a certain writer, much esteemed for his graceful style of saying silly things, informed us that the poor remained poor because they show no efficient desire to be anything else. Is that true? Are only the idle poor? Come with me, and I will show you where men and women work from morning till night, from week to week, from year to year, at the full stretch of their powers, in dim and fetid dens, and yet are poor, ay, destitute—have for their wages a crust of bread and rags. I will show you where men work in dirt and heat, using the strength of brutes, for a dozen hours a day, and sleep at night in styes, until brain and muscle are exhausted; and fresh slaves are yoked to the golden car of commerce, and the broken drudges filter through the union or the prison, to a felon's or a pauper's grave! And I will show you how men and women thus work and suffer, and faint and die, generation after generation, and I will show you how the longer and harder these wretches toil, the worse their lot becomes; and I will show you the graves and find witnesses to the histories of brave and noble industrious poor men, whose lives were lives of toil *and* poverty, and whose deaths were tragedies. And all these things are due to *sin;* but it is to the sin of the smug hypocrites who grow rich upon the robbery and the ruin of their fellow-creatures."

These extracts are from a small but weighty book called *Merrie England*, by Nunquam. In the form of a series of letters on Socialism to a working man, it contains more important facts, more acute reasoning, more conclusive argument, and more good writing, than are to be found in any English work on the subject I am acquainted with. When such men—and there are many of them—are returned to Parliament, and are able to influence the government of the country, the dawn of a new era, bright with hope for the long-suffering workers, will be at hand.

CHAPTER XXV

RALAHINE AND ITS TEACHINGS [1]

THE successful and most instructive experiment made at Ralahine in 1831-33 is very little appreciated except by a few advanced reformers; but it attracted great attention at the time, and deserves to be better known, because it affords a practical and conclusive answer to many of the objections now made to the possibility of successful co-operation, especially in agriculture. The adviser and organizer of this great experiment, Mr. E. T. Craig, exhibited a marvellous tact and knowledge of human nature, and has shown us how to avoid the rocks and pitfalls which have led to failure in many other cases, and this was especially remarkable in so young a man (then under thirty), and shows him to have been a born organizer and leader of men. Yet his great powers, which might have benefited the nation and the human race, were forbidden their full expansion through the influence of the money interests and religious prejudices of the ruling and landed classes. A brief sketch will now be given of the difficulties he overcame, the results he achieved, and the lessons to be learnt from his experience.

Ireland in 1830.

In 1830 the state of Ireland, especially in the south and west, was deplorable. The potato crop had failed and

---

[1] *The Irish Land and Labour Question illustrated in the History of Ralahine and Co-operative Farming.* By E. T. Craig. London: Trübner and Co. 1882.

200,000 people were starving. Agrarian outrages, murder, robbery, and intimidation were prevalent. Houses were broken open to obtain arms; midnight meetings were held; and neither the armed police nor the military could cope with the situation. County Clare, where Ralahine is situated, was in the very centre of the disturbed area. Many of the landlords left their houses in charge of the police, and went to Dublin or to England. Rents at this time were enormously high, often £10 or £12 an acre for small plots of land of average quality, so that all the produce, except potatoes enough to keep life in the cultivator's family, went into the landlords' pockets. The crisis had been aggravated by extensive evictions of the peasantry in order to form large grazing farms, the rents of which were more easy to collect and less likely to fail than those of small holdings. The excitement was intense, and hatred and suspicion of landlords and of all agents and stewards was at its height; and it was at this inopportune moment that Mr. Craig first came to Ralahine, on the invitation of its owner, Mr. Vandeleur, to see if he could establish a co-operative farm and thus restore peace to this one estate in a time of general anarchy.

The steward who managed the farm had just been murdered, and the owner's family had gone away for safety; and it was under these adverse circumstances, that a stranger from England, a Saxon who knew not a word of Irish, and a Protestant who, it was thought, would probably interfere with their religion, was brought over by the landlord, presumably in his own interest and to get all that was possible out of themselves, the labourers. The former steward had been a tyrant, a cruel and unfeeling one, and they naturally supposed that the new man from England would be as bad or even worse, and that the talk about their working for themselves was merely a pretence to get more work out of them and to rob them more completely than before. Within the first six weeks after Mr. Craig's arrival at Ralahine there were four murders in the immediate neighbourhood, and he himself received a letter with a sketch of a death's head and cross-bones and a coffin on which was written, " Death to the Saxon." He lodged in a poor cottage, which was sometimes in the middle of the night surrounded by a howling mob, which kept him in expectation of violence or death. Once he was warned to return home after dark by a different route from that he was following, and once a stone was thrown at him from behind and struck him on the head. In addition to his other troubles, the proprietor's family and most of the gentry around were entirely opposed to the new system he was preparing to introduce, and their servants made jests upon him to his face, and still further prejudiced the people against him.

Mr. Vandeleur, who had been struck by the example of co-operation he had seen at New Lanark under Robert Owen, had already made some preparations for the scheme by building several cottages, sheds, and a large building suitable for a dining hall, with a lecture or reading room above, as well as a store-room and some dormitories, and Mr. Craig was at first engaged in superintending the completion of these, getting in necessary stores, and making the acquaintance and endeavouring to gain the confidence of such of the mechanics and labourers as could speak English. He also arranged with Mr. Vandeleur the terms on which the farm, buildings, implements and stock should be taken, and drew up the rules and regulations which seemed to him most suitable for the success of the undertaking. The people who had been hitherto working on the farm lived scattered about the country, some of them three or four miles away, so that a long walk was added to their daily labour. But so wedded are the Irish peasantry to their homes that it was difficult to get them to come to live on the farm in the new houses, and still more difficult for them to agree to take their meals together. But when at length the more intelligent among them were satisfied that under the new plan they would have all surplus profits to divide among themselves, they saw that to live together and to have their meals in common would be a great saving, and would enable them to give more work to the farm; and as the benefit of all economies of this kind would not as heretofore go to the landlord but be really all their own, they soon persuaded the others to agree, and thus one great initial difficulty was overcome.

Description of Ralahine.

The farm of Ralahine contained 618 acres, only 268 of which were cultivated, the rest being pasture, some of it very stony or rough, and 63 acres of bog from which peat for fuel was obtained. There was also live stock and some farm implements, to the estimated value of £1,500, on which six per cent. interest was to be paid, the total of rent and interest being £900, which the landlord himself admitted was too high, but which was nevertheless punctually paid for the three years that the experiment lasted. Mr. Craig showed great judgment in stipulating that this rent should be paid entirely in produce, estimated at the average market prices at Limerick of the last two years, the grain to be delivered at Limerick, the stock at Dublin or Liverpool, a plan which saved all the inconvenience of fluctuations of price, as well as the loss due to forced sales to meet the rent at fixed dates. This arrangement was one of the causes of success, and it is of great importance to all peasant-cultivators and especially in barbarous or thinly populated districts. Yet so prejudiced are our rulers that they continue to insist on money rents in India and hut-taxes paid in money in our African possessions, entirely regardless of the wishes, habits, or convenience of the inhabitants. In Ralahine, unfortunately, this tenancy was a yearly one, and the experiment was thus dependent on the will or the life of the landlord. Had it been a secure permanent tenure, the whole subsequent history of Ireland might have been changed, while legislation would certainly have been beneficially influenced by it.

The Organization of the Ralahine Society.

Mr. Craig also drew up a constitution and rules of the association under forty-four separate heads, dealing with the purpose of the society, the modes and hours of work and nominal rates of wages, the arrangements for food, clothing, &c., rules for education and conduct, methods of government, accounts, &c. A few of the more important of these may be given. Any member wishing to leave the society could do so at a week's notice. The landlord had power, during the first year, to discharge any member for misconduct. If more labour-power were required new members could be introduced on being proposed and seconded. They first came for a week on trial (afterwards changed to a month), and were then balloted for and chosen by a majority of votes. The landlord chose the secretary, treasurer, and store-keeper. All were to work, and to assist in agriculture when specially wanted. All youths, male and female, were to learn some useful trade, as well as farming and gardening. All work usually done by domestic servants was to be performed by the boys and girls under seventeen years of age. Meals could be taken in the public rooms or not as desired, but those cooking in their own houses must pay for fuel. No spirituous liquor of any kind was to be kept at the stores or be brought to the premises. Holidays were to be arranged so that each of the members could pay occasional visits to their friends.

The whole business of the society was managed by a committee of nine members, chosen half-yearly by ballot by all adult members male and female. This committee met every evening to decide upon the work of the following day and any other matters of importance; and here Mr. Craig introduced an ingenious arrangement to prevent friction between the committee and the rest of the members, which was strictly carried out and was found to work admirably. In all such societies every person must know what work he or she has to do the next day. Now, if the members of the committee who have decided this have to tell each one individually, all kinds of difficulties are sure to arise. Many persons cannot give instructions simply and clearly, but are so verbose and explanatory that their meaning may be easily mistaken, from which endless disputes would result. Others speak too abruptly, and when asked to explain refuse or make disparaging remarks, hence more quarrels. In fact the giving and receiving orders among persons who look upon each other, and who really are, equals, is one of the most fertile sources of discord. To avoid this the names of all the members were arranged alphabetically and consecutively numbered, so that every one knew his or her number, and every horse and implement had also a number. A series of slates were hung up in the dining room at the end of each committee meeting, with the numbers and names of all the members in their proper order and an exact statement of the work they were to do the next day. Every one looked at these slates either before going to bed or early in the morning, and went straight to their work without any need of instructions and without any possibility of mistake. The members of the committee were divided into sub-committees dealing with special departments, and any alterations needed during the day on account of changes in the weather or other causes were settled by one of them. If any of the arrangements or allotments of work were thought to be injudicious by any of the members, they could state their objections in a "suggestion book" which was always open for the purpose. The remarks in this book were read by Mr. Craig as secretary, at the evening meetings of the committee, and the decision upon each point was noted therein by him.

Even more important, for the harmonious working of the society, was the weekly meeting of the whole body, at which the various suggestions during the week, with the decisions of the committee upon them, were read and subjected to remarks and criticism. It was thus seen that attention was given to all these remarks, and that many of them had been acted upon; and Mr. Craig tells us that—"sometimes very judicious suggestions would be made by men who, all their lives previously, had been treated as utterly unworthy of a moment's consideration." A healthy public opinion was thus formed, and every one gave his best thought as to how the affairs of the society could be improved, and the work carried on in the most economical and effective manner.

Self-government at Ralahine.

But although these admirable rules and methods were suggested by Mr. Craig it must not be thought that any of them were forced upon the people. At the very commencement each of them was put to the vote by ballot of the adult members, and were only adopted by them, after the purport and use of them had been explained by Mr. Craig and fully considered among themselves. Even the rule against drink and tobacco was accepted and strictly obeyed during the whole existence of the society, apparently because they believed that drinking would interfere with the general harmony, but no doubt chiefly because they knew that it would lead to neglect and bad work, and as all returns after paying the fixed rent were to be their own property, they all wanted to work as much and as effectively as possible.

Mr. Craig remarks on the change produced in the workers by this system of associated work and common benefit as compared with the old system, almost universal on large estates in Ireland, of badly paid uninterested labourers under the absolute rule of a tyrannical steward, who despised them and treated them as inferiors. The orders they received were often accompanied by oaths or personal insults, and they did as little work as they could without being discharged. They were then almost universally dissatisfied, and had the character of being lazy, untrustworthy and vicious. As the steward could not possibly be in all parts of the estate at once, and had often to be away a considerable part of the day, the loss to all parties must have been very great.

Many persons now came to see Ralahine, not being able to credit the accounts they heard of it. One of these, a large Irish farmer, found a single man repairing the masonry of a tunnel under a road, which had partly given way, and to do it he was standing up to his middle in water. The visitor was surprised, and the following conversation took place:—

Visitor.—" Are you working by yourself?"
Man.—" Yes, sir."

Visitor.—" Where is your steward ? "
Man.—" We have no steward."
Visitor.—" Then who sent you to do this work ? "
Man.—" The Committee."
Visitor.—" What Committee ? Who are the Committee ? "
Man.—" Some of the members, Sir."
Visitor.—" What members do you mean ? "
Man.—" The members of the new system—the ploughmen and labourers."

This gentleman afterwards expressed his astonishment at finding a solitary workman so industrious and doing the work so well. Another visitor, Mr. John Finch of Liverpool, who remained three days at Ralahine and published the results of his inquiry in fourteen letters to a Liverpool newspaper, makes the following statement :

"A sensible labourer with whom I conversed, when at Ralahine, in contrasting their present with their former condition under a steward, said to me,—' We formerly had no interest, either in doing a great deal of work, doing it well, or in suggesting improvements, as all the advantages and all the praise were given to a tyrannical taskmaster, for his attention and watchfulness. We were looked upon as merely machines, and his business was to keep us in motion ; for this reason it took the time of three or four of us to watch him, and when he was fairly out of sight, you may depend upon it we did not hurt ourselves by too much labour ; but now that our interest and our duty are made to be the same we have no need of a steward at all."

The first members of the society were forty men and women who had before worked on the estate, and twelve children, but as there had been much difference of opinion and some quarrels among them, and as it was desired to start with a set of people who would work harmoniously together, it was decided that each one should be balloted for by the rest, and Mr. Craig insisted that he too, should be balloted for. Before the ballot, in each case, a personal criticism took place, and as a result none were rejected, nor were any expelled afterwards for idleness or bad conduct ; and this was really a striking example either of the inherent goodness of the Irish peasant under reasonably fair conditions, or of the wonderful effect of associated labour for the common good in improving the character and conduct. For it must be remembered that these people were not selected at all, but were the very same who had before worked on the estate, many of them having been among what were considered the worst characters, while some of them were almost certainly the associates and abettors of the murderer of the former steward. Mr. Craig assures us that their characters *seemed* to be wholly changed; for whereas under the despotic rule of the steward they had been sullen, quarrelsome, and dissatisfied, when working under the men they had elected to manage the farm in which all had an equal interest, they became cheerful and contented.

Education and Sanitation at Ralahine.

Mr. Craig was a thorough educationist of the most advanced type. He was one of the first, if not the very first, to introduce the kinder-garten system, not only in the training of infants but throughout all education. He therefore at once established a school at Ralahine, in which this system was carried out under a trained teacher and his personal supervision. His own observation, after long experience, assures him, he tells us, that children can only attend with pleasure and profit to a purely intellectual lesson for a very limited period, which he puts at fifteen minutes for children of six to seven, increasing to thirty minutes for those of fifteen to sixteen. He therefore advocates a constant succession of subjects at each such interval, alternating with some mechanical, and, if possible, outdoor work, or with the experimental illustration of natural laws and phenomena. He was also a specialist in the general laws of health, and particularly as regards the need of pure air; and the result of his arrangements was visible when an epidemic of cholera and fever was raging all around the colony, attaining the proportions of a plague in many of the towns. Ralahine had not a single case, nor was there any illness during the three years in a population of eighty and upwards.

A striking illustration of the value of the prohibition

of spirituous liquors occurred during a time when Mr. Craig was on a visit to Manchester. Three of the members attended a "wake," where there was, as usual, abundance of whisky, resulting in the not unusual faction-fight, during which stones were thrown and a man was killed. The Ralahine blacksmith was accused of throwing the stone which killed the man, was tried and sentenced to seven years' transportation, and as the two others were mixed up in the same affair they were dismissed from the society. Yet there was plenty of enjoyment without drink, for once or twice every week there was dancing in the evening, which both men and women seemed to enjoy, notwithstanding their ten or twelve hours work in the fields. On other evenings Mr. Craig gave simple lectures on natural phenomena or the laws of health, illustrated by such experiments as were adapted to the intelligence of his audience.

All the people, at Ralahine, with the exception of Mr. and Mrs. Craig had been accustomed to live almost entirely on potatoes, and often not enough of these. They did not therefore expect or want meat; but they had a variety of vegetables and as much new milk as they wished at every meal, with sometimes a little pork and bread and butter as a luxury. Mr. Craig considered that abundance of new milk with vegetables, constituted a perfectly healthy and sufficient diet, on which the hardest work could be and was done. There was also a large orchard, which yielded so abundantly that although every one of the eighty members had as much fruit as they wished, in one year two cartloads rotted for want of consumers. It was no doubt owing to this wholesome food, abundance of fresh air, and cleanliness in all the houses and surroundings, together with the contented cheerfulness resulting from their improved condition and prospects, that the perfect healthiness of the Ralahine community is to be attributed.

An interesting result of the experiment was the change it produced in the peasant's attitude towards machinery. Hitherto it had been impossible to use agricultural machinery in Ireland, because as it clearly reduced the demand for labour and thus tended to lower wages, it benefited only the landlord or farmer, while it injured the labourers. But so soon as these labourers were working for themselves and any surplus profits were their own, labour-saving machinery became a blessing instead of a curse. The Ralahine people, therefore, invested their first savings in a reaping machine, which, in the third year of their work, enabled them to harvest economically a splendid crop of wheat which they had grown on some poor rocky pasture by trenching it eighteen inches deep and getting out all the rock with crowbars.

This third harvest reaped by the society was an abundant one, and they celebrated it by a harvest-home which the landlord attended and in a congratulatory speech summarized the work and success of the Association. He expressed the great satisfaction he felt at the progress the Society had made during the short time it had been in existence; at the harmony which prevailed in the social arrangements of the members; their evident comfort, prosperity, and contentment; contrasting the present happy state of Ralahine and the quiet condition of the county with what it was when his family were compelled, from the dread of outrages and murders, to leave their home in the care of an armed police force. He congratulated them on the operation of the new system, which had accomplished a success greater than he had expected; and he hoped that other landlords would appreciate the advantages of giving those they employed a share in the profits realized by mutual co-operation, as might be seen in the evidence given by the large crops raised on hitherto waste land, made richly productive by deep cultivation, producing a heavy crop of potatoes the first year, followed by the splendid crop of wheat the last load of which they were now carrying home.

It seems almost incredible that this unexampled success, material, social, and moral, did *not* lead to any general adoption of a similar plan, which was so well calculated to banish famine from the country and to bring about that peace, contentment, and general happiness which from that day to this has been constantly absent from this fertile and beautiful but sadly misgoverned island. The landlords seem to have had no real wish to benefit the workers, even though at the same time they would benefit themselves. Many no doubt were influenced by their stewards and agents, who, if the new system prevailed, would lose their often profitable employment, and, more important still, their power and influence. Others would not trust the peasantry with so much independence, and others again would insist upon using the schools on their estates as a means of influencing the children against the religion of their parents. One landlord appears to have tried the experiment on a small farm of a hundred acres, and though this was fairly successful he did not carry it any further. In another case the law itself intervened adversely. Mr. William Thompson, a disciple of Bentham, had large estates in county Cork, and having visited Ralahine he determined to adopt a system so beneficial to all parties. He died unfortunately soon afterwards, but left his property to trustees to have his wishes carried into effect. The will, however, was disputed by his relatives, the nature of the bequest being held to be a proof of insanity! The suit was carried from the Irish Probate Court to the Court of Chancery, and finally settled in favour of the claimants, thus stopping for ever an extension of the principles and methods that had been so successful at Ralahine!

The Last Days of the Great Experiment.

We now approach the sad ending of a social experiment which had been altogether successful and altogether beneficial, things that do not always go together. Not only had the admittedly too high rent been punctually paid for three years, but the estate itself had been greatly improved, by the bringing into cultivation of twenty acres of almost worthless land; by the erection of six additional cottages; by the purchase of a reaping machine and other tools, and by some increase in the stock. The men, women and children employed on the farm had increased in number from 52 to 81. All these entered upon it half-starved and in rags. At the end of the third year they were all strong and well fed, and had at least two suits of clothes each, and many of them had saved money in the form of labour-notes represented by the net increase in stock and crops above what was required for payment of the exorbitant rent. There had been no deaths, no illness, no quarrels, and no secessions from the little community. All were contented and happy, and were looking forward with confidence to a still greater prosperity in the coming year, and the certainty in a few years more of being able to pay off the value of the stock and implements and thus become the owners of everything but the land.

There was abundance of water-power on the property; and it was contemplated some day to utilize it for the purpose of establishing home-manufactures, which would give profitable employment to many of the members during bad weather, or at seasons of less pressure in agricultural work, and thus add still further to the productiveness and self-supporting character of the Association.

If the rent had been a fair one, that is at least £200 a year less than was actually paid, there is no reasonable doubt that there would have been a continuous increase of prosperity, and that ultimately double the number of persons could have been easily supported on the land. Never perhaps in the history of our country was there a more important social experiment tried, or one that was so completely successful and so thoroughly beneficial.

But suddenly a terrible misfortune fell upon them and shattered their prosperity and their hopes. The landlord and president of the Association, not long after the harvest-home suddenly disappeared. During the time that Mr. Craig and the other members of the Association had been working hard to ensure their own and his prosperity, he had spent much of his time from home, had been gambling in Dublin, and had got into such difficulties that he felt them to be overwhelming. He therefore escaped to America without a word of explanation to his family. A banker to whom he owed money obtained a decree of bankruptcy against the estate, and all the stock and movable property were appropriated and sold by auction. The Association was held to have no legal claim, the agreement was declared to be invalid, and the members were treated as mere labourers having no rights whatever beyond their weekly wages. Property which they had themselves bought, as well as the surplus stock and crops against which several of the members held labour-notes to the amount of £50, were all confiscated; and this oasis in the desert of Irish misery, this little "heaven upon earth" as the people around were accustomed to call it, became a thing of the past. Mr. Craig, however, determined that none but himself should lose their savings, and by selling his own personal effects and borrowing the balance from friends, succeeded in redeeming all the outstanding notes, and, so far as he was concerned, leaving no stain on the honour of Ralahine and the New System, which he had so judiciously inaugurated and so successfully supervised during the three years of its existence.

The Teachings of Ralahine.

Having thus given a brief history of this notable experiment, from the various fragmentary indications in Mr. Craig's interesting but very excursive and rather confusing little volume; supplemented by the more connected account by Mr. W. Pare, I wish to call special attention to some of the lessons to be learnt from it, which are of very great importance at this time, when writers of authority assert positively, that any general system of co-operative industry *must* fail on account of certain deficiencies in the character of workers as a class.[1]

1. It is said, again and again, that the majority of men will not work without either the dread of starvation or

---

[1] Mr. Craig's book is called *The Irish Land and Labour Question, illustrated in the History of Ralahine and Co-operative Farming*, Trübner and Co. 1882. Another work, *Co-operative Agriculture in Ireland*, by William Pare, F.S.S., Longmans, 1870, gives a more connected account from personal observation of the same experiment, and of some others.

the incitement of individual gain. The common good, the well-being of the community of which they form a part, and on the economical success of which their own well-being depends, it is said, is not sufficient. There will be numbers of men and women who are constitutionally lazy, and there will always be more or less of loafers, who will thus live upon the labour of their fellows.

The answer to this general proposition is, that such persons, who are perhaps not really numerous, do as little work as they can *now*, while great numbers do none at all but live by the plunder of society in various ways, some criminal, some quite respectable; whereas, in a co-operative or socialistic community of any kind, all these people would do *some* work, or they would be expelled from the community or treated in some way that would be far more disagreeable to them than working. It is further urged that the influences impelling them to work would be far stronger than now. At present the working classes of all grades, from the common labourer up to the engineer, architect, parson, or doctor, do not in any way *look down* upon the person who does no work whatever, and only lives to enjoy his life as best he can on inherited property. They do not for the most part *see* that such a person lives upon their labour just as much as the most thorough loafer among themselves, and with just as little real right or justice.[1] But in a co-operative state of society the very reverse would be the case. The lazy man who shirked his work in any degree, who did not do that fair share of work which he had both strength and ability to do, would be despised as a mean and dishonest individual, and if he persisted in his idleness would be so treated that he would feel like a detected cheat, liar, or thief, in a society of gentlemen; and, it is alleged, that under this moral compulsion every man would do a fair share of work.

But both these arguments are purely academic, and they may continue to be urged by each side, to their own satisfaction but without convincing the other, till doomsday. An experimental test, on however small a scale, is

[1] This point has been elaborated and demonstrated in Chapters XV. and XXIV. of this volume.

therefore of value, and at Ralahine we had such a test. Mr. Craig was a very close observer, but he gives no hint that any of the members shirked their work, while he declares that all were industrious, and that without any supervision men would work hard and well for the benefit of the community and of themselves. There was no doubt some considerable inequality in the work done by individuals, because men's capacity for work differs; but there is no indication whatever that systematic idleness formed a difficulty at Ralahine. Yet under the old system of work under a steward for daily wages only, the Irish peasants were always alleged to be incorrigibly idle; and the same thing is said of the cotters who worked their own land, when every increase of productiveness and every appearance of improvement in the house, or the food or clothing of the family, was the sure precursor of an increase of rent. Mr. Finch, however, who made a personal study of Ralahine, and who gave evidence before a Committee of the House of Commons in 1834, deals especially with this subject in his letters to the *Liverpool Chronicle* in 1838; he says, as quoted in Mr. Pare's book (p. 61):—

"There were at first two or three fellows inclined to be idle, and they were cured in the way wild elephants are tamed. The committee who fixed the labour knew their characters, and appointed one of these idlers to work between two others who were industrious — at digging, for instance; he was obliged to keep up with them, or he became the subject of laughter and ridicule to the whole society. This is what no man could stand. By these means they were soon cured: and when I was there, there was not an idle man, woman, or child in the whole society. Indeed, public opinion was found sufficient for the cure of every vice and folly."

And Mr. Craig tells us the result in the following passage:—

"At harvest-time the whole Society used voluntarily to work longer than the time specified, and I have seen the whole body occasionally, at these seasons, act with such energy, and accomplish such great results by their united exertions, that each and all seemed as if fired by a wild enthusiastic determination to achieve some glorious enterprise—and that too without any additional stimulus in the shape of extra pecuniary reward."

Here we have an indication of what will happen when labour is organized as we now organize our armies, and when all the ardour and enthusiasm now devoted to the destruction of life and property is excited in the interest of labour exerted for the preservation, well-being, and happiness of all.

2. It is said that working men will not submit to the orders of their equals even when chosen by themselves to be foremen of the work, and that quarrels will result, and the community be soon broken up. That this has sometimes happened is no doubt true, but that it will always happen or need happen is disproved by the example now before us. The Irish are perhaps rather more quarrelsome and more quick to take offence than many other races; yet with a few common-sense rules as to the management and the overpowering influence of self-interest, we find them at Ralahine living and working together for three years in the most perfect harmony. Of course there was the power of expulsion of any individual who could not or would not live at peace with the rest; but that power can be exerted by any community, and the dread of expulsion will probably be a sufficient deterrent in most other places, as it was at Ralahine.

3. One of the most serious allegations (if it were true) against socialism or any complete system of co-operative society is, that there would be no incentive to invention or improvement, and that civilization, instead of advancing, would either stand still or retrograde and ultimately fall back to barbarism. But all history shows that this supposed objection is utterly unfounded, and that the joy of the inventor, like that of the artist, arises first from the exercise of his special talent, next from the interest and admiration it excites in his fellows, and last of all and least of all, from any hope of exceptional money reward. The lives of such men as Kepler, Galileo, Palissy, Watt, Herschel, Faraday, and a hundred others, show the truth of this; and at Ralahine it was found that the most ignorant of the labourers were sometimes able to make suggestions of value to the community. Quite recently, Mr. Preece, in a lecture before the Liverpool Engineering

Society, stated, that since the telegraphs had been worked by the Post Office four times as much work was done by the same length of wire as could be done at first, and that this was mainly due to improvements made by officers of the postal service. These improvements, he said, had never been patented, and the inventors of them received no money reward. The objection that invention would cease, may therefore be dismissed as purely imaginary, and quite unsupported by an appeal to facts.

4. An objection made much of by Mr. Mallock and others is, that the power of organization, or business capacity in its higher forms, is the almost exclusive possession of the capitalist class, and will only be exercised under the stimulus of a very high salary or great prospective gain. The associated workers must therefore necessarily fail for want of this capacity. It may be admitted that great organizing power *is* rare, and that, under our present social arrangements and struggle for wealth it commands a high price, and often brings wealth to its possessor. But the assumption that it exists in one class only, and the supposition that, under a different state of society, it will not be utilized because it is not so highly paid, is unproved as fact and unsound as reasoning. The exercise of this faculty is the exercise of power; and this is *always* enjoyed for its own sake, or for the sake of the benefits it confers on humanity, and is still further enjoyed on account of the admiration and esteem of his fellow men which it usually brings to its possessor. The idea that the man who has this great faculty, and is asked by his fellow citizens to use it for the common good, will refuse because he will not be exceptionally paid, is about as absurd, and as contrary to all experience, as to maintain that the great leader of armies or fleets will not lead till he is assured of higher pay than all other leaders. Two of the greatest organizers of modern times, Count von Moltke and General Booth, were certainly *not* incited to exertion by the hope of a money reward.

And the experience at Ralahine shows that a sufficient business capacity does exist among very humble men so soon as they have an opportunity of exercising it. It would certainly be deemed by most persons to require considerable business talent, in addition to agricultural knowledge, to manage successfully a farm of over six hundred acres, employing eighty people and subject to an exorbitant rent, so as to be in a much better position at the end of three years than at the beginning. And yet this was done wholly by a committee of common Irish ploughmen and labourers. It may be said that they had an organizer in Mr. Craig, and no doubt much of the social success was due to him; but he was not a farmer, and though he no doubt was largely responsible for the good health and general harmony of the little community, the farm, as a business concern, was wholly managed by the farm labourers themselves. Here again the imaginary objections of the critics are fully answered by facts.

5. Perhaps one of the commonest, and at the same time wildest and least grounded of these allegations is, that any kind of socialism is slavery—is a despotism so rigid and so cruel that people will not long submit to it; and that the system will necessarily break down and men will gladly return again to the old, wise, perfect and wholly-beneficial-to-everybody system of competition, starvation, and slums!

There is not a particle of evidence adduced for these statements, and experience is wholly against them. Whenever there have been associations for the common good, of people *in a similar grade of education and of social advancement*, they have usually succeeded. The Shakers and some other communistic societies have succeeded marvellously. Owen's mills at New Lanark were a perfect success as long as he was allowed to carry on the experiment; and the case of Ralahine is particularly striking. There we had the direct comparison of co-operation against individualism under exactly the same external conditions and surroundings, and we have the opinion expressed both by the members who experienced its benefits, and the surrounding population which looked on with a wondering surprise. And what was their unanimous judgment? Did they say that the New

System was slavery, and that though it fed and clothed its members they would have none of it? Not one of the Irish workers left it voluntarily. Numbers applied for admittance for whom there was no room; and the only words they could find strong enough to express their approval were, that Ralahine had become "a little heaven upon earth" under the new system, while under the old one "it was a hell!" Now the Ralahine experiment was pure socialism. It was voluntary co-operation for the good of all. All benefited equally; all worked to the best of their ability; all fared alike. They lived together, worked together, and played together; and even those who would have been idlers and loafers if they could, did not go away, did not declare they could not stand the slavery, but remained, and worked on happily with the rest. The purely academic objection, the critic's idea of what he *thinks* would happen, is directly contradicted by the appeal to facts.

6. We come now to the last refuge of the individualist, an objection urged even by many who can find nothing but praise for the ideals of socialism. It is, that we are not good enough for such an ideal system; that we must alter human nature before a co-operative commonwealth (which is the brief definition of socialism) is possible. Here again we have bold assertion without any attempt, or the shadow of an attempt, at proof. All moralists, students of human nature, and social workers know well that a very large proportion of crime is *not* due to any exceptional badness of the criminals, but either to exceptional temptation or the character of the surroundings in childhood and youth. They know, and we all know, that the men and women who pass through life unstained by conspicuous vice or actual crime owe their immunity in a large number, perhaps in a majority of cases, to their happy surroundings and freedom from temptation, and *not* to any superiority of character to many of those who, after a first offence, are irresistibly driven by our vile systems of punishment and our still viler social environment to a life of crime. Look at the cases of what is termed kleptomania now and then occurring among men

and women almost wholly removed from the temptation to theft. For each one of these cases that comes before the public there must be scores and hundreds of persons in whom the impulse exists in a less pronounced degree, but who gratify it in various harmless ways—becoming collectors, picking up bargains, &c., or by exerting all their energy in the practice of those various devices, concealments, or adulterations, which in manufacture or trade soon lead to honourable fortune. Had all the people with these dispositions been born and brought up in the slums, they would certainly have gone to swell the ranks of thieves or burglars, their "human nature" being exactly the same as that which, under more favourable conditions, caused them to remain respected members of society. Again, look at the most suggestive history of the Pitcairn Islanders, the descendants of the mutineers of the *Bounty*, men brutalized by subjection to the cruelty of superiors perhaps no better than themselves, but given absolute power over them. After years of riot and fighting which made a very pandemonium of this tropic isle, all were killed but one, John Adams, whose influence then brought peace and contentment among the population of half-breeds, the descendants of the original mutineers, and for many years they remained a model community. We cannot suppose there was any great change of character, always for the better, in the descendants of these rough men and savage women, but the better conditions brought about by the influence of the one survivor, *appeared* to effect a radical change in their nature

And this view is strikingly supported by the case of Ralahine. The people there were considered to be among the worst of the Irish peasantry. They mostly belonged to the White Boys, Moonlighters, and other organizations for intimidating landlords and agents and carrying on the agrarian war then at its height. They were universally declared by their employers to be idle, wasteful, quarrelsome and vindictive. They had just connived at, perhaps helped in, the murder of the agent of the estate, and were universally considered to be about as bad a lot as could be found. They were, moreover, among the lowest and most ignorant class of Irish labourers. Apparently no more unpromising material could be found in the three kingdoms. Yet in a few months their whole natures *appeared* to be entirely changed. Their idleness became untiring industry; their wastefulness a most careful economy; their quarrelsomeness a cheerful good-temper and joyousness which lasted for three years! Here was a marvellous change of *conduct* under conditions of simple justice, sympathy, and self-interest; but there could have been no change whatever in the *nature* of these poor people. It was simply the result of a change from bad conditions to good conditions, from injustice, tyranny, contemptuous abuse and oppression by their immediate superiors and employers, to one of fairness, freedom, civility and mutual self-interest. And yet our would-be teachers, who claim to be of the "superior" classes, can find no remedy for the countless and terrible evils of our existing social system, but a vague appeal for a higher "human nature" in some distant future! May we not properly say to such people— "Physician! cure thyself." It may be true that *some* human natures need elevating; but it is quite as likely to be the nature of those who believe that the piling up of wealth for themselves and others of their class is the great object of life, as of those simple Irish ploughmen and labourers who only asked to be allowed to work hard under moderately fair conditions, and so long as they were permitted to do so lived joyous, contented, and blameless lives.

We are again and again told, that attempts at realizing socialism have all failed, and that they have failed in consequence of deficiencies in the character of the workers. History, however, tells a different story. The three experiments which in various degrees best illustrate the advantages of socialistic co-operation are those of Robert Owen at New Lanark, E. T. Craig at Ralahine, and more recently, of the Willimantic Thread Company of Connecticut under the management of Colonel Barrows.[1] All of these succeeded perfectly, so far as the conduct, contentment and

[1] See Mr. D. Pidgeon's account of this place in his *Old World Questions and New World Answers*, Chapter XIII.

improvement of the workers were concerned. They were all alike broken up by the misconduct or greed of the capitalistic owners—the self-styled "superior classes." But we do not therefore jump to the conclusion that these classes are worse than the workers, and that it is *their* " human nature " that wants altering. On the contrary we believe, as does Herbert Spencer, that on the average, both classes are equal in moral worth as well as in intellectual power, and that, when the wealthy and educated classes are once freed from the debasing influence of cut-throat competition and the worship of wealth, they will be amenable to the influences of a more elevating environment, and will be quite as well able to make the co-operative commonwealth a success, as were the humble Irish labourers and the enthusiastic young Englishman at Ralahine.

CHAPTER XXVI.

REOCCUPATION OF THE LAND: THE ONLY IMMEDIATE
SOLUTION OF THE PROBLEM OF THE UNEMPLOYED.

WE have now just completed a century which has far surpassed all preceding centuries in the increase of man's power over natural forces, and consequent enormous increase in the production of wealth. The amount of this increase may be judged from the fact, that fifteen years ago the steam-power in Great Britain was about ten times the labour power of the whole population. It is now certainly much greater, and by the use of enormously improved labour-saving machinery, this steam-power is again increased at least tenfold in efficiency—often very much more; so that our people now perform at least a hundred times as much productive work as during the preceding centuries, when steam-power and labour-saving machines were little used or almost unknown.

Yet with this hundred-fold capacity for producing the food, clothing and other commodities needed for the satisfaction of all the wants of human nature, and for obtaining all the comforts and enjoyments of life, what do we find? Huge masses of people suffering untold want, misery, and degradation in all our great cities, while in the country villages they are often surrounded by game preserves and untilled fields; an ever-increasing number dying of actual starvation, cold, or want; insanity and suicide increasing more rapidly than the population; and according to a very competent authority—a prison chaplain, who has studied the statistics of crime for thirty years—an equally large increase in the prison population and in crime itself.[1]

As confirming and illustrating all these terrible facts, we find, in the annual reports of the Registrar-General, proof that for the last forty years there has been a continuous increase in the proportion of deaths occurring in workhouses, hospitals, asylums, and other public charitable institutions, from 16 per cent. of the total deaths (in London) in the five years 1856-60, to 26·9 per cent. in the years 1892-96; while a similar increase, though not quite so great, is shown for the whole of England.

Coincident with all these facts, and to some extent explaining them, is the continuous depopulation of the rural districts and increase of town and city populations, certainly largely due, and I believe wholly due, to the monopoly of the land in the hands of a limited class, which has always forbidden, and still forbids to the workers, the free use of their native soil except in a very small number of cases, and usually on exorbitant terms. Hence has arisen the phenomenon of an ever-increasing lack of permanent employment; the flocking of large numbers of rural labourers to the towns; the increase of want, suicide, insanity, and crime; millions of acres of land going out of cultivation; and the cry of agricultural depression, now raised the more loudly because the pockets of the landlords themselves are affected by it.

Most of the aspects of the "problem of poverty" above adverted to, I have dealt with more or less fully elsewhere, and also to some extent in the chapters of this volume on the Land Question, the Social Quagmire, &c. My present object is to suggest an immediate practical remedy for some of the worst features of the present state of things, by withdrawing from the labour market the superabundant workers and rendering them wholly self-supporting on the land. This once effected, every other worker in the kingdom will be benefited, and the movement for a greatly improved organisation of society will be

[1] See "Increase of Crime," by Rev. W. D. Morrison, in the *Nineteenth Century* of June, 1892.

advanced by a practical illustration of the enormous waste involved in the capitalistic and competitive system that now prevails.

The Problem of the Unemployed.

The problem of general unemployment is well stated by Mr. J. Hobson in the *Contemporary Review*, April, 1898. He says :

"Why is it that, with a wheat-growing area so huge and so productive that in good years whole crops are left to rot in the ground, thousands of English labourers, millions of Russian peasants cannot get enough bread to eat ? Why is it that, with so many cotton-mills in Lancashire, that they cannot all be kept working for any length of time together, thousands of people in Manchester cannot get a decent shirt to their backs ? Why is it that, with a growing glut of mines and miners, myriads of people are shivering for lack of coals ? "

Now, not one of our authorized teachers of political economy, not one of our most experienced legislators can give any clear answer to these questions, except by vague reference to the immutable laws of supply and demand, and by the altogether false statement that things are not so bad as they were, and that in course of time they will improve of themselves. Mr. H. V. Mills had his attention directed to this subject by an individual instance of the same phenomenon. He found in Liverpool, next door to each other, a baker, a shoemaker, and a tailor, all out of work, all wanting the bread, clothes and shoes which they could produce, all willing and anxious to work, and yet all compelled to remain idle and half starving. His book has been before the world several years; it contains a practical and efficient remedy for this state of things; yet no attempt whatever has been made to give his plan a fair trial. Let us therefore see if we can throw a little more light on the problem, and thus help to force it upon the attention of those who have the power, but who believe that nothing can be done.

The answer to the question so well put by Mr. Hobson, and which Mr. Stead, in the *Review of Reviews*, considers to be the modern problem of the Sphinx which it needs a modern Œdipus to solve, is nevertheless perfectly easy. To put it in its simplest form it is as follows:—Unemployment exists, and must increase, *because, under the conditions of modern society, production of every kind is carried on, not at all for the purpose of supplying the wants of the producers, but solely with the object of creating wealth for the capitalist employer.*

Now, I believe that this statement contains the absolute root of the whole matter, and indicates the true and only lines of the complete remedy. But to many it will be a hard saying; let us therefore examine it a little in detail.

The capitalist cotton-spinner, cloth or boot-manufacturer, colliery-owner, or iron-master, care not the least *who* buys their goods or *who* uses them, so long as *they* can get a good price for them. The cotton, the boots, the coals, or the iron, may be exported to India or Australia, to America or to Timbuctoo, while millions are insufficiently clad or warmed in the very places where all these things are made. Even the very people who make them may thus suffer, through insufficient wages or irregular employment; yet the upholders of the present system will not admit that anything is fundamentally wrong. The lowness of wages and irregularity of employment, are, they tell us, due to general causes over which they have no control—such as foreign competition, insufficient markets, &c., which injure the capitalists as well as the workers. The unemployed exist, they say, on account of the improvements in machinery and in mechanical processes in all civilised countries, which economise labour and thus render production cheaper. The surplus labour, therefore, is not wanted; and that portion of it which cannot be absorbed in administering to the luxury of the rich must be supported by charity or starve. That is the last word of the capitalists and of the majority of the politicians. But though capitalists and politicians are satisfied to let things go on as they are, with ever-increasing wealth and luxury on the one hand, ever-increasing misery and discontent on the other, thinking men and women all over the world are *not* satisfied, and *will not* be satisfied, without a complete solution of the problem: which, though they are not yet able to see clearly, they firmly believe can be found.

Governments in modern times have gone on the principle that they have nothing whatever to do with the employment or want of employment of the people,—with high wages or low wages, with luxury or starvation, except inasmuch as the latter calamity may be prevented by the poor-law guardians. A great change has, however, occurred in the last few years. Both the local and imperial Governments have admitted the principle of a reasonable subsistence wage, and are acting upon it, in flagrant opposition to the principles of the old political economy. Now too, I observe, the buying of Government stores abroad because they can be obtained a fraction per cent. cheaper than at home, is being given up, though only three or four years ago the practice was defended as being in accordance with true economical principles, and also because it was the *duty* of the Government to buy as cheaply as possible in the interest of the tax-payer. I only mention these facts to show that new ideas are permeating modern society and are compelling Governments, however reluctantly, to act upon them. We may, therefore, hope to compel our rulers to acknowledge, that it is their duty also to provide the conditions necessary to enable those who are idle and destitute—from no fault of their own, but solely through the failure of our competitive and monopolist system—to support themselves by their own labour. Hitherto they have told us that it cannot be done, that it would disorganise society, that it would injure other workers. We must, therefore, show them *how* it can be done, and insist that at all events the experiment shall be tried. I will now give my ideas of how this great result can be brought about, and the reasons which I believe demonstrate that the method will be successful.

Hitherto there has been no organisation of communities or of society at large for purposes of production, except so far as it has arisen incidentally in the interest of the capitalist employers and the monopolist land-owners.

The result is the terrible social quagmire in which we now find ourselves. But it is certain that organisation in the interest of the producers, who constitute the bulk of the community, is possible; and as, under existing conditions, the millions who are wholly destitute of land or capital cannot organise themselves, it becomes the duty of the State, by means of the local authorities, to undertake this organisation; and if it is undertaken on the principle that all production is to be, in the first place, for consumption by the producers themselves, and only when the necessary wants of all are satisfied, for exchange in order to procure luxuries, such organisation cannot fail to be a success.

Why Organised Industry must be a Success.

My confidence in its success is founded on three considerations, which I will briefly enumerate. The first is, the enormous productive power of labour when aided by modern labour-saving machinery. Mr. Edward Atkinson, admitted to be the greatest American authority on the statistics of production and commerce, has calculated that two men's labour for a year in the wheat-growing States of America will produce, ready for consumption, 1,000 barrels of flour, barrels included; and this quantity will produce bread for 1,000 persons for a year. Now as we can grow more bushels of wheat an acre than are grown in America, we could also produce the bread for 1,000 persons by the labour of say four or five men including the baking. Again, he tells us that, with the best machinery, one workman can produce cotton cloth for 250 people, woollen goods for 300, or boots and shoes for 1,000. And as other necessaries will require an equally moderate amount of labour, we see how easily a community of workers could produce, at all events all the necessaries of life, by the expenditure of but a small portion of their total labour-power.

The next consideration is, that in the Labour Colonies of Holland, the unemployed *are* so organised as to produce all that they consume, or its value, *without the use of any labour-saving machinery.* The reason they have none, as the director told Mr. Mills, is, that it would lead to a difficulty in finding work for the people of the colony, and it would then be less easy to manage them. The difficulty in this case seems to be to provide against the possibility of a too great success![1]

The third consideration which points to the certainty of success is, the demonstrable enormous waste of the present capitalistic and competitive system; and the corresponding enormous economies of a community in which all production would be carried on primarily for consumption by the producers themselves. This economy will be illustrated as we consider the organisation of such a community.

The Economies of an Industrial Community.

A careful consideration of the whole problem by experts will determine the minimum size of a colony calculated to ensure the most economical production of all the chief necessaries of life. Let us take it at about 5,000 persons, including men, women and children, which is Mr. Mills' estimate. Enough land will be required to grow all the kinds of produce needed, both vegetable and animal—say two to three thousand acres—and a skilled manager will be engaged to superintend each separate department of industry. Not only will bread, vegetables, fruit and meat of all kinds be grown on the land, but the whole of the needful manufactures will be carried on, aided by steam, water, or wind power as may be found most convenient and economical. To provide clothes, tools, furniture, utensils, and conveniences of all kinds for 5,000 people, workshops and factories of suitable dimensions will be provided, and skilled workers in each department will be selected from among the unemployed or partially employed. A village with separate cottages or lodgings for families and individuals, with central cooking and eating-rooms for all who desire to use them, would form an essential part of the colony. The village would be built on a high yet central position, so that all the sewage could be applied by gravitation to the lower and more distant

[1] See, *Poverty and the State*, by H. V. Mills, Chapter X

portions of the land, while all the solid refuse and manurial matter would be applied to the higher portions. Here would be the first great economy, both in wealth and health. Every particle of sewage and refuse would be immediately returned to the land, where, under the beneficent action of the chemistry of nature, it would be again converted into wholesome food and other products.

Another economy, of vast amount but difficult to estimate, would arise from the whole effective population being available to secure the crops when at their maximum productiveness. Who has not seen, during wet seasons, hay lying in the fields week after week till greatly deteriorated or completely spoilt; shocks of wheat sprouting and ruined; fruit rotting on the ground; growing crops choked with weeds, all involving loss to the amount of many millions annually, and all due to the capitalistic system which has led to the overcrowding of the towns and the depopulation of the rural districts. But this is only a portion of the loss from deficiency of labour at the critical moment. Agricultural chemists know that, even in good seasons, a considerable portion of the nutritious qualities of hay is lost by the cutting of the grass being delayed a few weeks, owing to uncertain weather, the pressure of other work, or a deficiency of labour. The critical moment is when the grass is in flower. Every day later it deteriorates; and in our self-supporting colonies the whole population would be available to supply whatever assistance the head farmer required to get the hay made in the best possible state. A single fine day utilised, with the aid of machinery and ample labour, would often save hundreds of pounds value to the colony. The same would be the case with wheat and other corn crops, as well as with fruit and vegetables.

In such a colony education could be carried on in a rational manner not possible under the present conditions of society, where the means of industrial training have to be specially provided. Ordinary school work would be at the most three or four hours daily; the remainder of the working day being devoted to various forms of industrial work. Every child would be taught to help in the simpler agricultural processes, as weeding, fruit gathering, &c., and besides this each person would learn at least two trades or occupations, more or less contrasted; one being light and sedentary, the other more active and laborious, and involving more or less out-door work. By this means not only would a pleasant and healthful variety of occupation be rendered possible for each worker, but the community would derive the benefit of being able to concentrate a large amount of skilled labour on any pressing work, such as buildings or machinery.

But perhaps the greatest economy of all would arise from such a community being almost wholly free from costs of transit, profits of the middleman, and need for advertising. The total amount of this kind of waste, on the present system, is something appalling, and can be best realised by considering the difference between the cost of manufacture and the retail price of a few typical articles. Wheat now varies from 22s. to 30s. a quarter, which quantity yields nearly six hundred pounds of bread. In our proposed community the labour of making the flour would be repaid by the value of the pollard and bran, while the bread-making would employ two or three men and women. The actual cost of their four-pound loaf, reckoning the labourers to receive present wages, would be about 2d., while it now costs 3½d. or 4d. a saving of at least 40 or 50 per cent. Again, the best Cork butter sells wholesale at 8d. a pound, the actual maker probably getting no more than 7d., while the retail consumer has to pay double—here would be a saving of at least 50 per cent. Milk is sold wholesale by the farmers at about 7d. a gallon, while it is retailed at 16d. a gallon— a saving of more than 60 per cent. In meat there would be, probably, about the same saving as in bread; in vegetables and fruit very much more; in coals bought wholesale from the pit, as compared with the rate at which it is sold by the hundredweight or the pennyworth to the poor in great cities, an equally large saving. And in addition to all this, there would be the economy in the cooking for a large community; in the freshness and good quality of all food and manufactured products; and, further, in the saving of labour by all those improvements in gas and water supply, in disposing of refuse, in warming and ventilation, which can be easily provided for a large community living in a compact and well arranged set of buildings.

The Waste of Competition.

Taking all these various economies into consideration, it is probably far below the mark to say that our present system of production on a huge scale for the benefit of capitalists and landlords only, on the average doubles the cost of everything to the consumer; that is to say, the cost of distribution is equal to, and often much greater than, the cost of production. And this is said to be an economical system! A system too perfect, and almost too sacred to be touched by the sacrilegious hands of the reformer! We are to go on for ever spending a pound to get every pound's worth of goods from the producer to the consumer; just as under our poor-law system it costs a shilling to give a starving man a shilling's worth of food and lodging.

But there is yet another economy, which I have not hitherto mentioned, and which may perhaps be said to be the greater in real value and importance than all the rest—and that is the economy to the actual producer, of time, of labour, of health, and the large increase in his means of recreation and happiness. Agricultural labourers now often have to walk two or three miles to their work; mill-hands, including women and children, walk long distances in all weathers to be at the mill-gates by six in the morning; workers by the million undergo a process of slow but certain destruction in unsanitary workshops, or in dangerous or unhealthy occupations, many of which (as making the enamelled iron advertising plates, for example, as well as poisonous matches) are quite unnecessary for the needs of a properly organised community; while in all cases it is only a question of expense to save the workers from any injury to health. In our self-supporting communities all these sources of waste and misery would be avoided. All work would be near at hand. No work permanently injurious to health would be permitted; while the alternations of out-door and indoor work, together with the fact that every worker would be working for himself, for his family, and for a community of which he formed an integral part on an equality with all his fellow-workers, would give a new interest to labour similar to that which every gardener feels in growing vegetables for his own table, and every mechanic in fitting up some useful article in his own house. Then again, while living in and surrounded by the country and enjoying all the advantages and pleasures of country life, a community of five thousand persons would possess in themselves the means of supplying most of the relaxations and enjoyments of the town, such as music, theatricals, clubs, reading rooms, and every form of healthy social intercourse.

How to Establish Co-operative Communities.

This is not the place to go into the minute details of the establishment of such communities, but a few words as to ways and means may be considered necessary.

Mr. Mills has estimated that the capital required to buy the land and start such a colony, would not exceed two years' poor-rates of a Union where there are an equal number of paupers. But there is really no necessity for buying the land. It might be taken where required at a fair valuation and paid for by means of a terminable rental, similar to that by which Irish tenants have been enabled to purchase their farms; but in this case the county would be the purchaser, not an individual, and after the first year, or perhaps two years, this rent-charge would be easily payable by the colony. The capital needed for buildings, machinery, and one year's partial subsistence, should be furnished, half by the County or Union, and half by the Government, free of interest, but to be repaid by instalments to commence after, say, five or ten years. It would really be to the advantage of the community at large to give this capital, since it would inevitably lead to the abolition of unemployment and of able-bodied pauperism which would more than repay the initial outlay.

In each colony there would be grown or manufactured a considerable amount of surplus produce, which would be sold in order to purchase food which cannot be produced at home—as tea, coffee, spices, &c., and also such raw materials as iron and coal. The things produced for sale would vary according to the facilities for its production and local demand. In some colonies it would be wheat or barley, in others butter or cheese, in others again, flax, vegetables, fruit, or poultry. And as *all* the products of our soil except milk are largely imported, there is ample range for producing articles for sale which would not in any way affect prices or interfere with outside labour.

At first, of course, such colonies must be organised and all the work done under general regulations, and the same discipline as is maintained in any farm or factory, but with no unnecessary interference with liberty out of working hours. Accounts would be strictly kept and audited, and all profits would go to increasing the comfort of the colonists in various ways, and in paying surplus wages to be spent, or saved, as the individual pleased. Under reasonable restrictions as to notice, every one would be at liberty to quit the colony; but with such favourable conditions of life as would prevail there it is probable that only a small proportion would do so, as was the case at Ralahine, and at the Dutch colony of Frederiksoord.

But as time went on, and a generation of workers grew up in the colony itself, a system of self-government might be established; and for this purpose I think Mr. Bellamy's method the only one likely to be a permanent success. It rests on the principle that, in an industrial community, those only are fit to be rulers who have for many years formed integral parts of it, who have passed through its various grades as workers or overseers, and who have thus acquired an intimate practical acquaintance with its needs, its capacities, and its possibilities of improvement. Persons who had themselves enjoyed the advantages of the system, who had suffered from whatever injudicious restrictions or want of organisation had prevailed, and who had nearly reached the age of retirement from the more laborious work, would be free from petty jealousies of their fellow-workers, and would have no objects to aim at except the continued success of the colony and the happiness of all its inmates. On this principle those who had worked in the colony for at least fifteen or twenty years, and who had reached some grade above that of simple workmen, should form the governing body, appointing the superintendents of the various departments, and making such general regulations as were needed to ensure the prosperity of the community and the happiness of all its members.

Why is not the Experiment Tried?

Now, I would ask, what valid reason can be given against trying this great experiment in every county in Great Britain and Ireland, so as at once to absorb the larger part of the unemployed as well as all paupers who are not past work? The only real objection, from the capitalist's point of view, that I can imagine, is, that colonies in which the whole of the produce went to the workers themselves, including of course their own sick and aged, would be so attractive that they would draw to them large numbers of workers of all kinds, and thus interfere with the capitalists' labour-supply. This, I believe, would, after a few years, inevitably occur; but, from my point of view, and from that probably of most thinkers and workers, that circumstance would afford the greatest argument in favour of the scheme. For it would show that, with a proper organisation of labour, production under individual capitalists was unnecessary; it would afford practical proof that labourers can successfully produce without the intervention of capitalist employers.

In this connection I will quote a passage from the writings of that remarkable observer and thinker, the late Richard Jeffrey. He says:—

> "I verily believe that the earth in one year can produce enough food to last for thirty. Why then have we not enough? Why do people die of starvation, or lead a miserable existence on the verge of it? Why have millions upon millions to toil from morning till

evening just to gain a mere crust of bread? Because of the absolute lack of organisation by which such labour should produce its effects, the absolute lack of distribution, the absolute lack even of the very idea that such things are possible. Nay, even to mention such things, to say that they are possible, is criminal with many. Madness could hardly go further."[1]

This was written a good many years ago. Now, we who hold such opinions are considered to be, not criminals but merely cranks; and it is even allowed that we have good ideas sometimes, if only we were more practical. But, surely, nothing can be more practical than the proposal here made, since the experiment has already been tried in Holland and in Ireland, both under unfavourable conditions, yet in both it succeeded. To produce any real effect, however, it must be brought into operation on a large scale, and this can only be done by the local authorities, to whom all necessary powers must be given, with the needful financial assistance from the Government.

When labour-colonies of the kind here suggested have been established for a few years, it is quite certain that the District Councils will no longer endure the old, bad, wasteful, and degrading system of the Union Workhouses, but will obtain land in the vicinity of existing workhouses where possible, and establish labour-colonies of the same type. The effects of the new system will soon become palpable to every ratepayer in the kingdom by the greatly diminished rates together with the abolition of paupers, wherever they have been established. Public opinion will then be all in favour of the new system, and legislation will be demanded and quickly obtained, enabling any sufficient number of persons who wish to form such a community by voluntary association, to have the land required in any part of the country on a permanent tenure and at a fair agricultural rental.

Numerous self-supporting co-operative labour-colonies being thus established all over the country, their connection by tramways where required, together with the arrangements they would soon make for mutual assistance and exchange of products, for the common use

[1] *The Story of My Heart*, p. 194.

of mills or other costly machinery, and through the healthy rivalry that would inevitably spring up—all these varied influences would still further increase the benefits to be derived from them and the pleasures of associated life. And these advantages would in a less degree spread to every inhabitant of our country. For, not only would the withdrawal of the whole surplus labour now represented by the unemployed or partially employed, inevitably cause a considerable increase in the rate of wages in all departments of industry, but the high standard of comfort and the complete absence of the anxiety now inseparable from capitalistic wage-labour, would draw more and more of the workers to such co-operative communities and thus compel capitalists to offer yet higher wages, with probably a share of profits and a voice in the management of their factories in order to obtain men to work for them. This would compel the capitalist manufacturer to be satisfied with an amount of profit sufficient to pay him as an organiser and superintendent, as the only alternative to the loss of his fixed capital.

The whole of the surplus profits in all our industries would then be distributed among the various classes of workers, manual and intellectual, and labour would, for the first time, receive its full and just reward.

CHAPTER XXVII

HUMAN PROGRESS: PAST AND FUTURE

THE word progress, as used above, has two distinct meanings, not always recognized, whence has arisen some confusion of ideas. It may mean either an advance in material civilization, or in the mental and moral nature of man, and these are far from being synonymous. Material civilization is essentially cumulative. Each generation benefits by the trials and failures of the preceding generation; and since the discovery of printing has facilitated the preservation and circulation of all new knowledge, progress of this kind has gone on at an ever accelerated pace. But this does not imply any general increase of mental power. Step by step the science of mathematics has advanced immensely since the time of Newton, but the advance does not prove that the mathematicians of to-day have a greater genius for mathematics—are really greater mathematicians—than Newton and his contemporaries, or even than the Greeks of the time of Euclid and Archimedes. Our modern steam engines and locomotives far surpass those of Watt and Robert Stephenson, but of the hundreds who have laboured to improve them perhaps none have surpassed those great men in mechanical genius. And so it is with every item which goes to form that which we term our civilization. We have risen, step by step, on the ladders and scaffolds erected by our predecessors, and if we can now mount higher and see further than they could, it does not in the least prove that we are, on the average, greater men, intellectually, than they were. The question I propose to discuss is one quite apart from that of civilization as usually understood. It is, whether mankind have advanced as intellectual and moral beings; and, if so, by what agencies and under what laws have they so advanced in the past, and what are the conditions under which that advance may be continued in the future.

Has Human Nature improved during Historic Times?

We have, first, to inquire whether there is any evidence of such an advance in human nature during historic times; and this is by no means so simple a problem and one so easily answered as is sometimes supposed. If there has been any cause constantly at work tending to elevate human nature, we should expect it to manifest itself by a perceptible rise in the culminating points reached by mankind, in the intellectual and moral spheres, at successive periods. But no such continuous rise of the high-water mark of humanity is perceptible. The earliest known architectural work, the great pyramid of Egypt, in the mathematical accuracy of its form and dimensions, in its precise orientation, and in the perfect workmanship shown by its internal structure, indicates an amount of astronomical, mathematical, and mechanical knowledge, and an amount of experience and practical skill, which could only have been attained at that early period of man's history by the exertion of mental ability no way inferior to that of our best modern engineers. In purely intellectual achievements the Vedas and the Mahabharata of ancient India, the Iliad of Homer, the book of Job, and the writings of Plato, will rank with the noblest works of modern authors. In sculpture and in architecture the ancient Greeks attained to a height of beauty, harmony and dignity, that has never been equalled in modern times; and taking account also of the great statesmen, commanders, philosophers, and poets of the age of Pericles, Mr. Francis Galton is of opinion "that the average ability of the Athenian race was, on the lowest possible estimate, very nearly two grades higher that our own—that is, about as much as our race is above that of the African negro."[1]

There is, therefore, some reason to think that the intellectual high-water level of humanity has sunk rather than risen during the last two thousand years; but this is not absolutely incompatible with the elevation of the mean level of the human ocean both intellectually and morally. We must, therefore, briefly consider the various agencies that have been at work, some tending to raise others to depress this level; and by balancing the one against the other, and taking account of certain modern developments of human nature in civilized societies, we may be able to arrive at an approximate conclusion as to the final result.

During the whole course of human history the struggle of tribe with tribe and race with race has inevitably caused the destruction of the weaker and lower, leaving the stronger and higher, whether physically or mentally stronger, to survive. Another, and perhaps not less potent cause of the destruction of lower tribes is the greater vital energy and more rapid increase of the higher races, which crowds the lower out of existence even when no violent destruction of life takes place. To this latter cause quite as much as to actual warfare must we ascribe the total disappearance of the Tasmanians, and the continuous diminution of population among the Maoris of New Zealand and the inhabitants of the Eastern Pacific Islands, as well as of the Red Indians of the North American continent. Here we see survival of the fittest among competing peoples necessarily leading to a continuous elevation of the human race as a whole, even though the higher portion of the higher races may remain stationary or may even deteriorate.

But a similar and even more complex process is ever going on within each race, by the survival of the more fit and the elimination of the less fit under the actual conditions of society. On the whole, we cannot doubt that the prudent, the sober, the healthy, and the virtuous, live longer lives than the reckless, the drunkards, the unhealthy, and the vicious; and also that the former, on

[1] *Hereditary Genius*, p. 342.

the average, leave more descendants than the latter. It is true that the latter not unfrequently marry earlier and have larger families; but many of these die young, and as, on the average, children resemble their parents, fewer of these will survive and leave offspring. Thus, accidents, violence, and the effects of a reckless and vicious life, are natural checks to the increase of population among these classes, and this inevitably gives an advantage to the more intellectual, the more prudent, and the more moral portion of each race. The latter will, therefore, increase at the expense of the former, and thus again tend to raise the mean level of humanity.

But society has always, in one way or another, interfered with these beneficent processes, and has thus retarded the general advance. The celibacy of the clergy and the refuge offered by monasteries and nunneries to many to whom the rude struggle of the world was distasteful, and whose gentle natures fitted them for deeds of charity or to excel in literature or art, prevented the increase of these nobler individuals; and thus, as Mr. Galton well remarks, "the Church, by a policy singularly unwise and suicidal, brutalized the breed of our forefathers." By a still more deplorable policy, independent thought and that true nobility which refuses to purchase life by a lifelong lie, was almost exterminated in Europe by religious persecution. It is calculated that for the three centuries between 1471 and 1781, a thousand persons annually were either executed or imprisoned by the Inquisition in Spain alone. In Italy it was even worse; while in France during the seventeenth century three or four hundred thousand Protestants perished in prison, at the galleys, or on the scaffold.

Another cause which has had a prejudicial effect at all times, and which continues in action in the civilized societies of to-day, is the system of inherited wealth, which often gives to the weak and vicious an undue advantage both in the certainty of subsistence without labour, and in the greater opportunity for early marriage and leaving a numerous offspring. We also interfere with the course of nature by preserving the weak, the sickly, or the malformed infants; but in this, probably, humanity gains rather than loses, since many who are in infancy weak or distorted exhibit superior mental or moral qualities which are a gain to civilization, while the cultivation of humane and sympathetic feelings in their care and nurture is itself of the greatest value.

Balancing, as well as we are able, these various opposing influences, it seems probable that there has been, on the whole, a decided gain. Health, perseverance, self-restraint, and intelligence have increased by slowly weeding out the unhealthy, the idle, the grossly vicious, the cruel, and the weak-minded; and it may be in part owing to the increased numbers of the higher and gentler natures thus brought about that we must impute the undoubted growth of humanity—of sympathy with the sufferings of men and animals, which is perhaps the most marked and most cheering of the characteristics of our age.

The Effect of Education.

But although the natural process of elimination does actually raise the mean level of humanity by the destruction of the worst and most degraded individuals, it can have little or no tendency to develop higher types in each successive age; and this agrees with the undoubted fact that the great men who appeared at the dawn of history and at the culminating epochs of the various ancient civilizations, were not, on the whole, inferior to those of our own age. It remains, therefore, a mystery how and why mankind reached to such lofty pinnacles of greatness in early times, when there seems to be no agency at work, then or now, calculated to do more than weed out the lower types. Leaving this great problem as, for the present, an insoluble one, we may turn to that aspect of the question which is of the most vital present day interest—whether any agencies are now at work or can be suggested as practicable, which will produce a steady advance, not only in the average of human nature, but in those higher developments which now, as in former ages, are the exceptions rather than the rule.

Till quite recently the answer to this question would have been an unhesitating affirmative. Education, it would have been said, is such an agency; and although hitherto it has done comparatively little, owing to the very partial and extremely unscientific way in which it has been applied, we have now acquired such a sound knowledge of its philosophy and have so greatly improved its methods, that it has become a power by which human nature may be indefinitely modified and improved. When every child is really well educated, when its moral as well as its intellectual faculties are trained and developed, some portion of the improvement effected in each generation will be transmitted to the next, and thus a continual advance both in the intellectual and moral nature will be brought about.

Almost all who have discussed the subject have held that this is the true and only method of improving human nature, because they believe that in the analogous case of the bodily structure the modification and improvement of all organisms has been effected by a similar process. Lamarck taught that the effects produced by use and exertion on the body of the individual animal were, wholly or in part, transmitted to the offspring; and although Darwin's theory of natural selection rendered this agency almost if not altogether unnecessary, yet it was so universally held to be a fact of nature that Darwin himself adopted it as playing a subsidiary but not unimportant part in the modification of species. So little doubt had he of this "transmission of acquired characters" that his celebrated theory of Pangenesis was framed so as to account for it. In order to explain, hypothetically, how it was that the increased size or strength given to a limb or an organ by constant exercise was transmitted to the progeny, he supposed that the male and female germ-cells were formed by the aggregation of inconceivably minute gemmules from every tissue and cell of every part of the body, that these gemmules were continually renewed and continually flowing towards the reproductive organs, and that they had the property of developing into cells and structures in the offspring which more or less closely resembled the corresponding cells and structures in the parents at that particular epoch of their lives. Thus was explained the transmission of disease, and the supposed transmission of the changes produced in the parents by use or disuse of organs or by other external conditions.

To illustrate this by an example: if two brothers, equally strong and healthy, became one a city clerk, the other a farmer, land-surveyor, or rural postman, living much in the open air and walking many miles every day of his life, and if they married two sisters equally alike in constitution, then the children of these two couples, especially those born when their parents approached middle age and the different conditions of their existence had had time to produce its full effect on their bodily structure, ought to show a decided difference, the one family being undergrown, pale, and rather weak in the lower limbs, the other the reverse; and this difference should be observable even if the children of the two families were brought up together under identical conditions. It may be here stated that no trustworthy observations have ever been made showing that such effects are really produced, but it has always been believed that they must be produced.

As Darwin's theory of Pangenesis led to considerable discussion, Mr. Francis Galton, who had at first accepted it provisionally, endeavoured to put it to the test of experiment. He obtained a number of specimens of two distinct varieties of domestic rabbits which breed true, and, by an ingenious and painless arrangement, caused a large quantity of the blood of one variety to be transfused into the blood-vessels of the other variety. This having been effected with a number of individuals without in any way injuring their health, they were separated and bred from. It was found that in every case the offspring resembled their parents and showed no trace of intermixture of the two varieties. It was also pointed out by another critic that if the theory of Pangenesis were true, the stock on which a fruit is grafted ought to change the character of the fruit produced by the graft, which, as a rule, it does not do.

The Non-inheritance of Acquired Character.

Doubt being thus thrown on the validity of the theory, Mr. Galton suggested another, in which the germs in the reproductive organs of each individual were supposed to be derived directly from the parental germs and not at all from the body itself during its growth and development. A very similar theory was proposed some years later by Professor Weismann under the now well-known term "the continuity of the germ-plasm." Both these theories imply that, except among the lower single-celled animals and in certain exceptional cases among the higher animals, no change produced in the individual during life, by exercise or other external conditions, can be transmitted to its offspring. What is transmitted is the capacity to develop into a form more or less closely resembling that of the parents or their direct ancestors, the characteristics of these appearing in the offspring in varying degrees and compounded in various ways, leading to that wonderful variety in details while preserving a certain unmistakable family resemblance. Thus are explained not only bodily but mental characteristics, even those peculiar tricks of motion or habits which are often adduced as proofs of the transmission of an acquired character, but which are really only the transmission of the minute peculiarities of physical structure and nervous or cerebral co-ordination, which led to the habit in question being acquired by the parent or ancestor, and, under similar conditions, by his descendant.

Finding that this theory, if true, did not allow of the hereditary transmission of the majority of individually *acquired* characters, Weismann was led to examine the evidence of such transmission, and found that hardly any real evidence existed, and that in most cases which appeared to prove it, either the facts were not accurately stated, or another interpretation could be given to them. The transmission had been assumed because it appeared so natural and probable; but in science we require as the foundation of our reasoning not probability only, but proof; or if we cannot get direct proof, then the probability which arises from all the phenomena being such as would occur if the theory in question were true, and this so completely as to give us the power of predicting what will occur under new and hitherto untried conditions. Of this nature is the probability in favour of the existence of an ethereal medium whose undulations produce light and heat; of atoms which combine to form the molecules of the various elements; and of the molecular theory of gases. The biologists of Europe, though usually slow to accept new theories in the place of old ones, have given to the theories of Weismann and Galton an amount of acceptance which was never accorded to Darwin's theory of Pangenesis, notwithstanding the weight of his great reputation; and they are now seeking earnestly for facts which shall serve as crucial tests of the rival theories, just as the phenomena of interference served as a test of the rival theories of light.

We have here only to deal with the theory of the non-inheritance of acquired characters as it affects mental and moral qualities; and in this department it has to encounter great opposition, because it seems to bar the way against any improvement of the race by means of education. If the theory is a true one, it certainly proves that it is not by the direct road of education, as usually understood, that humanity has advanced and must advance, although education may, in an indirect manner, be an important factor of progress. Let us, however, look at the problem as presented by the rival theories, and see what light is thrown upon it by the history of those great men who have most contributed to the advance of civilization, and who serve well to illustrate the successive high-water marks attained by human genius.

Illustrations of the Non-inheritance of Culture.

If progress is in any important degree dependent on the hereditary transmission of the effects of culture, as distinguished from the transmission of innate genius, or of the various talents and aptitudes with which men and women are born, then we should expect to see indications of such transmission in the continuous increase of mental power wherever any family or group of families have for several generations been subjected to culture or training of any particular kind. It has, in fact, been claimed that this is the case, for in his presidential address to the Biological Society of Washington, in January, 1891, Mr. Lester F. Ward argues that not only is Professor Weismann's great ability a result of the rigid methods of training in the German universities, but that

"those rigid methods themselves have been the product of a series of generations of such training, transmitted in small increments and diffused in increasing effectiveness to the whole German people. . . . And the fact, that out of the barbaric German hordes of the Middle Ages there has been developed the great modern race of German specialists is one of the most convincing proofs of the transmission of acquired characters, as well as of the far-reaching value to the future development of the race of such an educational system as that which Germany has had for the last two or three centuries."

It will, I think, be admitted that, if this is " one of the most convincing proofs " of the transmission of the effects of culture, the theory of its transmissibility has but a weak foundation ; for not only may the facts be explained in another way, but there is another body of facts which point with at least equal clearness in an exactly opposite direction. It may be said, for instance, that the eminence of German specialists in science is due primarily to special mental qualities which have always been characteristic of the German race, and to the facilities afforded for the culture of those faculties throughout life, by the large numbers of professorships in their numerous universities, and by the comparative simplicity of German habits, which renders the position of professor attractive to the highest intellects. And when we turn to other countries we find facts which tend in the opposite direction. In England, for example, during many centuries, Oxford and Cambridge Universities were closed to nonconformists, and their honours and rewards were reserved for members of the Established Church, and very largely for the families of the landed aristocracy. Yet in the short period that has elapsed since they were opened to dissenters, these latter have shown themselves fully equal to the hereditarily trained churchmen, and have carried off the highest honours in as great, and perhaps even in greater proportion than their comparative numbers in the universities would render probable.

Again, it is a remarkable fact, that almost all our greatest inventors and scientific discoverers, the men whose originality and mental power have created landmarks in the history of human progress, have been self-taught, and have certainly derived nothing from the training of their ancestors in their several departments of knowledge. Brindley, one of the earliest of our modern engineers, was the son of a dissipated small freeholder; Telford, our greatest road and bridge builder, was the son of a shepherd, and apprenticed to a rough country mason; George Stephenson, the inventor of the locomotive engine, was a self-taught collier; Bramah, the inventor of the hydraulic press, of improved locks, and· almost the originator of machine tools, was the son of a farmer, and at seventeen years of age was apprenticed to the village carpenter; Smeaton, who designed and built the Eddystone lighthouse, was the son of a lawyer, and a wholly self-taught engineer; Harrison, the inventor of the modern chronometer, was a joiner and the son of a joiner; the elder Brunel was the son of a French peasant farmer, and was educated for a priest, yet he became a great self-taught engineer, designed and executed the first Thames tunnel, and at the beginning of this century designed the block-making machinery in Portsmouth dock-yard which was so complete both in plan and execution that it is still in use.

Coming now to higher departments of industry, science, and art, we find that Dolland, the inventor of the achromatic telescope, was a working silk-weaver, and a wholly self-taught optician; Faraday was the son of a blacksmith, and apprenticed to a bookbinder at the age of thirteen; Sir Christopher Wren, the son of a clergyman and educated at Oxford, was a self-taught architect, yet he designed and executed St. Paul's Cathedral, which will certainly rank among the finest modern buildings in the world; Ray, the son of a blacksmith, became a good mathematician, and one of the greatest of our early naturalists; John Hunter, the great anatomist, was the son of a small Scotch landholder; Sir William Herschel was a professional musician, the son of a German bandmaster; Rembrandt was the son of a miller; the great linguists and oriental scholars, Alexander Murray and Dr. Leyden, were both sons of poor Scotch shepherds; while Shelley, whose poetic genius has rarely been surpassed, was the son of an altogether unpoetic and unsympathetic country squire.

These few examples, which might be easily increased so as to fill a volume, serve to show, what is indeed seldom denied, that genius or superexcellence in any department of human faculty tends to be sporadic, that is, it appears suddenly without any proportionate development in the parents or immediate ancestors of the gifted individual. No doubt there is usually, or perhaps always, a considerable amount of the same mental qualities dispersed through the diverging ancestral line of all these men of genius, and their appearance seems to be well explained by a fortunate intermingling of the germ-plasms of several ancestors calculated to produce or to intensify the various mental peculiarities on which the exceptional faculties depend. This is rendered probable, also, by the fact that, although genius is often inherited it rarely or never intensifies after its first appearance, which it certainly should do if not only the genius itself, but the increased mental power due to its exercise were also inherited. Brunel, Stephenson, Dollond, and Herschel, all had sons who followed in the steps of their fathers; but it will be generally admitted that in no case did the sons exceed or even equal their parents in originality and mental power. So, if we look through the copious roll of names of great poets, and painters, sculptors, architects, engineers, or scientific discoverers, we shall hardly ever find even two of the same name and profession, and never three or four, rising progressively to loftier heights of genius and fame. Yet this is what we ought to find if not only the innate faculty, but the increased development given to that faculty by continuous exercise, tends to be inherited.[1]

If it is thought that this non-inheritance of the results of education and training is prejudicial to human progress, we must remember that, on the other hand, it also prevents the continuous degradation of humanity by the inheritance of those vicious practices and degrading habits which the deplorable conditions of our modern social system undoubtedly foster in the bulk of mankind. Throughout all trade and commerce lying and deceit abound to such an extent that it has come to be considered essential to success. No dealer ever tells the exact truth about the goods he advertises or offers for sale, and the grossly absurd misrepresentations of material and quality we everywhere meet with have, from their very commonness, ceased to shock us. Now it is surely a great blessing if we can believe that this widespread system of fraud and falsehood does not produce any inherited deterioration in the next generation. And it is equally satisfactory to believe that the physical deterioration produced on the thousands who annually exchange country for town life will have no permanent effect on their offspring if they return at any time to more healthy conditions. And we have direct evidence that this is so, in the fact that the street arabs of our great cities, when brought up under healthy and elevating conditions in the colonies, usually improve both physically, intellectually, and morally, so as to be fully equal to the average of their fellow-countrymen.

It appears, then, that the non-inheritance of the effects of training, of habits, and of general surroundings, whether these be good or bad, is by no means a hindrance to

[1] The only prominent example that looks like a progressive increase of faculty for three generations is that of Dr. Erasmus Darwin and his grandson Charles Darwin. But in this case the special faculties displayed by the grandson were quite distinct from those of the grandfather and father; while if we consider the different state of knowledge at the time when Erasmus Darwin lived, his occupation in a laborious profession, and the absence of that stimulus to thought which the five years' voyage round the world gave to his grandson, it is not at all certain that in originality and mental powers, the former was not fully the equal of the latter.

human progress, if, as seems not improbable, the results on the individual of our present social arrangements are, on the whole, evil. It may be fairly argued that the rich suffer, morally and intellectually, from these conditions quite as much as do the poor; and that the lives of idleness, of pleasure, of excitement, or of debauchery, which so many of the wealthy lead, is as soul-deadening and degrading in its effects as the sordid struggle for existence to which the bulk of the workers are condemned. It is, therefore, a relief to feel that all this evil and degradation will leave no permanent effects whenever a more rational and more elevating system of social organization is brought about.

How Progress will be Effected.

If, then, education, training, and surrounding conditions can do nothing to affect permanently the march of human progress, how, it may be asked, is that progress to be brought about; or are we to be condemned to remain stationary in that average condition which, in some unknown way, the civilized nations of the world have now reached? We reply, that progress is still possible, nay, is certain, by the continuous and perhaps increasing action of two general principles, both forms of selection. The one is that process of elimination already referred to, by which vice, violence, and recklessness so often bring about the early destruction of those addicted to them. The other, and by far the more important for the future, is that mode of selection which will inevitably come into action through the ever-increasing freedom, joined with the higher education of women.

There have already been ample indications in literature that the women of America, no less than those of other civilized countries, are determined to secure their personal, social, and political freedom, and are beginning to see the great part they have to play in the future of humanity. When such social changes have been effected that no woman will be compelled, either by hunger, isolation, or social compulsion, to sell herself whether in or out of wedlock, and when all women alike shall feel the refining influence of a true humanizing education, of beautiful and elevating surroundings, and of a public opinion which shall be founded on the highest aspirations of their age and country, the result will be a form of human selection which will bring about a continuous advance in the average status of the race. Under such conditions, all who are deformed either in body or mind, though they may be able to lead happy and contented lives, will, as a rule, leave no children to inherit their deformity. Even now we find many women who never marry because they have never found the man of their ideal. When no woman will be compelled to marry for a bare living or for a comfortable home, those who remain unmarried from their own free choice will certainly increase, while many others, having no inducement to an early marriage, will wait till they meet with a partner who is really congenial to them.

In such a reformed society the vicious man, the man of degraded taste or of feeble intellect, will have little chance of finding a wife, and his bad qualities will die out with himself. The most perfect and beautiful in body and mind will, on the other hand, be most sought and therefore be most likely to marry early, the less highly endowed later, and the least gifted in any way the latest of all, and this will be the case with both sexes. From this varying age of marriage, as Mr. Galton has shown, there will result a more rapid increase of the former than of the latter, and this cause continuing at work for successive generations will at length bring the average man to be the equal of those who are now among the more advanced of the race.

When this average rise has been brought about there must result a corresponding rise in the high-water mark of humanity; in other words, the great men of that era will be as much above those of the last two thousand years as the average man will have risen above the average of that period. For, those fortunate combinations of germs which, on the theory we are discussing, have brought into existence the great men of all ages, will have a far higher average of material to work with, and we may reasonably expect that the most distinguished among the poets and philosophers of the future will decidedly surpass the Homers and Shakespeares, the Newtons, the Goethes, and the Humboldts of our era.

Mr. Lester F. Ward has indeed urged, in his article on "The Transmission of Culture" (*Forum*, May, 1891), that, if Weismann's theory is true, then "education has no value for the future of mankind, and its benefits are confined exclusively to the generation receiving it." Another eminent scientist, Professor Joseph Le Conte, in his article on "The Factors of Evolution" (*The Monist*, Vol I., p. 334), is still more desponding. He says:—

> "If it be true that reason must direct the course of human evolution, and if it be also true that selection of the fittest is the only method available for that purpose; then, if we are to have any race-improvement at all, the dreadful law of *destruction of the weak and helpless* must with Spartan firmness be carried out voluntarily and deliberately. Against such a course all that is best in us revolts."

These passages show that the supposed consequences of the theories of Weismann and Galton, have, very naturally, excited some antagonism, because they appear, if true, to limit or even to destroy all power of further evolution of mankind, except by methods which are revolting to our higher nature.

But I have endeavoured to show, in the present article, that we are not limited to the depressing alternatives above set forth,—that education *has* the greatest value for the improvement of mankind,—and that selection of the fittest may be ensured by more powerful and more effective agencies than the destruction of the weak and helpless. From a consideration of historical facts bearing upon the origin and development of human faculty I have shown reason for believing that it is only by a true and perfect system of education and the public opinion which such a system will create, that the special mode of selection on which the future of humanity depends can be brought into general action. Education and environment, which have so often stunted and debased human nature instead of improving it, are powerless to transmit by heredity either their good or their evil effects; and for this limitation of their power we ought to be thankful.

It follows, that when we are wise enough to reform our social economy and give to our youth a truer, a broader, and a more philosophical training, we shall find their minds free from any hereditary taint derived from the evil customs and mistaken teaching of the past, and ready to respond at once to that higher ideal of life and of the responsibilities of marriage which will, indirectly yet surely, become the greatest factor in human progress.

CHAPTER XXVIII

TRUE INDIVIDUALISM—THE ESSENTIAL PRELIMINARY OF A
REAL SOCIAL ADVANCE

Now that we have entered the last year of this our Nineteenth Century, in many respects the most eventful century for good and evil the world has witnessed, most thinking men are looking forward with anxious hope as to what of real good the Twentieth Century may have in store for humanity. Any words of hopeful guidance as to how we may help to bring about such good; any indication of the true path to such social regeneration as may not only enable the middle classes to reach a still higher pitch of refinement, but may raise up the masses from the deadly slough of want, misery, starvation and crime in which so many millions are now floundering, often from no fault of their own and in the midst of the most wealthy and most civilized countries in the world,—will certainly be welcome to the humane and thoughtful in all modern societies.

It is clear, that if we wish to do any real good, we must cease to deal in generalities, or to suggest mere palliatives. We must seek for the fundamental error in our social system which has led to the damning result, that, in the latter half of the nineteenth century there has been a far greater mass of human misery arising directly from want —and an equal, perhaps greater, amount in proportion to population—than in the preceding century. This is clearly indicated by the figures given by the statistician Mulhall in 1883, in a paper read at the British Association.

In the period 1774 to 1800—which may be taken as representing the latter half of the eighteenth century—he gives the wealth per head of the population of Great Britain at £110, and for 1860 to 1882, representing a corresponding period of the nineteenth century, at £216. But the purchasing power of money is estimated to have been so much greater in the earlier period that Mr. Mulhall calculates the effective income per head to have then been £227, or actually higher than in our own time. This apparent paradox can only be explained by the *proportion* of the very poor to the whole population being now exceptionally large, so that, although there has been such an enormous increase of total wealth and a considerable increase of very rich men, yet the great army of workers who produce this wealth has increased so much more largely that the proportion coming to them is smaller than ever. And this is quite in accordance with the evidence of Mr. Charles Booth, who has shown that about 1,300,000 of the population of London live " below the margin of poverty ; " and if we add to these the inmates of the workhouses, hospitals, &c., we shall find that close upon *one-third* of the whole population are in this miserable condition; and we may be sure that in all our great manufacturing towns and cities, the proportion of the very poor is not much less.

Again, we must remember that in the last century the majority of the workers lived in the rural districts or in the smaller towns, and possessed many additions to their means of living which they have now lost:—such as gardens, common-rights, wood for fuel, gleaning after harvest, pig and poultry-keeping, and often skim-milk or butter milk from the farms where they worked.

It thus appears that the conclusions arrived at by myself from the statistics of poverty, suicide, insanity, physical deterioration, and crime, during the last forty years,[1] are supported by a quite different set of facts, extending over a much longer period, and set forth by a statistical authority of the first rank, who has no special views to support. Let us therefore now consider the main

[1] See my *Wonderful Century*, Chapter XX.

problem. What is the fundamental error in our social system that has allowed this state of things to persist, notwithstanding all our increase in wealth, and how we may most certainly and most safely bring about the desired change to a social state in which none who are willing to work shall ever suffer the extreme of want?

The Society of the Future.

I am myself convinced that the society of the future will be some form of socialism, which may be briefly defined as *the organization of labour for the good of all*. Just as the Post Office is organized labour in one department for the benefit of all alike; just as the railways might be organized as a whole for the equal benefit of the whole community; just as extensive industries over a whole country are now organized for the exclusive benefit of combinations of capitalists; so all necessary and useful labour might be organized for the equal benefit of all. When a combination or trust deals with the whole of one industry over an extensive area, there are two enormous economies; advertising, which under the system of competition among thousands of manufacturers and dealers wastes millions annually, is all saved; and distribution, when only the exact number of stores and assistants needful for the work are employed, effects an almost unimaginable saving over the scores of shops and stores in every small town, competing with each other for a bare living. What then would be the economy when *all* the industries of a whole country were similarly organized for the common good; and when all absolutely useless and unnecessary employments were abolished—such as gold and diamond mining except to the extent needed for science, and art; nine-tenths of the lawyers, and all the financiers and stock-gamblers? It is clear that under such an organized system three or four hours work for five days a week by all persons between the ages of twenty and fifty would produce abundance of necessaries and comforts, as well as all the refinements and wholesome luxuries of life, for the whole population.

But although I feel sure that some such system as this will be adopted in the future, yet it may be only in a somewhat distant future, and the coming century may only witness a step towards it; it is important that this step should be one in the right direction. The majority of our people dislike the very idea of socialism, because they think it can only be founded by compulsion. If that were the case it would be equally repulsive to myself. I believe only in *voluntary* organization for the common good, and I think it quite possible that we require a period of true individualism—of competition under strictly equal conditions—to develop all the forces and all the best qualities of humanity, in order to prepare us for that voluntary organization which will be adopted when we are ready for it, but which cannot be profitably forced on before we are thus prepared.

In our present society the bulk of the people have no opportunity for the full development of all their powers and capacities, while others who have the opportunity have no sufficient inducement to do so. The accumulation of wealth is now mainly effected by the misdirected energy of competing individuals; and the power that wealth so obtained gives them is often used for purposes which are hurtful to the nation. There can be no *true* individualism, no *fair* competition, without equality of opportunity for all. This alone is social justice, and by this alone can the best that is in each nation be developed and utilized for the benefit of all its citizens. I propose, therefore, to state briefly what is the ethical foundation for this principle, and what its practical application implies.

The Law of Social Justice.

In Herbert Spencer's volume on "Justice," forming Part IV. of his *Principles of Ethics*, he gives as the foundation of social justice the following:—

"Of man, as of all inferior creatures, the law by conformity to which the species is preserved, is that among adults the individuals best adapted to the conditions of their existence shall prosper most, and that the individuals least adapted to the conditions of their existence shall prosper least—a law which, if uninterfered with, entails survival of the fittest, and spread of the most adapted varieties. And, as before, so here, we see that, ethically considered, this law implies, *that each individual ought to receive the benefits and evils of his own nature and consequent conduct: neither being prevented from having whatever good his actions normally bring him, nor allowed to shoulder off on to other persons whatever ill is brought to him by his actions.*"

The passage printed in italics is the "law of social justice" deduced from the law of the survival of the fittest, and it is appealed to again and again throughout the volume, but is usually indicated by the shorter formula —"each shall receive the benefits and evils due to his own nature and consequent conduct."[1] In all our sports and trials of skill or endurance, we aim at equality of conditions for the competitors, who are all of nearly equal age and in good health, while their preliminary training has been nearly the same; and it is universally recognized that the skill or endurance of each can only be ascertained by such equal or fair conditions.

But when it is a question, not of mere sports or amusements, but of the real battle of life, failure in which often means continuous hardship, want, or premature death, with the loss to friends and to the community of all those higher qualities or talents which were undeveloped through want of leisure or opportunity, we make no attempt whatever to give fair play to all alike. How much we lose by this unfairness no one can tell, but the poets have always recognized that there is such a loss. Gray tells of the "village Hampdens" and the "mute inglorious Miltons" that may have passed away unknown, and of the hearts "once pregnant with celestial fire"—

> "But knowledge to their eyes her ample page
> Rich with the spoils of time did ne'er unroll;
> Chill Penury repress'd their noble rage
> And froze the genial current of the soul":

[1] It would operate, not as among the lower animals and plants by the actual destruction of the unfit, but by their less rapid increase, since, under equal conditions of education and mode of life, it is certain that marriage would be delayed till some industrial success had been reached by both parties.

while even the refined and critical Tennyson could say—
"Plowmen, shepherds, have I found, and more than once, and still could find,
Sons of God and kings of men in utter nobleness of mind."

And everywhere we see illustrations of the same fact in the fortunate accidents that have here and there rescued some great mind from a life of obscure drudgery. If Watt, the mathematical instrument maker, had not lived in Glasgow, where he had the model of a steam-engine sent him from the University for repair, the advent of the modern steam-engine might have been delayed half a century. If Faraday had not had a ticket given him to Sir Humphry Davy's lectures on chemistry at the Royal Institution, he might have always remained a working bookbinder, and the progress of electrical science might have been seriously checked. Numbers of our inventors and original thinkers have sprung from the ranks of peasants and mechanics, and we may be sure that many more who were equally gifted have been wholly lost to the world owing to the absence of favourable conditions at the right period of their lives, or to some inherent modesty or timidity that prevented them from forcing their way in spite of all obstacles. What we need in order to profit by all the skill, and talent, and genius that may exist in our whole population, is that all should have the education and the opportunities for developing whatever abilities they may possess, which are now accessible only to the higher and the wealthier classes; and when we find that this is also the teaching of philosophy, and that only in this way can we apply the fundamental principle of organic evolution to the development of the social organism, we have both experience and theory in favour of adopting it as a sure guide.

Equality of Opportunity.

While discussing Herbert Spencer's "Justice" in an address to the Land Nationalization Society in 1892, I remarked:

"It is strange that Mr. Spencer did not perceive that if this law of the connection between individual character and conduct and their economical results, is to be allowed free play, some social arrangement must be made by which all may start in life with an approach to *equality of opportunities.*"

Two years later this term was used and popularized by Mr. Benjamin Kidd in his " Social Evolution," and is now often used with approval by political and social writers, most of whom, however, do not appear to see all that it implies. The term includes all that is contained in Spencer's principle of social justice, and as it is much shorter and more expressive, it is well adapted to become the watchword of social reformers. Let us then see what its full application would really mean.

Equality of Opportunity is absolute fair play as between man and man in the struggle for existence. It means that all shall have the best education they are capable of receiving; that their faculties shall all be well trained, and their whole nature obtain the fullest moral, intellectual, and physical development. This does not mean that all shall have the *same* education, that all shall be made to learn the same things and go through the same training, but that all shall be so trained as to develop fully all that is best in them. It must be an adaptive education, modified in accordance with the peculiar mental and physical nature of the pupils, not a rigid routine applied to all alike, as is too often the case now.

It further implies that during this period of thorough education every endeavour shall be made to ascertain how the special faculties of each can be best utilized for the good of society and for his own happiness, and thus will be determined the particular work, both bodily and mental, to which each youth shall be trained, subject always to the demand for workers in the various industries or occupations.

Yet further, equality of opportunity requires that all shall have an endowment to support them during the transition period between education and profitable employment, and to furnish them with such an outfit as their special avocations require.

Inheritance of Wealth causes Inequality.

But even this is not all. We must also take care that inequality is not introduced by private gifts or bequests to individuals which might enable them to live permanently in idleness and luxury, since every one who so lives must necessarily be supported by the labour of others, and is in all essentials a pauper, as has been so forcibly urged in the remarkable work of Mr. A. J. Ferris—" Pauperizing the Rich." It is here that most people (including Herbert Spencer and Mr. Kidd) object to the application of the principle that every man shall receive the results of his own nature and conduct, or, in other words, shall have " equality of opportunity," as being unjust or injurious. But if this principle is the essential feature of social justice, its full application cannot be unjust; while if it is the true correlative in human society of survival of the fittest among the lower forms of life, it cannot be injurious.

The difficulty seems to arise from the fact that if the accumulation of property either by the labour, the foresight, or the good fortune of an individual, is right, and is for the benefit of society as a whole, as is generally assumed, it is also assumed that the power of transferring this property to another must be also both right and beneficial. This, however, does not logically follow. If equality of opportunity is a true and just principle, then the society that gives to every man that equality, and protects him in his work throughout his life, may fairly claim to inherit any surplus wealth that he leaves behind him, *in order to ensure similar advantages to all.* And it is still more obvious, that a society which has adopted the principle of equality of opportunity as the only means of securing true individualism and competition under fair and equal conditions, may justly prevent individuals from introducing inequality by their injudicious gifts or bequests. From either point of view it follows, that society should protect itself by a strict regulation of the transmission or inheritance of wealth.

Public Debts Impolitic and Immoral.

There is another consideration that is usually overlooked in this connection, and thus helps to obscure the real issue. Under our highly artificial and complex monetary system, the "property" left by rich men is seldom real wealth, but consists almost wholly of claims upon, or tribute exacted from society at large. Real wealth is highly perishable—food, clothing, houses, tools, machinery, &c.— and if such wealth were given to another in large quantities it would rapidly deteriorate or require a considerable annual expenditure to preserve its value. But by our money-market system of funds, stocks, shares, and rents, permanent incomes are derived from perishable wealth, to the injury of all who are forced to pay these incomes. Money has been diverted from its original and beneficial purpose of facilitating the mutual exchange of commodities—"a tool of exchange," as some economists have termed it—into a means of enabling large numbers of wealthy individuals to live permanently at the expense of their working fellow-citizens. This is the real reason of the objection of the ancient law-givers to usury, that it enables men to live without doing any useful work; and the objection of modern socialists to interest is, not that to take interest for the use of money is morally wrong, but that the general application of the principle of national or municipal interest-bearing debts, railway shares, &c., afford the conditions by which perishable wealth is changed into permanent property, and offers facilities for the most gigantic and harmful system of gambling the world has ever seen.

All wealth so acquired is a means of impoverishing those whose work produces all the real wealth that is consumed annually. Adam Smith again and again states this fact, that the annual consumption of the whole population, including all the idle rich, is produced annually by the workers; and it is because the system of interest enables false wealth, which is really tribute exacted from the people, to go on increasing indefinitely, and thus tends continually to impoverish the workers and to increase the numbers of the idle, that it has been condemned as both impolitic and evil. And we now see that, as it leads to results which are opposed to "equality of opportunity," it is also ethically unjust.

Hereditary Wealth bad for its Recipients.

There is yet another consideration which leads to the same conclusion as to the evil of hereditary or unearned wealth—its injurious effects to those who receive it, and through them to the whole community. It is only the strongest and most evenly balanced natures that can pass unscathed through the ordeal of knowing that enormous wealth is to be theirs on the death of a parent or relative. The worst vices of our rotten civilization are fostered by this class of prodigals, surrounded by a crowd of gamblers and other parasites, who assist in their debaucheries and seek every opportunity of obtaining a share of the plunder. This class of evils is too well known and comes too frequently and too prominently before the public to need dwelling upon here; but it serves to complete the proof of the evil effects of private inheritance, and to demonstrate in a practical way the need for the adoption of the just principle of equality of opportunity.

Conclusion.

Under such a system of society as is here suggested, when all were well educated and well trained and were all given an equal start in life, and when every one knew that however great an amount of wealth he might accumulate he would not be allowed to give or bequeath it to others in order that they might be free to live lives of idleness or pleasure, the mad race for wealth and luxury would be greatly diminished in intensity, and most men would be content with such a competence as would secure to them an enjoyable old age. And as work of every kind would have to be done by men who were as well educated and as refined as their employers, while only a small minority could possibly become employers, the greatest incentive would exist towards the voluntary association of workers for their common good, thus leading by a gradual transition to various forms of co-operation adapted to the conditions of each case. With such equality of education and endowment none would consent to engage in unhealthy occupations which were not absolutely necessary for the well-being of the community, and when such work was necessary they would see that every possible precautions were taken against injury. All the most difficult labour-problems of our day would thus receive an easy solution.

I submit, therefore, that the adoption of the principle of Equality of Opportunity as our guide in all future legislation, should be acceptable to every social reformer who believes in the supremacy of Justice. To the individualist it would mean the fullest application of his principle of individual freedom limited only by the like freedom of others, since this principle is a mere mockery under the present negation of fair and equal conditions to the bulk of the citizens of all civilized states. And it should be equally acceptable to the socialist, because the greatest obstacle to his teachings would be removed by the abolition of ignorance and of that grinding poverty and want which leaves no time or energy for any struggle but that for bare existence. Equality of Opportunity, founded as it is upon simple Justice between man and man, is therefore well fitted to become the watchword of the social reformers of the Twentieth Century.

CHAPTER XXIX

JUSTICE, NOT CHARITY, AS THE FUNDAMENTAL PRINCIPLE
OF SOCIAL REFORM. AN APPEAL TO MY READERS [1]

OUR conceptions of social duty—of what constitutes justice in social life—will be to a considerable extent dependent upon the views we hold as to man's spiritual nature, and more especially upon the relation believed to exist between the present life and that which is to follow it. On this subject there has been a great change of opinion during the last forty years.

The old doctrine as to the nature of the future life was based upon the idea of rewards and punishments, which were supposed to be dependent upon *dogmatic beliefs* and ceremonial *observances*. The atheist, the agnostic, even the Unitarian, were for centuries held to be certain of future punishment; and, with the unbaptised infant, the Sabbath-breaker, and the abstainer from church-going, were alike condemned to hell-fire. Beliefs and observances were then held to be of the first importance; disposition, conduct, health, and happiness were of little or no account.

The new doctrines—founded almost wholly on the teachings of Modern Spiritualism, though now widely accepted even among non-Spiritualists—are the very reverse of all this. They are based upon the conception of mental and moral continuity; that there are no imposed

[1] The following pages (with verbal modifications) constitute the main portion of an Address to the International Congress of Spiritualists, at St. James's Hall, in June, 1898.

punishments; that dogmatic beliefs are absolutely unimportant, except so far as they affect our relations with our fellows; and that forms and ceremonies, and the complex observances of most religions, are equally unimportant. On the other hand, what are of the most vital importance are motives with the actions that result from them, and everything that develops and exercises the whole mental, moral, and physical nature, resulting in happy and healthy lives for every human being. The future life will be simply a continuation of the present, under new conditions; and its happiness or misery will be dependent upon how we have developed all that is best in our nature here.

Under the old theory the soul could be saved by a mere change of beliefs and the performance of certain ceremonial observances. The body was nothing; happiness was nothing; pleasure was often held to be a sin; hence any amount of punishment, torture, and even death were considered justifiable in order to produce this change and save the soul.

On the new theory it is the body that develops, and to some extent saves, the soul. Disease, pain, and all that shortens and impoverishes life, are injurious to the soul as well as to the body. Not only is a healthy body necessary for a sound mind, but equally so for a fully-developed soul—a soul that is best fitted to commence its new era of life and progress in the spirit world. Inasmuch as we have fully utilized and developed all our faculties—bodily, mental, and spiritual—and have done all in our power to aid others in a similar development, so have we prepared future well-being for ourselves and for them.

All this is the common knowledge and belief of Spiritualists; and I should not have thought it necessary to restate it were it not that their creed is often misunderstood and misrepresented by outsiders, and also because it is preliminary to certain conclusions which, I think, logically follow from it, but which are not so generally accepted among us.

It seems to me that, holding these beliefs as to the future life and what is the proper and only preparation for it, not only Spiritualists, but all those to whom these beliefs are acceptable, must feel themselves bound to work strenuously for such improved social conditions as may render it possible for *all* to live a full and happy life, for *all* to develop and utilize the various faculties they possess, and thus be prepared to enter at once on the progressive higher life of the spirit-world. We *know* that a life of continuous and grinding bodily labour, in order to obtain a bare existence; a life almost necessarily devoid of beauty, of refinement, of communion with Nature; a life without adequate relaxation, and with no opportunity for the higher culture; a life of temptation and with no cheering hope of a happy and a peaceful old age, is as bad for the welfare of the soul as it is for that of the body.

If the accounts we get of the spirit world have any truth in them, the reclamation and education of the millions of undeveloped or degraded spirits which annually quit this earth, is a sore though cheerfully accepted burden, a source of trouble and sorrow to those more advanced spirits who have charge of them. This burden must, for a long time to come at all events, *necessarily* be great, on account of the numbers of the less advanced races and peoples still upon the earth; but that *we*, who call ourselves *civilized*, who have learnt so much of the secret powers and mysteries of the universe, who by means of those powers could easily provide a decent and rational and happy life for our whole population—that *we* should send to the spirit world, day by day and year by year, millions of men and women, of children and of infants, all destroyed before their time through want of the necessary means of a healthy life, or by the various diseases and accidents forced upon them by the vile conditions under which alone we give them the opportunity of living at all—this is a disgrace and a crime!

I firmly believe—and the fact is supported by abundant evidence—that the very poorest class of our great cities, those that live constantly below the margin of poverty, who are without the comforts, the necessaries, and even the decencies of life, are, nevertheless, as a class, quite as good morally, and often as high intellectually, as the middle and upper classes who look down upon them as in every way their inferiors. Their degraded condition, socially and morally, is the work of society; and in so far as they appear worse than others they are made so by society. What should we ourselves have been if we had had no education, no repose, no refined or decent homes, no means of cleanliness, which is not only next to, but is a source of, godliness; surrounded by every kind of temptation, and not unfrequently forced into crime? And a direct consequence of the millions who are compelled to lead such lives are the millions of infants who die prematurely—a slaughter a thousand times worse than that of Herod, going on year by year in our midst; surely their innocent blood cries out against our rulers, against all of *us*, who choose such rulers; and more especially against those Spiritualists and Christians, who know the higher law, if they do not work with all their strength for a radical reform.

I ask you to think over this question; and above all things, I ask you to consider the necessity for real and fundamental remedies, not mere palliatives, which have been tried with ever-increasing energy and good-will throughout the whole of the nineteenth century, and have absolutely failed. The evil has grown, just as if no such remedies had been applied at all. Charity has increased enormously, and has completely failed. Now it is time for us to try Justice.

A few years since a talented writer used, and at once popularized, a new term—"equality of opportunity." It expresses, briefly and forcibly, what may be termed the minimum of social justice. The same idea had been urged by other writers, especially by Herbert Spencer in his volume on "Justice," when he declared that justice requires every man to receive "the results of his own nature and consequent actions"—this and this only. Fundamentally, the two ideas are the same, but "equality of opportunity" is the more simple and intelligible expression of it.

To Christians and Spiritualists, who realize that every

child born into this world is a living soul, which has come here to prepare itself for the higher life of the spirit world, it must appear a crime against that world and against humanity not to see that every such child has the best possible nurture and training, at the very least till it arrives at the adult age and becomes an independent unit of the social organism. And if to each is due the best, then none can have more than the best, and we come thus again to EQUALITY OF OPPORTUNITY.

Of course, many of my readers will say, "This is impossible. How can we possibly give this equality of nurture and education to every child?" I admit that it is difficult—by no means impossible. It must, of course, be brought about gradually; and where there is a will there is a way. As Herbert Spencer said of another matter—the nationalization of the land—"justice sternly demands that it be done"; and if we, boasting of our civilization, declare that it cannot be done, then so much the worse for us and for our false civilization. But it wants only the will. And it is our duty to help to create that will.

But, again, you will say, "Where are the means of doing this? We are already taxed as much as we can bear." True, we are shamefully over-taxed for all kinds of unnecessary and hurtful expenses, some of which have been exposed in preceding chapters; but, instead of increasing the taxes, there is a necessary corollary of "equality of opportunity" which will not only give us ample funds to bring it about, but will at the same time greatly reduce taxation and ultimately abolish it altogether. For, if every child is given equality of opportunity, and every man and woman receives only "the results of their own nature and consequent actions," then it is evident that there must be no *inequality* of inheritance; and to give equality of inheritance, the State, that is, the community, must be the universal inheritor of all wealth. At first, of course, it would only be needful to take surplus wealth above a fixed maximum; and, so far from this being an injury to the heirs of a millionaire, it would be a great benefit; for it is admitted that nothing has so demoralizing an effect on the young as the certainty of inheriting great wealth; and examples of this come before us every year and almost every month. This is the real teaching of the parable of Dives and Lazarus; this gives us the true meaning of Christ's saying that a rich man shall hardly enter into the kingdom of heaven.

Now, many who dislike the idea of Socialism—chiefly, I think, through not understanding what it really implies—will perhaps look more favourably on this great principle of "equality of opportunity," since it would leave individualism untouched, would in fact render it far more complete and effective than it is now. For our present state of society is *not* true individualism, because the inequalities of opportunity in early life are so great that often the worst are forced to the top, while many of the best struggle throughout life without a chance of using their highest faculties, or developing the best part of their nature. Even Tennyson, whose mind was of an aristocratic bent, could say—

"Plowmen, shepherds, have I found, and more than once, and still could find,
Sons of God and kings of men in utter nobleness of mind;
Truthful, trustful, looking upward to the practised hustings-liar;
So the Higher wields the Lower, while the Lower is the Higher.
Here and there a cotter's babe is royal-born by right divine;
Here and there my lord is lower than his oxen or his swine."

Equality of opportunity would put all this right; every one would be able to show what power for good he possessed, and society would be enormously benefited in consequence. At the same time, there would be all the stimulus to be derived from individual effort. The man who could surpass his fellows under such equal and fair conditions would be truly great. Some would achieve honour, some would acquire wealth; but it would be all due to their own "nature and consequent actions," and neither the honour nor the wealth would be handed on to individuals who might not be worthy of the one or be able to acquire, or properly to use, the other.

I believe myself that such a perfectly fair competition,

in which all started on equal terms materially and socially, would be an admirable training, and would be sure to lead, ultimately, to a voluntary co-operation and organization of labour which would produce most of the best results of Socialism itself. But whether it would or not, I claim that it embodies a great and true principle—Social Justice; and that it affords the only non-socialistic escape from the horrible social quagmire in which we find ourselves. All who believe in a moral law as a guide to conduct must uphold justice; and equality of opportunity for all is but bare justice. Believing that the life here is the school for the development of the spirit, we must feel it our duty to see that the nascent spirit in each infant has the fullest and freest opportunity of developing all its faculties and powers under the best conditions we can provide for it. And I have ventured to bring this subject before the reading and thinking world, because it is the one hope nearest to my heart; and I am sure that if the great and rapidly-increasing body of true philanthropists and all who are spiritually-minded, can be brought to consider it, and to feel that the misery and degradation around them *must* be and *can* be got rid of, and further, that it is especially *their* business and *their* duty to help to get rid of it, the great work will soon be taken in hand.

What we want, above all things, is to educate the people and create a public opinion adequate for the work. In this movement for justice and right, Christians and Spiritualists should take the lead, because they, more than any others, know its vital importance both for this world and the next. The various religious sects are all working, according to their lights, in the social field; but their forces are almost exclusively directed to the alleviation of individual cases of want and misery by means of charity in various forms. But this method has utterly failed even to diminish the mass of human misery everywhere around us, because it deals with symptoms only and leaves the causes untouched. I would not say a word against even this form of charity, for those who see no higher law; but we want more of the true charity of

St. Paul—the charity that thinketh no evil, that suffereth long and is kind, that rejoiceth in the truth—not only the lesser and easier charity which feeds the poor out of its superfluity, an action which St. Paul does not allow to be charity at all.

To all advanced thinkers, to all earnest reformers, to all humanitarians—and especially to Christians and Spiritualists, I urge, that it is now time to sever ourselves from these old and utterly useless methods and to take higher ground. Let us unflinchingly demand Social Justice, as developed in the preceding chapter. This will be a work of such grandeur, of such far-reaching influence, so deeply founded in Right, so absolutely impregnable against the attacks of logic, or theology, or expediency, that it must succeed in a not far distant future. Knowing that we are striking at the very roots of our social evils, that every step we take will make the next step easier, let us work strenuously for the elevation and permanent well-being of our fellows, and let our watchword be—not Charity, but JUSTICE.

INDEX

INDEX

A.

Agassiz, Prof. L., and the Harvard Museum, 19
    Alex., on defects of museums, 21
Agnostics as moral as believers, 376
    have no adequate motive for morality, 377
America, depression of trade in, 207
America, eastern forests of, 83
American cities, condition of, 413
American museums, 16
Ancient ash-pits and cemeteries, 54
    mounds in N. America, 54
Animal sculptures from the mounds, 51
Animals, how to exhibit in a museum, 8
Aquired characters not inherited, 500
Argyll, Duke of, on economic use of land, 424
Armies should be industrial, 387
Arrow-heads, twisted, 43
Atkinson, Mr. E., on economy of production, 483
Aveling, Dr. E., on American workers, 412

B.

Bad habits not inherited, 505
Bankruptcies showing depression, 193
Bequest, limitation of right of, 341
    unlimited is doubly injurious, 342
Botany in a museum, 6
Bread, cheap, not always a blessing, 407
Breathing in language, 121
British Museum of Nat. History, defects of, 17
Broderick, Hon. G. C., on results of peasant-husbandry, 321
Business capacity of peasants at Ralahine, 472

C.

Cairnes, Prof., on landlordism, 301
California, forests of, 86
Capitalism causes poverty, 414
Carrington, Lord, on peasant cultivation, 217
Cave-dwellings in Arizona, 56

Character, apparent improvement of, at Ralahine, 474
Child-language, no guide to its origin, 135
Christianity, modern, a slow growth, 108
Church, a native, in South Africa, 109
Church, a really national, 235
Church property is national property, 236
Clarke, Mr. Hyde, on mouth, tongue, and tooth words, 134
Coal, a national trust, 138
    unlimited export of, injurious, 141
Concave globe, advantages of, 68
    how to construct, 71
Conquests desired by governing classes, 385
Commission, the Devon, on evictions, 311
Commons enclosed under false pretences 449
Competition, the waste of, 487
Co-operative colonies, how to establish, 488
    why not tried, 490
Craig, Mr., comes to Ralahine, 456
    his excellence as an organiser, 458
Credit at creditor's risk, 163, 165
Culture, effects of, not inherited, 501
Cups and bowls in stone, 48

D.

Dead men should not dictate to the living, 157
Debt, how to extinguish the national, 257
Debtor and creditor, 162
Debts should not be collected by the State, 163
Depopulation of rural districts, 210
    causes of, 214
    effects of, 212
Depression of trade, causes of, 188
    its main features, 190
    remedies for, 220
    true causes of, 194
Destitution, evidence of increase of, 213
Differences of forests, how caused, 94
Disestablishment and disendowment, 235

Disputes about work, how avoided, 459
Dives and Lazarus, real teaching of the parable of, 526
Draining injurious in a natural forest, 92
Dutch people healthy in the Moluccas, 104

E.

Earth, how best to model the, 59
Eastern Asia and Japan forests of, 88
Europe, forests of, 88
Economics of an industrial community, 484
Economy of health and happiness, 487
Education at Ralahine, 463
    needed to abolish militarism, 390
    the effect of, 497
Effort in pronouncing words, significant, 129
Enclosures under false pretences, 449
Endowed Schools Commission, 156
Epping Forest, 74
    how to treat, 91
Equality of opportunity, 515, 524
    a guide to legislation, 520
Equatorial zone generally healthy, 101
    a Scotchman's work in, 103
Essential equality in nature of rich and poor, 524, 526
Estates, origin of great, 449
Ethical principles applied to land tenure, 410
Ethnology in a museum, 10
European forests, 82
Europeans treat natives badly, 113
Eviction should be illegal, 444
Experiments in socialism have not all failed, 476
Exports and imports showing depression, 192
Extinct animals in Harvard Museum, 35

F.

Fair play in life, absence of, 514
Farmers, remedy for poverty of, 407
    suggested remedies for poverty of, 402
Farms, large versus small, 406
Farrar, Dean, on transference of meanings, 132
    and Wedgwood on imitative words, 127
    on origin of language, 116
Fawcett, Prof., against land-nationalisation, 296
    on free trade and protection, 175
    unsound argument of, 176
Finch, Mr. J., visits Ralahine, 462
Foreign loans a cause of depression, 195
Forest regions, proposed illustrations of, 82
    the temperate, 74

Food at Ralahine, 464
Free access to land the remedy for want, 417
Free contract a mockery, 416
Freeholders, system of small, unstable, 349
Free trade, acknowledged exceptions to, 169
    and reciprocity, 167
    in limited natural products wrong, 138
    the whole programme of, 173
Froude, J. A., on property in land, 265

G.

Galton, F., on superiority of Athenians, 494
Genius rarely inherited, 503
    no successive increase of, 504
Geography and geology, a museum of, 4
Geographical distribution in Harvard Museum, 26
    suggested museum of, 33
George, Henry, an illustration of land-lordism, 302
    on state of American workers, 412
Gestures and words used in earliest speech, 135
Giffen, Mr., on depression of trade, 191
Globe, gigantic, alternative plan of, 65
    objections to the plan of, 63
    proposed by M. Roclus, 59
Gold not a permanent standard of value 145
Gould Mr. Baring, on German peasants, 319
Government tested by happy homes, 407
Governments all seek enlarged territory, 485
Gray, Prof. Asa, on forest regions, 82
Great and small in Malay languages, 130
Guernsey market-notes, 200

H.

Harvard Museum, origin of, 19
Harvest-home at Ralahine, 464
Health at Ralahine, 464, 467
Hereditary wealth injures its receivers, 519
Himalayas, temperate forests of, 90
Home and household goods inviolable, 164
    the inviolability of, 443
Homes, to form secure, 361
Hooker, Sir Jos., on museums, 12
Howitt, Wm., on German peasants, 319
House of Lords, a representative, 223
    conditions of its reform, 226
Human nature advances by survival of the fittest, 495
    adverse influences, 496
    has it improved in historic times, 494
Human progress, past and future, 498

INDEX

I.

Idlers, how cured at Ralahine, 470
Increased war expenditure and depression of trade, 197
Individualism, true, 510
Industrial colony, economics of, 484
    organisation needed, 388
Inheritance causes inequality, 517
Inherited wealth injurious, 446
Interest-bearing funds injurious and unjust, 254
Invention among co-operators, 471
Ireland, chronic starvation in, 310
    in 1830, 455
Irish cottars, industry of, 323

J.

Jeffrey, Richard, on organisation of labour, 490
Jevons, Prof. W. S., objections to inconvertible paper money, 147
    on changes in value of gold, 146
Justice, economic and social, 432
    for all an elementary right, 150
    not charity, 521
    Spencer's formula of, 341
    the law of social, 513

K.

Kaffir clergyman in England, 110
Kavanagh on myths and language, 122
Kay, Mr. Joseph, on pauperised England, 305
    on farming in Switzerland, 318

L.

Labour colonies of Holland a success, 483
    results of organisation of, 471
Land, benefit of free access to, 288
    evils of free trade in, 234
    evil of speculation in, 409
    free access to, a cure for want, 417
    freedom of choice of, 330
    how to be distributed, 328
    how to deal with the, 259
    how to nationalise, 265
    is private property in, just, 298
    Lord Carrington on benefits of, 419
    monopoly, effects of, 347
    nationalisation, objections answered, 345
        probable results of, 232
        State tenants versus freeholders, 345
        why and how of, 296
    question and Herbert Spencer, 333
    re-occupation of the, 478
    robbery of, by landlords, 440
    special iniquity of, 450
    tenure, proposed system of, 274
    transfer, a general principle of, 268
    value not due to improvement of soil, 336

Landlordism, Froude, on evils of, 285
    in Bengal, 286
    results of, in Ireland, 309
        in Scotland, 313
    the effects of, 304
    the remedy for, 323
    the right or wrong of, 290
Language, a factor in origin of, 115
Large and small holdings, results compared, 423
Laveleye, on results of great properties, 321
Law, how to simplify, 152
    system, antiquated and inefficient, 151
Legislation of 19th century, nature of, 436
    its outcome, 437
Legislators, the impotence of our, 440
Lewis, poverty in island of, 315
Life at Ralahine free and happy, 473, 476
    why live a moral, 375
Limitations of free trade, 171
Limitation of State functions, 150
Limited Liability Act, evil results of, 205, 221
Lord Tollemache's peasant-farms, 322
Lords, a representative House of, 223
    constitution of new House of, 226
    mode of election of representative, 229
    representative, advantages of, 231
Lowe, Mr., a reply to, 183

M.

Macdonald, Dr., on Highland clearances, 315
Machinery at Ralahine, 464
Malay words for great and small, 130
Manning, Cardinal, on the land question, 452
Martius, on lip-pointing, 117
M'Culloch on Flemish agriculture, 320
Mercy, works of, on the Sabbath, 369
"Merrie England," extracts from, 453
Mill, J. S., on the incidence of a rent-tax, 352
Mills, Mr. H. V., on failure of our civilisation, 415
    on willing workers idle, 480
Militarism can only be banished by the people, 390
    its good and evil effects, 386
Millionaires causing depression of trade, 201
    increase of, 400
    supposed uses of, 254
Mineralogy, museum of, 5
Money-fines unjust, 151
Moral qualities, their expressive names, 132
More, Sir T., on wrongs of the workers, 433
Motions, and their expressive names, 124

INDEX

Mounds and earthworks in N. America, 54
Mouth-gesture, 115, 117
Mouth-words, 121
Museums, American, 16
　for the people, 1
Museum, position and plan of, 19

N.

National Church, advantages of the new, 249
　the proposed, 240
　debt, how to extinguish, 257
Natural history, a typical museum of, 4
Nineteenth-century legislation, character of, 436
Nose-words, 122

O.

Objections to reciprocity answered, 181
Ogilvie on right of property in land, 435
Opportunity, equality of, 515
Opportunities equalised by rent, 404
Organisation not injurious to character, 388
　the good side of militarism, 387
Organised industry must succeed, 483

P.

Pare, Mr. W., book on Ralahine, 468
Parish rectors, duties of, 241
Paper money as a standard of value, 145
　Jevons's objections to inconvertible, 147
Pauperism not really diminishing, 306
Peasant cultivation, results of, 217
　farms in Annandale, 322
　in Cheshire, 322
　farmers co-operate, 425
　good agriculture of, 426
Pim, Mr. J., on industry of Irish cottars, 323
Pope, expressive verses of, 130
Possession, results of secure, 320
Post obits should be illegal, 164
Pottery from mounds, 53
Poverty due to capitalism, 414
　in great cities of United States, 396
　proof of its increase in America, 399
　in England, 398
　proof of increasing, 203
Prehistoric archaeology, museums of, 37
　at Cambridge, Mass., 39
　at Washington, D.C., 40
Priests, the teaching of, as to the land, 452
Produce of land more important than profit, 424
Progress, how it will be effected, 506
Public debts immoral, 518
　rooms in the Harvard Museum, 23

Q.

Qualities indicated by sounds, 128
Queensland, white men work in, 104

R.

Ralahine a striking success, 470, 477
　and its teachings, 455, 468
　confiscation of workers' savings at, 468
　description of, 458
　education and sanitation at, 463
　end of great experiment at, 466
　enthusiastic industry at, 470
　organisation of the society at, 458
　self-government at, 461
　why no imitators, 465
Real reforms, suggestions for, 442
Rebellion not a crime, 392
　is always justifiable, 392
Reciprocity the essence of free trade, 167
　what it means, 177
Reclus on proposed gigantic globe, 59
Rectors, new, advantages of, 249
　may be introduced gradually, 244
　new parish, 241
　objections to and answers, 245
Remedies for depression of trade, 220
Rent of small holdings, how to fix, 358
　paid in produce, advantages of, 458
　the social importance of, 404
Reports, agricultural, on results of small holdings, 420
Republican simplicity now vanished, 411
Ricardo on effects of a rent-tax, 351

S.

Sabbatarians, a suggestion to, 364
Sabbath, effects of a true observance of, 370
　the command to keep, 365
Sabbath breaking by sabbatarians, 365
　excuses for, 366
Salvation, old and new theories of, 522
Sanitation at Ralahine, 463
Savages, how to civilise, 107
Scotland, clearances of the crofters, 314
　landlordism in, 313
Selection of the fittest will act on mankind in the future, 508
Self-government at Ralahine, 461
Shell-mounds and cave-dwellings, 56
Sismondi on peasant cultivators, 318
Small holdings, beneficial results of, 420
　good results of, 316
　to secure success of, 354
Smith, Adam, on effects of a land-tax, 351
Socialistic experiments successful, 476
Social justice, the law of, 513
　quagmire, the, 394
　reform, the fundamental principle of, 521
Society of the future, 512

INDEX

Sounds, continuous and abrupt, 123
 which represent motions, 127
South Africa, educated natives in, 111
South temperate zone, forests of, 90
Spades and knives in stone, 46
Speculation causing depression of trade, 206
 increase of, 204
Speech, how it originated, 135
 the expressiveness of, 115
Spencer, Herbert, misstatements of, 335, 337
 on iniquitous power of landlords, 435
 on social justice, 513
 on the land question, 333
Spiritualism and justice, 527
 higher teachings of, 380
 is rationalism, 381
 the teachings of, 521, 525
Spiritualists are impelled to live morally, 382
Sportsmen healthy in the tropics, 103
Standard of value, a, 145
State-functions, limitation of, 150
 tenants have rights of freeholders, 348
Steam-power, estimate of, 309
Stone implements at Washington, 41
 objects of unknown use, 49
 yokes from Porto Rico, 52
Stubbs, Dean, on peasant cultivation, 217
Success of small holdings, how to secure, 354
Sunday service an introduction to useful work, 373

T.

Taxation of future generations immoral, 447
Taxation of land, who will pay it? 351
Teeth and palate words, 122
Tennyson, expressive verses of, 130
Tollemache, Lord, his peasant farmers, 322
Tropics, can white men work in the, 99
 unhealthiness of parts of, 100
Trusts should be personal and voluntary, 155
 should not be recognised by the law, 154
Twentieth century, the work of, 440
Tylor, Mr. E. B., on ingenious contrivances in language, 120

U.

Unborn, the, not to inherit property, 445
Unemployed, political economists have no remedy for the, 480

Unemployed, solution of problem of, 478
 workers a man-created evil, 415
Unemployment, why it exists, 481
United States, condition of, in early years of the 19th century, 394
 present condition of, 396
 supposed causes of absence of poverty in, 395

V.

Value, paper money as a standard of, 145
Vandeleur, Mr., disappearance of, 467
 invites Mr. Craig to organise Ralahine, 456

W.

Wage-workers in America, 411
War, evil results of, 198
 its causes and the remedies, 384
 traders in materials of, are enemies of their country, 391
 teachers of, to lower races are enemies of civilisation, 391
War-expenditure and depression of trade, 197
Water not bullets to disperse mobs, 392
Wealth, how to abolish fictitious, 257
 how to deal with accumulated, 448
 real and fictitious, 256
White men in the tropics, 99
Wickstead, Mr. C., on proposed land tax, 354
Words expressing disgust, 129
 imitating sounds, 124
 more widely useful than gestures, 136
 not originally conventions, 118
 specially expressive, 131
 with changed meanings, 134
Work, meaning of in Fourth Commandment, 367
Workers in harmony at Ralahine, 471
Wyld's globe in Leicester Square, 67

Y.

Young, Arthur, on secure possession of land, 320

Z.

Zoology in a museum, 7

THE END

www.ingramcontent.com/pod-product-compliance
Lightning Source LLC
Chambersburg PA
CBHW021131230426
43667CB00005B/81